Road Extraction and Distress Assessment by Spaceborne, Airborne and Terrestrial Platforms

Road Extraction and Distress Assessment by Spaceborne, Airborne and Terrestrial Platforms

Guest Editors

Alessandro Mei

Xianfeng Zhang

Valerio Baiocchi

Basel • Beijing • Wuhan • Barcelona • Belgrade • Novi Sad • Cluj • Manchester

Guest Editors

Alessandro Mei
Institute of Atmospheric
Pollution Research (CNR-IIA)
National Research Council of
Italy
Monterotondo Stazione
(Rome)
Italy

Xianfeng Zhang
Institute of Remote Sensing
and Geographic Information
Systems
School of Earth and Space
Sciences
Peking University
Beijing
China

Valerio Baiocchi
Department of Civil
Construction and
Environmental Engineering
Sapienza University of Rome
Rome
Italy

Editorial Office
MDPI AG
Grosspeteranlage 5
4052 Basel, Switzerland

This is a reprint of the Special Issue, published open access by the journal *Remote Sensing* (ISSN 2072-4292), freely accessible at: https://www.mdpi.com/journal/remotesensing/special_issues/Road_Extraction.

For citation purposes, cite each article independently as indicated on the article page online and as indicated below:

Lastname, A.A.; Lastname, B.B. Article Title. *Journal Name* **Year**, *Volume Number*, Page Range.

ISBN 978-3-7258-2831-9 (Hbk)
ISBN 978-3-7258-2832-6 (PDF)
https://doi.org/10.3390/books978-3-7258-2832-6

Cover image courtesy of Alessandro Mei

Contents

About the Editors

Alessandro Mei

Alessandro Mei, Technologist at Institute of Atmospheric Pollution Research - National Research Council of Italy (CNR-IIA), has a graduate degree in Geological Sciences (2005) and has obtained a PhD in Infrastructure Engineering in 2017. He is an expert in the fields of remote sensing, Geographic Information Systems, Thematic Cartography, Unmanned Aircraft Systems (UAS), and Geomatic. He has great experience in the use of remote sensing and GIS systems applied to various environmental topics and has held positions of responsibility within several research projects related to remote sensing and GIS. Finally, he cooperates with research groups at the national and international level. Some of the thematic areas covered by him include the following: development of multi-platform methodologies for environmental monitoring through the integration of analytical techniques and remote sensing sensors; geostatistical analysis of environmental data (e.g., atmospheric pollutants, soil characteristics); analysis of road infrastructures by hyperspectral/multispectral remote sensing and UAS data; UAS acquisitions for 3D reconstructions of artifacts and territorial areas and multispectral/thermographic investigations; laboratory and field spectroscopy of natural and man-made materials; and construction of environmental geo-databases and spectral libraries starting from field and remote sensing data.

He is the Scientific Responsible of UAS, Geomatic and Spectroscopy laboratories and coordinates the "Soil Laboratory for Environmental Protection (SAVE)" of CNR-IIA Laboratory registered in the Global Soil Laboratory Network (GLOSOLAN) of Food and Agriculture Organization of the United Nations.

Xianfeng Zhang

Xianfeng Zhang is currently a full professor and deputy director at the Institute of Remote Sensing and Geographical Information Systems, School of Earth and Space Science, Peking University, Beijing, China. He obtained his first Ph.D. in Cartography and GIS from the Institute of Remote Sensing Applications, Chinese Academy of Sciences, in 2000 and his second Ph.D. in Geography from the University of Western Ontario, Canada, in 2005. His research interests include the remote sensing of ecology, geospatial techniques for natural disaster management, and geospatial data handling and modeling. He has received three grants from the National Natural Science Foundation of China, and four National Key Research and Development Program projects have been conducted under his leadership. At present, he is the co-editor-in-chief of the *International Journal of Applied Earth Observation and Geoinformation*. He is also a referee for several international academic journals such as *ISPRS Journal*, *IEEE TGRS*, *Applied Remote Sensing*, *JSTARS*, *Remote Sensing*, *PE & RS*, *CJRS*, *IJRS*, *IJDE*, *CEUS*, etc. He has published more than 120 papers, edited/co-edited three books, and has an H-index of 22 on Scopus.

Valerio Baiocchi

Valerio Baiocchi is a professor at La Sapienza University of Rome, Italy, and is an Engineer and a Geologist, with both being a full graduation from Sapienza University of Rome, one of the oldest universities of Europe, with top marks. He has obtained a Ph.D. in Geodesy and Survey, form 'Parthenope' University, Napoli, Italy (1996–1999), a master's in Environmental Sciences (Scuola di specializzazione), from Urbino University, Italy (1995–1997), and a second Ph.D. in Infrastructures and Transports, from 'Sapienza' University of Rome (2006–2009).

He has authored more than 250 scientific papers and has an H-index of 18 on Scopus, 16 on WOS, and 19 on Google scholar.

In 2020, he obtained the Associate Professor Affiliation DICEA (Department of Civil, Constructional and Environmental Engineering), 'La Sapienza' University, Roma, Italy; in 2008–2020, he obtained the Assistant Professor Affiliation DICEA; in 2000–2008, he became a Professor on contract in Department of Hydraulic, Transports and Roads (DITS), 'La Sapienza' University, Roma, Italy. He specializes in Geodesy, geomatics, Teledection, GIS, and GPS/GNSS.

The main courses taught by him as a professor are related to survey (2008.06–2013.06), geometric use of high-resolution satellite imagery (2003.06–2008.06), geodetic survey (2014.06–present), geomatics (2005.06–present), and survey and cartography (2005.06–present).

 *remote sensing*

Editorial

Road Extraction and Distress Assessment by Spaceborne, Airborne, and Terrestrial Platforms

Valerio Baiocchi [1], **Xianfeng Zhang** [2] **and Alessandro Mei** [3,*]

[1] Department of Civil Construction and Environmental Engineering, Sapienza University of Rome, 00184 Rome, Italy; valerio.baiocchi@uniroma1.it

[2] Institute of Remote Sensing and Geographic Information Systems, School of Earth and Space Sciences, Peking University, Beijing 100871, China; xfzhang@pku.edu.cn

[3] Institute of Atmospheric Pollution Research (CNR-IIA), National Research Council of Italy, Monterotondo, 00185 Rome, Italy

* Correspondence: alessandro.mei@cnr.it

Citation: Baiocchi, V.; Zhang, X.; Mei, A. Road Extraction and Distress Assessment by Spaceborne, Airborne, and Terrestrial Platforms. *Remote Sens.* **2024**, *16*, 1416. https://doi.org/10.3390/rs16081416

Received: 27 March 2024
Accepted: 7 April 2024
Published: 17 April 2024

1. Introduction

The road systems connecting villages, cities, and countries stand as a pivotal transportation infrastructure in modern society [1], and road maps are widely used in navigation, intelligent transportation, location-based services, emergency rescue, and urban design [2]. Road extraction from remotely sensed imagery is one of the early-stage applications in the traffic industry. A road is usually seen as linear features in medium- to low-resolution satellite imagery, or the central line of a road is extracted from high-resolution imagery [1,2]. With the increasing availability of high-resolution remote sensing, roads are no longer just extracted as linear features from images, but can be used to evaluate the health conditions of road pavements [3,4]. In the fields of computer vision and autonomous driving, the focus is mainly on the recognition and extraction of targets such as cracks, curbs, pedestrians, and cars [5]. In addition to conventional shallow machine learning and mathematical morphology methods, deep neural networks have carried out a significant amount of work in road pavement distress and road target extraction in recent years [6,7]. The main remotely sensed data used in these studies are high-resolution RGB images captured by vehicle-mounted and handheld cameras such as DeepCrack [8] and RDD2022 [9]. The methods for assessing pavement aging and distress conditions can be categorized into three types: image classification [10,11], object detection [5,6], and semantic segmentation [7]. The pavement management system (PMS) often consists of mounted sensors including CCD cameras and LiDAR, as well as ground-penetrating radar (GPR) and thermal infrared sensors. At the same time, many researchers in the field of remote sensing have attempted to use sub-meter satellite image data for pavement aging assessment [4,11], and apply unmanned aerial vehicle (UAV)-captured RGB, multispectral, and hyperspectral data for road distress detection and semantic segmentation [3,12]. In addition, navigation street view images are also used for road distress identification in urban areas [3]. From this point of view, the various remote sensor data with different resolutions obtained by spaceborne, UAV, and terrestrial remote sensing systems offer a new possibility for road aging and distress assessment. A research direction that is becoming a hot topic is how to integrate remote sensing data from multiple modalities to enhance sensing capability for pavement health conditions. For example, cracks may be difficult to distinguish from gasoline stains, shadows, etc., in RGB and multispectral images, but can be easily differentiated if high-resolution thermal infrared images can be obtained simultaneously [13]. Closely related to multimodal remote sensing applications, it is necessary to study new deep learning models that fully utilize the spatial, spectral, depth, and thermal characteristics of road pavement distresses to construct deep artificial neural networks with strong generalization ability in order to provide more reliable technologies for road maintenance. It is undeniable that remote sensing technology has become a new tool for assessing road pavement health conditions, and it is worthy of further study. Therefore, we compiled a Special Issue for the journal *Remote Sensing* in 2022: "Road Extraction and Distress Assessment by Spaceborne, Airborne, and Territorial Platforms", which received contributions from several scholars. The 12 papers published in this Special Issue will be introduced briefly in the following section.

2. An Overview of Published Articles

In the first analysis of the contributions presented, it can be observed that the majority of them consistently utilize imagery from ground-based vehicle systems, developing sensors as well as algorithms to automate the extraction of pavement damage in either fully automatic or semi-automatic modes [2,3,5,7,9]. A growing number of papers, on the other hand, employ ground-penetrating radar (GPR) techniques either alone [8,10,11] or in combination with optical techniques [12]. Surveys from unmanned aerial vehicles (UAVs) are also beginning to proliferate [4,6], while the use of true remote sensing techniques appears to be more limited [1]. This evidence may reflect the fact that satellites and UAVs are not globally integrated in terms of technical requirements for road management procedures. Nonetheless, this kind of technology could be efficient and promising, especially when used with AI techniques, to examine large road networks and to extract valuable parameters to establish intervention priorities or to set up preventive maintenance programs.

In Liu et al. (contribution 1), the authors introduce a lightweight dynamic addition network (LDANet) tailored for rural road extraction. To address the unique characteristics of rural roads—narrowness, complexity, and diversity—they propose an enhanced Asymmetric Convolution Block (ACB)-based Inception structure to augment low-level features in the feature extraction layer. In the deep feature association module, they leverage depth-wise separable convolution (DSC) to reduce computational complexity and design an adaptation-weighted overlay to capture salient features effectively. Additionally, they curate a rural road dataset based on the Deep Globe Land Cover Classification Challenge dataset. Hence, LDANet exhibits promise for the rapid extraction and monitoring of rural roads from remote sensing imagery.

The article by Song et al. (contribution 2) presents the creation of the ISTD-PDS7 dataset, the first of its kind aimed at multi-type pavement distress segmentation. This dataset comprises natural charge-coupled device (CCD) images and encompasses seven types of pavement distress across nine different scenarios, including negative samples with texture similarity noise, resulting in a total of 18,527 annotated images, surpassing previous benchmarks in scale. Additionally, the authors explore the efficacy of negative samples in mitigating false positive predictions in complex scenes and propose two potential data augmentation methods to enhance segmentation accuracy. The authors think that these efforts will catalyze advancements in both academic research and industrial applications within the field.

In Inácio at al. (contribution 3), the authors introduce a straightforward system aimed at expediting road pavement surface inspection and analysis to facilitate maintenance decision-making. Leveraging a low-cost video camera mounted on a vehicle, pavement imagery was captured and processed through an automatic crack detection and classification system based on deep neural networks. The system offers a cracking percentage per road segment, alerting experts to areas requiring attention, as well as a segmentation map highlighting cracked areas on the road pavement surface. The system seems to exhibit promising performance in highway pavement analysis, and its automation and low processing time make it a valuable tool for experts engaged in road pavement maintenance activities.

In the fourth text (contribution 4), the authors present a methodology for real-time road extraction and condition detection using video footage captured by UAV multispectral cameras or pre-downloaded multispectral images from satellites. The primary objective is to detect road conditions and identify emergencies to provide timely assistance to individuals in the wild. By leveraging a normalized difference vegetation index (NDVI), the UAV effectively distinguishes between bare soil roads and gravel roads, enhancing the accuracy of route planning data. In the context of low-altitude human–machine interaction, the authors utilized media-pipe hand landmarks and machine learning techniques to develop a dataset comprising four fundamental hand gestures for dynamic gesture recognition. The experimental results demonstrate that the model achieves very high accuracy on the testing set. Through this proof-of-concept study, the authors claim that the described approach fulfills the expected tasks of UAV rescue and route planning effectively.

In Wang et al. (contribution 5), the authors propose an improved version of the You Only Look Once version three (YOLOv3) object detection model, integrating data augmentation and structure optimization, to achieve the intelligent and accurate measurement of pavement surface potholes. Initially, color adjustment techniques were employed to enhance the image contrast, followed by data augmentation through geometric transformations. Pothole categories were further categorized into P1 and P2 based on the presence of water. Subsequently, the structure of the YOLOv3 model was optimized using the Residual Network (ResNet101) and complete IoU (CIoU) loss, while the multiscale anchor sizes were refined through clustering and modification using the K-Means++ algorithm. Lastly, the robustness of the proposed model was evaluated through the generation of adversarial examples. The experimental results indicate a significant improvement over the original YOLOv3 model.

The article by Qiu et al. (contribution 6) proposes an Adaptive Spatial Feature Fusion YOLOv5 Network (ASFF-YOLOv5) for the automatic recognition and detection of multiple multiscale road traffic elements. Initially, the K-means++ algorithm is utilized for clustering statistics on the range of multiscale road traffic elements, facilitating the determination of suitable candidate box sizes for the dataset. Subsequently, a Spatial Pyramid Pooling Fast (SPPF) structure is integrated to enhance the classification accuracy and speed, enabling richer feature information extraction. An ASFF strategy based on a Receptive Field Block (RFB) is then introduced to improve the feature scale invariance and enhance the detection of small objects. Finally, the experimental effectiveness is evaluated through mean average precision (mAP) calculations. The results demonstrate that the proposed method achieves a significant improvement over the original YOLOv5 model.

Zhang et al. (contribution 7) evaluated mainstream CNN structures for road crack segmentation and propose a novel method, termed a Recurrent Adaptive Network (RAN), inspired by the second law of thermodynamics. The RAN dynamically assesses the imbalance degree, adjusts sampling rates, and modifies loss weights during training to maintain a balanced flow between precision and recall, akin to temperature conduction. The authors realized a dataset of high-resolution road crack images with pixel-level annotations (HRRC) from real inspection scenes, enabling the comprehensive evaluation of CNN performance in highway patrol scenarios. The primary contribution lies in addressing data imbalance and guiding model training by analyzing precision and recall. The experimental results seem to demonstrate the effectiveness of the RAN, achieving state-of-the-art performance on the HRRC dataset.

Qi's text (contribution 8) concerns a specific problem of some road infrastructure: the block-stone embankment that is vital for stabilizing underlying warm and ice-rich permafrost. It faces various damages over time, potentially compromising its cooling function and exacerbating issues along the Qinghai–Tibet Highway (QTH). Ground-penetrating radar (GPR), a nondestructive testing technique, was employed to assess damage properties in the embankment. An analysis of GPR imagery alongside other data and methodologies revealed several damage categories: loosening of the upper sand–gravel layer, loosening of the block-stone layer, settlement of the block-stone layer, and dense filling of the block-stone layer. While the first two conditions were widespread, settlement and dense filling of the block-stone layer were less common, with occurrences of combined damages also noted. The observed correlations among different damages suggest underlying causal relationships.

Chen's study (contribution 9) introduces LeViT, a novel Transformer method for automatic asphalt pavement image classification. LeViT comprises convolutional layers, transformer stages alternating between Multi-layer Perception (MLP) and multi-head self-attention blocks using residual connections, and two classifier heads. Leveraging three different sources of pavement image datasets and pre-trained weights from ImageNet, the authors compare the performance of LeViT with six state-of-the-art (SOTA) deep learning models trained using transfer learning. The experimental results demonstrate that after training for 100 epochs with a batch size of 16, LeViT achieves good results on the Chinese asphalt pavement dataset as well as on the German asphalt pavement dataset, outperforming all tested SOTA models. Moreover, LeViT exhibits superior inference speed compared to the original ViT method as well as prominent CNN-based models like DenseNet, VGG, and ResNet. Furthermore, the authors propose a visualization method combining Grad-CAM and Attention Rollout to enhance the interpretability of LeViT, facilitating the analysis of the classification results and insights into the learned features in each MLP and attention block.

In the tenth study (contribution 10), the authors developed a method for the rapid target identification and comparison of time-lapse GPR profiles. A field experiment was conducted to monitor a backfill pit using three-dimensional GPR (3D GPR), with time-lapse data collected over four months. A U-Net, a fast neural network based on convolutional neural networks (CNNs), was trained using the collected data. The trained model effectively segmented the backfill pit from inline profiles, achieving an Intersection over Union (IoU) of 0.83 on the test dataset. Additionally, the comparison of segmentation masks revealed potential changes in the southwest side of the backfill pit.

The article by Ling et al. (contribution 11) proposes a road subgrade monitoring method based on the time-lapse full-coverage (TLFC) 3D GPR technique. The approach focuses on resolving key challenges related to time and spatial position mismatches in experimental data. By employing time-zero consistency correction, 3D data combination, and spatial-position-matching methods, the approach seems to significantly enhance the 3D imaging quality of underground spaces. Furthermore, the authors utilized time-lapse attribute analysis on TLFC 3D GPR data to extract detailed characteristics and overall patterns of dynamic subgrade changes.

The last text (contribution 12) introduces a novel approach for the inverse calculation of material parameters to determine the mechanical response of asphalt pavements. Initially, a modulus correction method is developed to minimize the error between the tested and simulated strains. Furthermore, a dual sinusoidal regression model effectively illustrates the relationship between temperature at various depths within the pavement structure and atmospheric temperature. An analysis of the pavement monitoring data reveals that increased loading weight and temperature, coupled with decreased loading speed, lead to elevated three-way strain in the asphalt layer. Consequently, a relationship model between loading conditions and three-way strain is established with high fidelity ($R^2 > 0.95$). This comprehensive methodology addresses reliability issues with pavement structure parameters and provides a quantitative assessment of structural conditions, supporting the performance prediction and maintenance analysis of asphalt pavements with a semi-rigid base.

The Guest Editors would like to extend their thanks to all of the contributors to this Special Issue.

Author Contributions: A.M., X.Z. and V.B. contributed equally to this work. All authors have read and agreed to the published version of the manuscript.

Funding: This research received no external funding.

Acknowledgments: The Guest Editors would like to thank all authors who have contributed to this Special Issue by sharing their research results and experiences; the reviewers who dedicated their time to provide constructive recommendations; and the journal editorial team for their continuous support.

Conflicts of Interest: The authors declare no conflicts of interest.

List of Contributions:

1. Liu, B.; Ding, J.; Zou, J.; Wang, J.; Huang, S. LDANet: A Lightweight Dynamic Addition Network for Rural Road Extraction from Remote Sensing Images. *Remote Sens.* **2023**, *15*, 1829. https://doi.org/10.3390/rs15071829.
2. Song, W.; Zhang, Z.; Zhang, B.; Jia, G.; Zhu, H.; Zhang, J. ISTD-PDS7: A Benchmark Dataset for Multi-Type Pavement Distress Segmentation from CCD Images in Complex Scenarios. *Remote Sens.* **2023**, *15*, 1750. https://doi.org/10.3390/rs15071750.
3. Inácio, D.; Oliveira, H.; Oliveira, P.; Correia, P. A Low-Cost Deep Learning System to Characterize Asphalt Surface Deterioration. *Remote Sens.* **2023**, *15*, 1701. https://doi.org/10.3390/rs15061701.
4. Liu, C.; Szirányi, T. Road Condition Detection and Emergency Rescue Recognition Using On-Board UAV in the Wildness. *Remote Sens.* **2022**, *14*, 4355. https://doi.org/10.3390/rs14174355.
5. Wang, D.; Liu, Z.; Gu, X.; Wu, W.; Chen, Y.; Wang, L. Automatic Detection of Pothole Distress in Asphalt Pavement Using Improved Convolutional Neural Networks. *Remote Sens.* **2022**, *14*, 3892. https://doi.org/10.3390/rs14163892.
6. Qiu, M.; Huang, L.; Tang, B.-H. ASFF-YOLOv5: Multielement Detection Method for Road Traffic in UAV Images Based on Multiscale Feature Fusion. *Remote Sens.* **2022**, *14*, 3498. https://doi.org/10.3390/rs14143498.
7. Zhang, Y.; Fan, J.; Zhang, M.; Shi, Z.; Liu, R.; Guo, B. A Recurrent Adaptive Network: Balanced Learning for Road Crack Segmentation with High-Resolution Images. *Remote Sens.* **2022**, *14*, 3275. https://doi.org/10.3390/rs14143275.
8. Qi, S.; Li, G.; Chen, D.; Chai, M.; Zhou, Y.; Du, Q.; Cao, Y.; Tang, L.; Jia, H. Damage Properties of the Block-Stone Embankment in the Qinghai–Tibet Highway Using Ground-Penetrating Radar Imagery. *Remote Sens.* **2022**, *14*, 2950. https://doi.org/10.3390/rs14122950.
9. Chen, Y.; Gu, X.; Liu, Z.; Liang, J. A Fast Inference Vision Transformer for Automatic Pavement Image Classification and Its Visual Interpretation Method. *Remote Sens.* **2022**, *14*, 1877. https://doi.org/10.3390/rs14081877.
10. Shang, K.; Zhang, F.; Song, A.; Ling, J.; Xiao, J.; Zhang, Z.; Qian, R. Fast Segmentation and Dynamic Monitoring of Time-Lapse 3D GPR Data Based on U-Net. *Remote Sens.* **2022**, *14*, 4190. https://doi.org/10.3390/rs14174190.
11. Ling, J.; Qian, R.; Shang, K.; Guo, L.; Zhao, Y.; Liu, D. Research on the Dynamic Monitoring Technology of Road Subgrades with Time-Lapse Full-Coverage 3D Ground Penetrating Radar (GPR). *Remote Sens.* **2022**, *14*, 1593. https://doi.org/10.3390/rs14071593.
12. Liu, Z.; Yang, Q.; Gu, X. Assessment of Pavement Structural Conditions and Remaining Life Combining Accelerated Pavement Testing and Ground-Penetrating Radar. *Remote Sens.* **2023**, *15*, 4620. https://doi.org/10.3390/rs15184620.

References

1. Alshehhi, R.; Marpu, P.R. Hierarchical graph-based segmentation for extracting road networks from high-resolution satellite images. *ISPRS J. Photogramm. Remote Sens.* **2017**, *126*, 245–260. [CrossRef]
2. Xu, Q.; Long, C. Road extraction with satellite images and partial road maps. *IEEE Trans. Geosci. Remote Sens.* **2023**, *61*, 4501214. [CrossRef]
3. Pan, Y.; Zhang, X.; Cervone, G.; Yang, L. Detection of asphalt pavement potholes and cracks based on the unmanned aerial vehicle multispectral imagery. *IEEE J. Sel. Top. Appl. Earth Obs. Remote Sens.* **2018**, *11*, 3701–3712. [CrossRef]
4. Mei, A.; Salvatori, R.; Fiore, N.; Allegrini, A.; D'Andrea, A. Integration of field and laboratory spectral data with multi-resolution remote sensed imagery for asphalt surface differentiation. *Remote Sens.* **2014**, *6*, 2765–2781. [CrossRef]
5. Ren, M.; Zhang, X.; Chen, X.; Zhou, B.; Feng, Z. YOLOv5s-M: A deep learning network model for road pavement damage detection from urban street-view imagery. *Int. J. Appl. Earth Obs. Geoinf.* **2023**, *120*, 103335. [CrossRef]
6. Lee, J.K.; Huh, Y. Multi-sensorial image dataset collected from mobile mapping system for asphalt pavement management. *Sens. Mater.* **2022**, *34*, 2615–2624. [CrossRef]
7. Maeda, H.; Kashiyama, T.; Sekimoto, Y.; Seto, T.; Omata, H. Generative adversarial network for road damage detection. *Comput.-Aided Civ. Infrastruct. Eng.* **2021**, *36*, 47–60. [CrossRef]
8. Liu, Y.; Yao, J.; Lu, X.; Xie, R.; Li, L. DeepCrack: A deep hierarchical feature learning architecture for crack segmentation. *Neurocomputing* **2019**, *338*, 139–153. [CrossRef]
9. Arya, D.; Maeda, H.; Ghosh, S.K.; Toshniwal, D.; Sekimoto, Y. RDD2022: A multi-national image dataset for automatic Road Damage Detection. *arXiv* **2022**, arXiv:2209.08538.

10. Nhat-Duc, H.; Nguyen, Q.-L.; Tran, V.-D. Automatic recognition of asphalt pavement cracks using metaheuristic optimized edge detection algorithms and convolution neural network. *Autom. Constr.* **2018**, *94*, 203–213. [CrossRef]

11. Chen, X.; Zhang, X.; Li, J.; Ren, M.; Zhou, B. A new method for automated monitoring of road pavement aging conditions based on recurrent neural network. *IEEE Trans. Intell. Transp. Syst.* **2022**, *23*, 24510–24523. [CrossRef]

12. Truong, L.N.H.; Mora, O.E.; Cheng, W.; Tang, H.; Singh, M. Deep learning to detect road distress from unmanned aerial system imagery. *Transp. Res. Rec.* **2021**, *2675*, 776–788. [CrossRef]

13. Pozzer, S.; De Souza, M.P.V.; Hena, B.; Hesam, S.; Rezayiye, R.K.; Azar, E.R.; Lopez, F.; Maldague, X. Effect of different imaging modalities on the performance of a CNN: An experimental study on damage segmentation in infrared, visible, and fused images of concrete structures. *NDT E Int.* **2022**, *132*, 102709. [CrossRef]

Article

Assessment of Pavement Structural Conditions and Remaining Life Combining Accelerated Pavement Testing and Ground-Penetrating Radar

Zhen Liu [†], Qifeng Yang [†] and Xingyu Gu *

Department of Roadway Engineering, School of Transportation, Southeast University, Nanjing 211189, China; 230208344@seu.edu.cn (Z.L.); 220215101@seu.edu.cn (Q.Y.)
* Correspondence: guxingyu1976@seu.edu.cn
[†] These authors contributed equally to this work.

Abstract: The inspection and monitoring of structural conditions are crucial for the maintenance of semi-rigid base pavement. To achieve the inverse calculation of material parameters and obtain the mechanical response of asphalt pavement, a method of modulus correction by reducing the error between tested and simulated strains was first developed. The relationship between the temperature at various depths within the pavement structure and atmospheric temperature was effectively demonstrated using a dual sinusoidal regression model. Subsequently, pavement monitoring data illustrated that as loading weight and temperature increased and loading speed decreased, the three-way strain of the asphalt layer increased. Thus, the relationship model between loading conditions and three-way strain was established with a good fitting degree ($R^2 > 0.95$). The corrected modulus was obtained by approximating the error between simulated and measured strains. Then, the finite element analysis was performed to calculate key mechanical index values under various working conditions and predict the fatigue life of asphalt and base layers. Finally, ground-penetrating radar (GPR) detection was performed, and the internal pavement condition index was defined for quantitative assessment of structure conditions. The results show that there is a good correlation between the internal pavement condition index (IPCI) and remaining life of pavement structure. Therefore, our works solve the problems of the parameter reliability of pavement structures and quantitative assessment for structural conditions, which could support the performance prediction and maintenance analysis on asphalt pavement with a semi-rigid base.

Keywords: pavement structure conditions; modulus inversion; accelerated pavement testing; ground penetrating radar; remaining life; finite element method (FEM)

Citation: Liu, Z.; Yang, Q.; Gu, X. Assessment of Pavement Structural Conditions and Remaining Life Combining Accelerated Pavement Testing and Ground-Penetrating Radar. *Remote Sens.* **2023**, *15*, 4620. https://doi.org/10.3390/rs15184620

Academic Editor: Lorenzo Capineri

Received: 30 June 2023
Revised: 6 September 2023
Accepted: 18 September 2023
Published: 20 September 2023

1. Introduction

The assessment of pavement structural conditions and remaining life has historically been a major research focus [1]. Among the fields of study, the two main research directions are real-time monitoring of structural health status and the periodic monitoring of structural conditions.

The focus of pavement structural health monitoring is the acquisition and analysis of dynamic response data. Strain is a critical index used to characterize pavement structural characteristics. Pavement fatigue damage is analyzed by measuring the tension and compression strain at the bottom of the asphalt layer [2]. The theoretical calculations based on the elastic layer system do not consider the dynamical behaviors of pavement structures under environmental and vehicle load factors. Thus, these calculations poorly reflect its actual mechanical state. On the other hand, although the dynamic pavement responses can be acquired through laboratory tests, they might not faithfully reflect the gradual deterioration of pavement mechanical properties because of the difference between indoor tests and the actual road under traffic loads [3]. Meanwhile, the long-term observation of operating roads is time-consuming and labor-intensive. Therefore, there are significant limitations to the monitoring of pavement dynamic responses.

The emergence of accelerated pavement testing (APT) has compensated for the shortcomings of theoretical calculations and laboratory tests [4,5]. The AASHO test road in America established the statistical relationship between pavement performance and the number of axial loads for the first time in 1962. After that, the field measurement of pavement strain has gradually become a research hotspot in road engineering during APT monitoring. The MnRoad test road in America has pioneered the application of strain sensors in road structure tests and achieved good results [6]; it established a modified pavement material model based on the local climate. Still, the effect of axle loads was not considered in the research results. Then, many studies [7,8] began analyzing the relationship between the field-measured stress or strain and loading speed, seasonal variation, and the depth of road structures, respectively. Among them, the heavy vehicle simulator (HVS) [9] is widely used by the University of California [10], Florida Department of Transportation, and Swedish Transport Research Institute [11], etc. However, the shortcoming of HVS is obvious: the loading speed is relatively low (the highest is only 14 km/h). Afterward, South Africa developed the new mobile load simulator (MLS) series accelerated loading equipment based on absorbing previous experience. The strain test of the asphalt structural layer was carried out in the first MLS10 loading test in Switzerland [12]. The Bundesanstalt für Straßenwesen (BASt) in Germany further used the MLS30 equipment to answer difficult questions regarding pavement engineering combined with indoor tests [13]. Additionally, some studies were conducted on the longitudinal and transverse strains of asphalt and base layers in flexible pavements. The Liaoning Transportation Research Institute used the most advanced MLS66 equipment in APT monitoring [14]. For similar layers, they found that asphalt pavement strain amplitudes were more minor than rubber asphalt pavement strain amplitudes.

Although several strain studies in APT have been carried out, almost all are aimed at flexible pavements [15]. However, most of China's operational pavements have semi-rigid bases [16]. This difference means that previous conclusions need to be further demonstrated. There are few full-scale APT at present, and no omnidirectional tests for strain response have been conducted. In addition to the outdoor tests, numerical simulations based on finite element (FE) analysis are also an essential adjuvant method to analyze and calculate the pavement structural dynamic responses [17]. Yang used the FE tire–pavement contact model to predict the pavement strain responses under conventional loading forms through APT monitoring [18]. They concluded that the numerically simulated results agree with the measured permanent deformations. Some scholars also applied a three-dimensional (3D) pavement model in FE software (ABAQUS 2016) by considering the nonlinear and heterogeneous features of mobile loads [19]. Chun measured the dynamic responses at the junction of the field pavement structures and combined the improved FE models with the full-scale APT [20]. The modulus parameters in the (ABAQUS 2016) software were obtained based on the inverse calculation of falling weight deflectometer (FWD) measured results. In these available studies, little work has been focused on adopting a more realistic distribution state of loads to obtain the distribution characteristics in 3D space for pavement dynamic responses. In addition, parameter reliability in (ABAQUS 2016) software also needs to be improved [21].

Furthermore, the temperature effect on dynamic responses is crucial in field monitoring and numerical simulation [22]. As a typical temperature-sensitive material [23–25], temperature affects both the mechanical properties and the service properties of asphalt mixtures [26]. The most common failure forms of asphalt pavement are fatigue cracking and rutting deformation. Temperature distribution on pavements is closely related to its mechanisms and development processes. Unfortunately, previous studies barely considered temperature differences at different road structural depths. Therefore, it is necessary to acquire temperature fields at the actual pavement structures. Current temperature field investigations include statistical analysis and theoretical analysis methods. A prediction model for road surface temperature was developed by Barber, making it possible to apply this temperature field in numerical analysis [27]. However, the statistical method does not have universality, and primarily relies on the comprehensiveness and the precision of measured results. Recently, simulations using FE have made a breakthrough in theoretical analysis methods. Using FE modeling, researchers [28,29] studied the impact of temperature on pavement structure responses combined with laboratory tests [30,31]. Although these studies used theoretical analyses and numerical simulations that are more applicable, numerous parameters must be acquired. Most of these parameters are difficult to obtain, making the application too complicated and not intense in practical engineering applications. Therefore, it is necessary to obtain more reliable parameters relating to asphalt pavement to analyze and predict the change law in materials' properties.

In addition, ground-penetrating radar (GPR) has been widely used for the rapid non-destructive testing of pavement structural integrity because of its advantages regarding it high precision and good continuity [32–35]. GPR is a geophysical method used to detect the physical properties and spatial distribution of a medium by transmitting an antenna signal and receiving a high-frequency electromagnetic wave reflection signal. Because the electrical characteristics of the underground medium change during the propagation of the electromagnetic wave, information about the internal structure of the road can be obtained by analyzing the signal [36]. Many researchers have analyzed the typical characteristic images of road GPR detection [37], and summarized the electromagnetic wave phase characteristics of common distresses in road structures [38,39], so as to provide support for road detection and maintenance. However, studies of non-destructive assessment and quantitative evaluation of pavement structures are lacking. Therefore, it is warranted to establish a quantitative evaluation method for pavement structural integrity and analyze its relationship with the pavement remaining life.

This study has four objectives, as follows:

(1) To investigate the relationship between pavement temperature and atmospheric temperature in the depth direction.
(2) To analyze the influencing factors of mechanical responses of pavement structure layer and reveal its influencing law.
(3) To explore the establishment of more reliable parameters in FE simulation for fatigue life prediction under more complex conditions.
(4) To reveal the relationship between the pavement structure conditions and remaining life of different structural layers.

2. Materials and Method

As shown in Figure 1, the representative combination form of asphalt surface of Jiangsu expressway is "SMA-13 + SUP-20 + SUP-25". Specifically, the material of the base layer is mainly crack-resistant cement-stabilized macadam (CSM), with a thickness of 36 cm to 40 cm. Therefore, the structure of the tested road in Zhenjiang, Jiangsu province was determined. The research procedure consisted of the following steps.

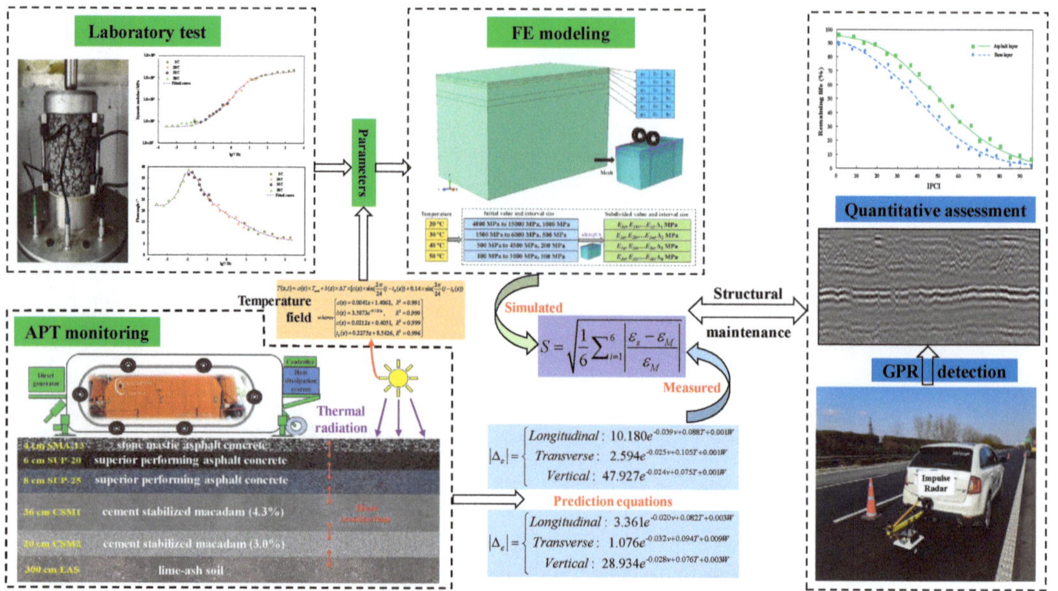

Figure 1. Workflow of the APT, laboratory tests, numerical simulations, and GPR detection on pavement.

2.1. Field Investigation

2.1.1. Dynamic Modulus Tests

According to the test standard (JTG E20−2011) [40], the test temperature was divided into 5 °C, 20 °C, 35 °C, and 50 °C. The dimensions of the core samples (cylindrical shapes) of the asphalt layer were 100 mm × 100 mm × 150 mm. The specimens were stored in the thermostat for 4 to 5 h at each temperature grade. The specimens were installed and tested in UTM-25 equipment. To obtain the time spectrum representing the temperature dependence of asphalt mixtures, the test results were fitted according to the nonlinear least square method. The master curves of *E** (20 °C) adhered to Equations (1)–(3) [41].

$$\lg(\alpha_T) = \frac{C_1(T - T_s)}{C_2 + T - T_s} \tag{1}$$

where α_T denotes the displacement factor, T denotes the test temperature, C_1 and C_2 represent the fitted parameters, and T_s denotes the reference temperature [42,43].

$$\lg|E^*| = \delta + \frac{\alpha}{1 + e^{\beta + \gamma \lg(f_r)}} \tag{2}$$

$$\lg(f_r) = \lg(f) + \lg(\alpha_T) \tag{3}$$

where f denotes testing frequency, f_r represents the adjusted frequency, and α_T, α, β, γ and δ are the regression constants, respectively.

2.1.2. Field Temperature Monitoring

The temperature of pavement structures at different depths z was collected from March 2020 to December 2020 using a resistive sensor with a frequency of 2 Hz. To eliminate the influence of noise on the testing results, the curves of temperature obtained by means of continuous collection were smoothed according to an interval of 10 min. According to the characteristic that the warming stage takes less time than the cooling stage, the hyperbolic sine function in Equation (4) was used to conduct regression analysis on the temperature field *T(z, t)* of asphalt pavement under different seasons (only non-rain weather was studied) and different depths.

$$T(z,t) = a \times T_{ave} + b \times \Delta T \times (c \times \sin(\omega(t - t_0)) + d \times \sin(2\omega(t - t_0)))$$ (4)

where T_{ave} denotes the average of the maximum and minimum temperatures during one day, ΔT is the difference between them, ω is the angular frequency, t_0 is the initial phase, and *a*, *b*, *c*, and *d* represent the regression coefficients.

After t_0 was obtained by fitting the temperature under different months and depths, the regression analysis was performed again to obtain the initial phase t_0 (*z*) at other *z*. Similarly, a and b were used to perform the regression analysis again after the first fitting to obtain the initial phase *a* (*z*) and *b* (*z*), respectively. After obtaining these regression relationships, the average values of *a* (*z*), *b* (*z*), *c* (*z*), and t_0 (*z*) were obtained by means of different *z* fitting in each month and were functionally fitted with *z* as an independent variable, respectively.

2.1.3. Accelerated Pavement Testing

As shown in Figure 2, the mobile load simulator (MLS66) equipment with internal twin-tires (305/70/R22.5) was performed to conduct the loading of the tested road under different temperature and speed conditions [42]. Figure 2 also depicts the cross-sectional layout scheme of monitoring sensors, considering the test coverage and sensor survival rate [44]. Temperature sensors were divided into two groups: one group was located in the middle and lower asphalt layers under the wheel track center for the purpose of monitoring temperature conditions, and the other group was situated on the edge of the road from the middle asphalt layer to the CMS2 layer for the purpose monitoring the natural temperature field of the road surface [45].

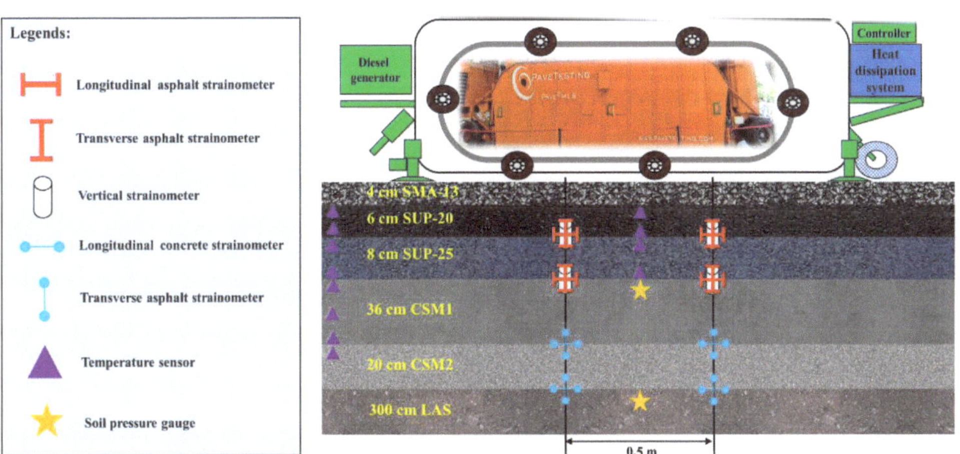

Figure 2. Sensor layout of accelerated loading test road section.

The strain sensors were all buried below the tire action point. The location of longitudinal, transverse, and vertical strain sensors was the same as that of the temperature sensors. At the bottom of the base layers, transverse and longitudinal horizontal-strain sensors were used [46].

Temperature (*T*), loading weight (*W*) and speed (*v*) were the three influencing factors [19]; thus, the loading scheme was as follows: 50, 60, and 75 kN for *W*; 10, 15, and 22 km/h for *v*; *T*: low (18 °C to 28 °C), middle (28 °C to 37 °C), and high temperatures (37 °C to 48 °C) for T. Then, the strain-temperature coefficient ε_T was defined in Equation (5) to calculate the strain transformation corresponding to the change in unit temperature [47].

$$\varepsilon_T = \frac{\varepsilon_2 - \varepsilon_1}{T_2 - T_1},\tag{5}$$

where T_1 and T_2 are the minimum and maximum values of the temperature range, and ε_1 and ε_2 are the corresponding strains, respectively. Combined with the measured strain values, the prediction equation of the strain amplitude (three-way: lateral, longitudinal, and vertical) $|\Delta_\varepsilon|$ for the bottom of the measured layers is described in Equation (6) [42].

$$\left|\Delta\varepsilon\right| = a \times e^{b \times v + c \times T + d \times W},\tag{6}$$

where a, b, c, and d denote the fitted parameters.

2.1.4. GPR Investigation

GPR investigation was performed on the road site with the same pavement structures in Figure 1 for its structural condition assessment. The detection equipment used was CO4080 vehicle-mounted GPR produced by Swedish Impulse, as shown in Figure 3. It is generally believed that the wheel track of the road is the most heavily loaded, and the structure of the wheel track can represent the internal condition of the overall road structure. Therefore, when selecting the vehicle-mounted GPR for the road survey, the measuring line should be situated on the wheel track of each lane. The detection parameter settings used by this GPR are shown in Table 1.

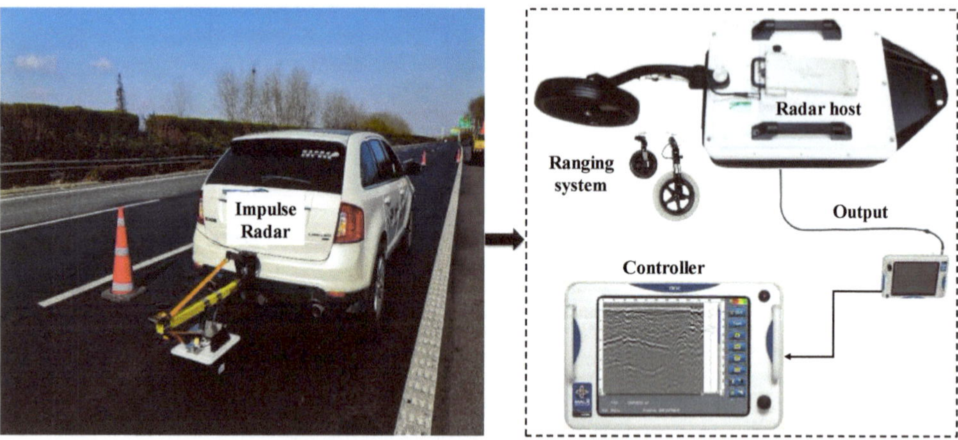

Figure 3. GPR detection for pavement structure conditions.

Table 1. Detection parameters of vehicle-mounted CO4080 GPR manufactured by Impluse Company.

Bandwidth (MHz)	Detection Depth (m)	Sampling Point in Horizontal Direction	Time Window (ns)	Ranging Method	Signal-To-Noise Ratio	Sampling Interval
400 MHz (channel 1) 800 MHz (channel 2)	4.5 m for channel 1 1.5 m for channel 2	400	26	DMI	>100 dB	5 cm

The distress characteristics were identified through GPR detection images of the longitudinal section of the road, according to the judgment results in previous studies [32–35,48]. In this study, the three most prevalent distresses were detected in this road section: cracks, looseness, and interlayer separation (IS). Then, the internal pavement condition index (IPCI) was used to evaluate the integrity of different structural layers of pavement, as shown in Equation (7).

$$IPCI = 100 - \frac{\sum_{i=1}^{n} S_i}{S_0} \times 100\%, \tag{7}$$

where S_i refers to the area of the ith distress (crack, looseness, and interlayer separation) in a structural layer (m^2). S_0 represent the area of the asphalt layer ($0.18 \times 100 = 18$ m^2) or the base layer ($0.36 \times 100 = 36$ m^2), as shown in Figure 4. In this study, only the structural conditions of asphalt surface and CSM base layers were analyzed. GPR investigation was performed for several sections of road on the G15 expressway of Jiangsu province, China, with different service lives. The length of the test road was 5 km, and the pavement structure was the same as in Figure 1. The IPCI index of the structural layers per 100 m of each road section was calculated to analyze the relationship between it and the remaining life of the structural layers.

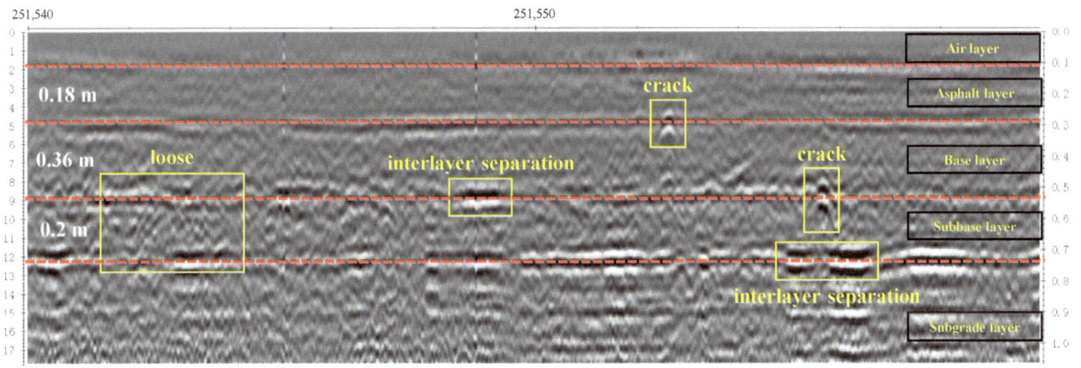

Figure 4. The division of road structure layers and distress labeling in the GPR image.

Through the investigation and statistical analysis of typical pavement structure maintenance projects in Jiangsu Province, the overhaul cycle of the semi-rigid base pavement structure under different traffic conditions was obtained according to previous studies [49,50], as shown in Table 2. The traffic grade was divided according to the cumulative number of N_e (million times) of a single lane under the standard axle load (100 kN). Then, the remaining life ratio (RLR, %) was taken as an evaluation index, which represented the ratio of the service life of existing pavement to the upper service life limit of newly built pavement.

Table 2. Service life of expressway structures with different traffic grades.

			Traffic grades			
	Performance classification		Light ($N_e < 300$)	Moderate ($300 \leq N_e < 1200$)	Slightly heavy ($1200 \leq N_e < 2500$)	Heavy ($N_e \geq 2500$)
RLR of pavement (%)	Excellent	Capital repair	23.5	31.0	25.3	29.7
		Partial repair	59.2	55.7	65.6	65.5
	Average	Capital repair	33.7	42.5	38.3	43.4
		Partial repair	69.4	67.2	78.6	79.3
	Poor	Capital repair	43.9	54.0	51.3	91.6
		Partial repair	79.6	78.7	91.6	93.1

Thus, the RLR index of the structure in each maintenance stage could be calculated by comparing the findings of capital repair cycle and the partial repair cycle with the pavement service life, respectively, under various operating conditions.

2.2. Numerical Simulation

2.2.1. FE Modelling

The FE simulations of the field APT monitoring were performed using ABAQUS 2016 software to investigate the distribution state of dynamic responses under moving loads. Structural parameters of the asphalt surface were inverted, and the key mechanical response indexes were analyzed. In the FE model of the tested road, the X-axis was the lateral direction of the pavement, the Y-axis was the vehicle driving direction, and the Z-axis was the pavement layer depth direction, with lengths of 6 m, 10 m and 3.74 m, respectively.

The fixed constraint was chosen in the model (U1 = U2 = U3 = 0). The displacement in the Z-direction (U3 = 0) and X-direction was limited (U1 = 0), respectively [51]. Considering computing accuracy and efficiency, the mesh density of the surface layer and the grid near the loading area was improved. From the surface layer and the part of the load to the edge of the road, the grid was gradually thinned by means of the principle of equal difference distribution. The model mesh was divided using C3D8R (a three-dimensional eight-node hexahedral element with a linear reduction integral isoperimetric).

In addition, the load was 100 kN, the grounding pressure was 0.4456 MPa, and the action area was 0.035712 m^2 (18.6 cm $\times$ 19.2 cm) in this FE model.

2.2.2. Viscoelastic Parameters

Based on the resting results of E^*, the viscoelastic parameters were obtained using Prony series. The detailed steps in this process were as follows.

First, E^* was converted to the relaxation modulus $E(t)$. The first step was to calculate the storage modulus $E'(f)$ at each frequency using Equation (8).

$$E'(f) = |E^*| \cos(\varphi), \tag{8}$$

where φ is the phase angle. The second step was to calculate log(f) of each frequency and perform a differential solution of lg(f) to calculate the slope n using Equation (9).

$$n = \frac{d\lg E'(f)}{d\lg(f)}, \tag{9}$$

The third step was to calculate the adjustment function λ' using Equation (10).

$$\lambda' = \Gamma(1 - n) \cos\left(\frac{n\pi}{2}\right)$$
$$where : \Gamma(s) = \int_0^{+\infty} x^{s-1} e^{-x} dx, s0 \tag{10}$$

The fourth step was to calculate $E(t)$ using Equation (11).

$$E(t) = \frac{E'(f)}{\lambda'}, \tag{11}$$

Then, the master curve of $E(t)$ was fitted. Subsequently, the shear modulus $G(t)$ and its ratio $g(t)$ were calculated by using Equations (12) and (13), respectively.

$$G(t) = \frac{E(t)}{2(1 + \mu)}, \tag{12}$$

$$G(0) = \frac{E(0)}{2(1+\mu)} = \frac{\Sigma E(t)}{2(1+\mu)} = \Sigma G(t)$$
$$g(t) = G(t)/G_0 \tag{13}$$

Finally, Equation (14) was used to obtain the Prony series of asphalt mixtures.

$$g(t) = 1 - \sum_{i=1}^{n} g_i \left(1 - e^{(-t/\tau_i)}\right), \tag{14}$$

where g_i is the material constant, n is the number of items, and τ_i is the delay time [42].

Considering that the load capacity of the base, subbase, and subgrade layers was always stable during the loading process, these three layers were all taken as linear elastic materials in the FE model. The median value of the inverse modulus based on the FWD test was rounded as the elastic parameters of these three layers.

2.2.3. Material Parameter Inversion

Based on the response data measured on the tested road, the dynamic modulus values under different working conditions were inverted using ABAQUS 2016 software. The master curve of E^* based on the measured dynamic responses was established and compared with that obtained from laboratory tests to achieve the E^* correction of the asphalt layer.

First, monitoring response data of tested road were obtained according to Equation (6) under different working conditions, including four different loading temperatures (20 °C, 30 °C, 40 °C, and 50 °C) and three different loading speeds under an axial load of 50 kN. The calculated response data were the three-way strain of two measured layers (six data types).

As described in Figure 5, the varying ranges of the modulus of the asphalt surface in the FM model at different temperatures were set with the interval values of 1000 MPa, 500 MPa, 200 MPa, and 100 MPa, respectively, based on the test results of the dynamic modulus at different temperatures in Section 3.2.1. For each modulus value, six types of response peaks at different speeds were obtained by means of FE simulation, and the error S between the simulated strains ε_s and the measured strains ε_M was obtained according to Equation (15).

$$S = \sqrt{\frac{1}{6}\sum_{i=1}^{6}\left|\frac{\varepsilon_s - \varepsilon_M}{\varepsilon_M}\right|}, \tag{15}$$

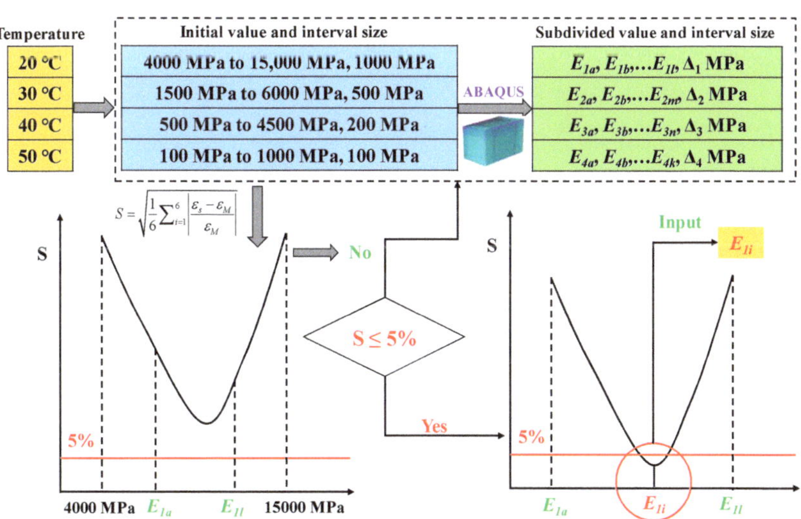

Figure 5. Diagram of material parameter inversion under different working conditions.

If $S \leq 5\%$, the inverse modulus could be determined. Otherwise, smaller interval values were further divided in the modulus range with the most minor error. The FE simulation was repeated until the error requirement was met. Thus, the inversion modulus values under twelve working conditions were obtained.

2.2.4. Fatigue Life Indexes

According to the design standard (JTG F50-2017) [52], fatigue cracking and permanent deformation of the asphalt layer and fatigue cracking of the base layer should be controlled. The fatigue life of the asphalt layer N_{f1} and base layer N_{f2} was calculated via Equations (16) and (17), respectively.

$$N_{f1} = 6.32 \times 10^{15.96-0.29\beta} k_a k_b k_{T1}^{-1} \left(\frac{1}{\varepsilon_a}\right)^{3.97} \left(\frac{1}{E_a}\right)^{1.58} (VFA)^{2.72}, \tag{16}$$

where k_a denotes the adjusted coefficient of seasonally frozen ground, k_b denotes the coefficient of fatigue loading modes, E_a represents a E^* at 20 °C, VFA represents the saturation of asphalt mixtures, k_{T1} is the adjusted coefficient of temperature, and β is the reliability coefficient.

$$N_{f2} = k_a k_{T2}^{-1} 10^{a-b\frac{\sigma_t}{R_S}+k_c-0.57\beta}, \tag{17}$$

where k_{T2} denotes the adjusted temperature coefficient, σ_t denotes the tension stress at the base layer bottom, R_S represents the tension strength at the base layer, k_c denotes the field comprehensive correction factor, and a and b denote fitted parameters.

Based on the predicted results of Equation (6), some indicators cannot be obtained via the APT monitoring. Therefore, the maximum shear strain, maximum compression stress of asphalt layer, and tension stress of base layer were calculated by means of FE simulation. Meanwhile, the most adverse and favorable conditions were selected to determine the reference scope for the values of the critical mechanical responses.

3. Results and Discussion

3.1. Field Temperature Monitoring

According to the measured temperature change curve of pavement structures along the z-direction, fitted results of the hyperbolic sine function in other representative months are shown in Figure 6.

The fitted results of $a(z)$, $b(z)$, $c(z)$, and $t_0(z)$ are presented in Figure 7. Then, the temperature-field prediction model was obtained using Equation (18). The relationship between the measured and fitted temperatures was further explored. The correlation coefficient was above 0.95. This illustrates that the correlation was highly significant, which could predict the pavement structural temperatures.

$$T(z,t) = a(z) \times T_{ave} + b(z) \times \Delta T \times \left[c(z) \times \sin\left(\frac{2\pi}{24}(t-t_0(z))\right) + 0.14 \times \sin\left(\frac{2\pi}{24}(t-t_0(z))\right)\right.$$

$$where \begin{cases} a(z) = 0.0041z + 1.4062, & R^2 = 0.991 \\ b(z) = 3.5873e^{-0.1101z}, & R^2 = 0.999 \\ c(z) = 0.0212z + 0.4051, & R^2 = 0.999 \\ t_0(z) = 0.2275z + 8.5426, & R^2 = 0.996 \end{cases}, \tag{18}$$

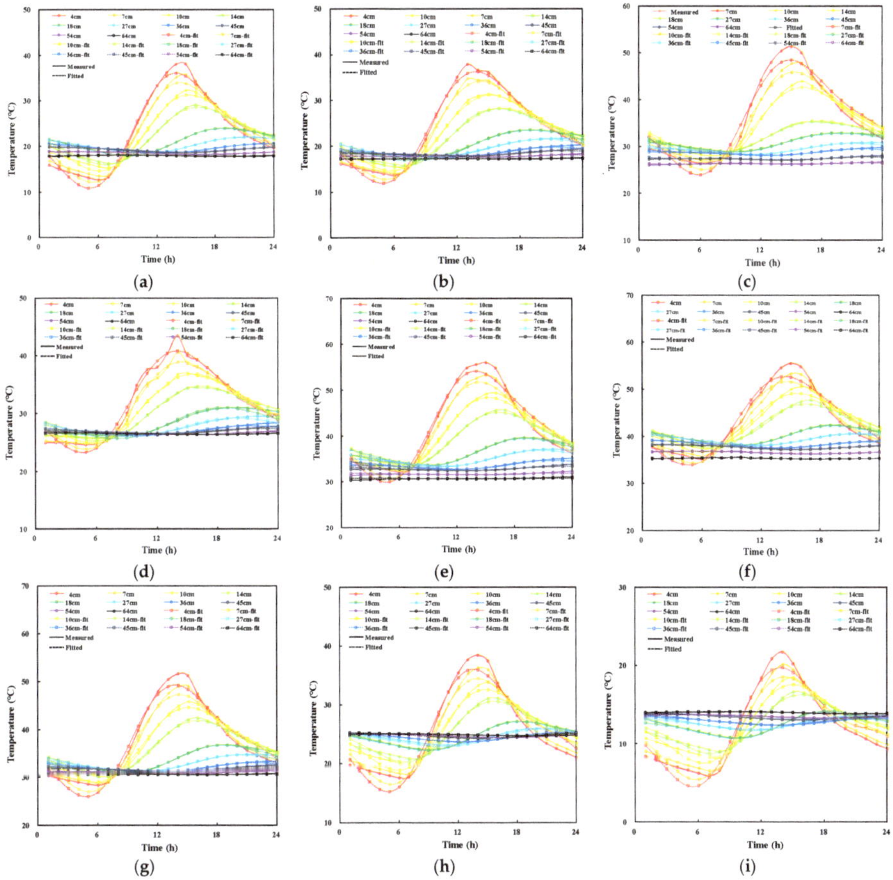

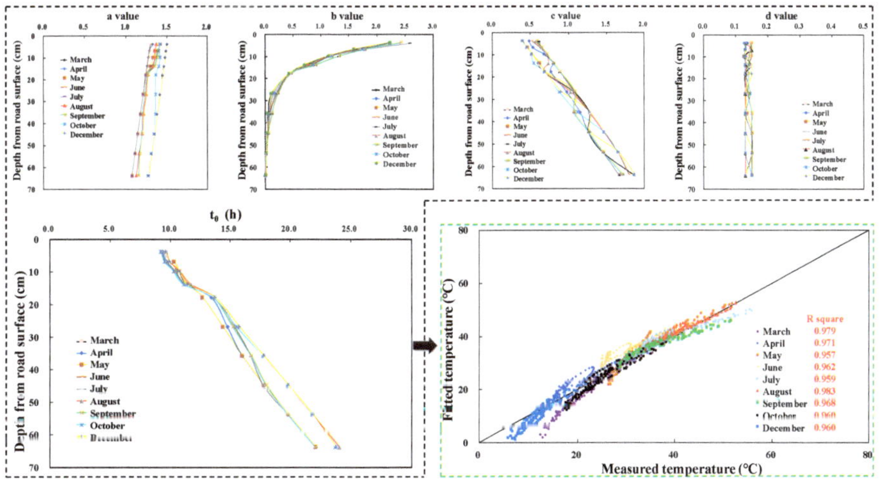

Figure 6. Measured and fitted temperature fields in different months: (**a**–**i**) March to December.

Figure 7. Fitting results of parameters in different months and the relationship between measured temperatures and fitted temperatures.

3.2. Dynamic Responses to Different Influencing Factors

In this section, the dynamic response to at different factors was analyzed, including the following points.

3.2.1. Field Temperature

The other two variables were a v of 22 km/h and a W of 50 kN. Figure 8 illustrates the measured strains at the bottom of the middle asphalt layer (z = 10 cm) at 1# to 5# temperature field. The peak strain increased with an increase in temperature. When the temperature was 47.9 °C (5# high-temperature field), the extreme values of longitudinal and vertical strains were 262.53 με and −979.70 με, 25.4 times and 21.6 times those of 18.9 °C, respectively. This shows that the strain at the bottom of the middle asphalt layer is greatly affected by the pavement temperature field.

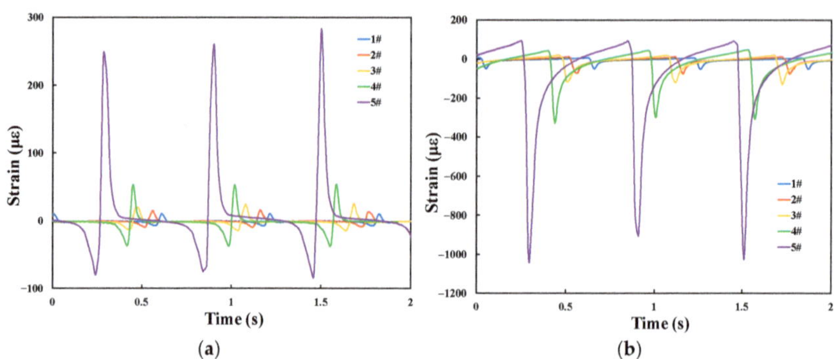

Figure 8. Strain–time-history curves at the bottom of the middle asphalt layer under different temperature fields: (**a**) longitudinal strain; (**b**) vertical strain.

Figure 9a further presents the calculation results of ε_T. This figure illustrates that the ε_T of vertical compression strain was the largest, indicating that it is most sensitive to temperature changes in different regions. Under the high-temperature condition, the vertical compression strain increased by 59.4 με for every 1 °C temperature increase, being 8.1 times that under the low-temperature condition. The second was the longitudinal tension-strain: it increased by 4.0 με and 18.7 με for each 1 °C increase in temperature under the middle- and high-temperature conditions, respectively. In addition, the longitudinal compression-strain and vertical tension-strain were little affected by temperature, and the ε_T was in the range of 1 to 4 με/°C under three temperature conditions.

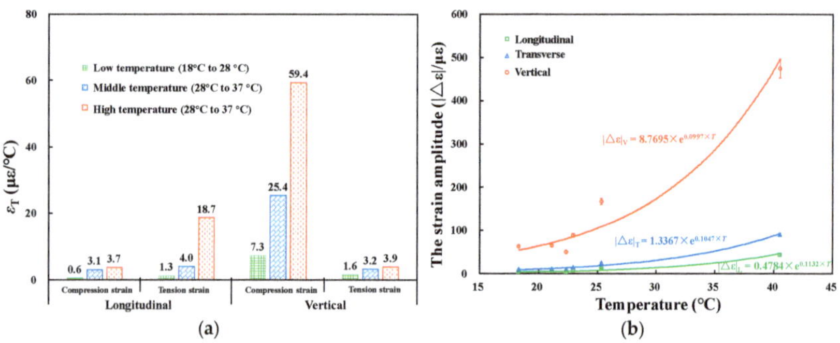

Figure 9. Index change laws with temperature conditions at the bottom of the middle asphalt layer: (**a**) three-way ε_T; (**b**) three-way strain amplitudes.

Figure 9b displays the curve of three-way strain amplitude changing with temperature. The three-way strain amplitudes all showed positive exponential growth with temperature. The vertical strain amplitudes were larger than the transverse and longitudinal strain amplitudes under different temperature fields. The results are consistent with previous studies [17].

3.2.2. Loading Weight

The other variables were a T of 37 °C (bottom of middle asphalt layer bottom) and a v of 22 km/h. Figure 10 demonstrates the measured strain curves under different W. W is a critical influencing factor affecting pavement dynamic responses, and strain increases linearly with the change in the loading weight. Specifically, when W was 50 kN, the longitudinal and vertical strain amplitudes were 32.7 με and 230.2 με, respectively. When W was 60 kN, the longitudinal strain and vertical strain amplitudes were 38.6 με and 255.9 με, respectively, which increased by 18% and 11.2% compared with those of 50 kN and was further increased by 48% and 39.8% when W was 75 kN. This study's relationship between loading weight and mechanical response agrees with earlier studies [53,54].

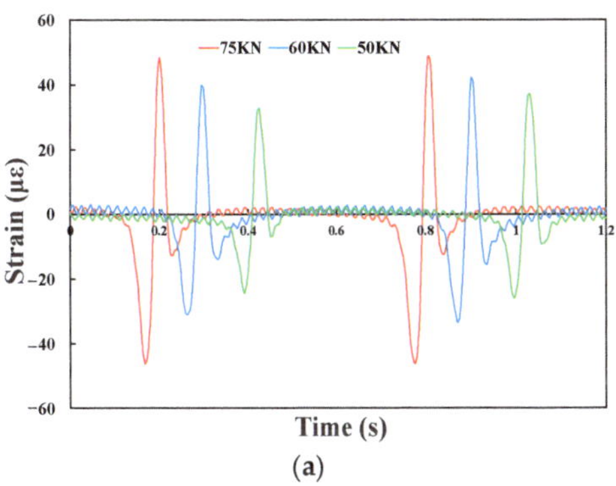
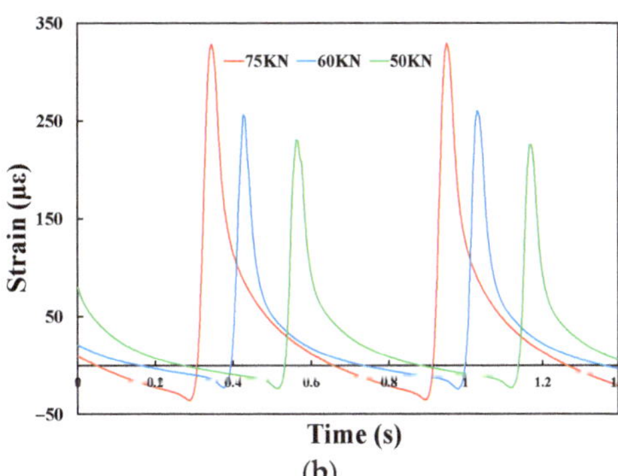

Figure 10. Strain curves at the bottom of the middle asphalt layer: (**a**) longitudinal and (**b**) vertical strain.

3.2.3. Loading Speed

The other two variables were a T of 37 °C (bottom of middle asphalt layer bottom) and a W of 50 kN. Three-way strain curves under different v are presented in Figure 11.

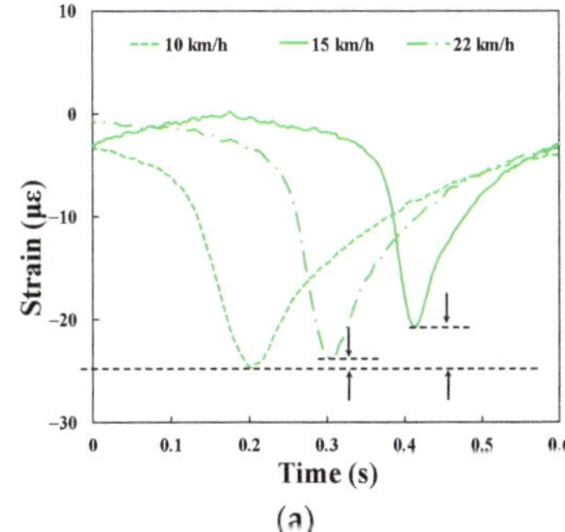
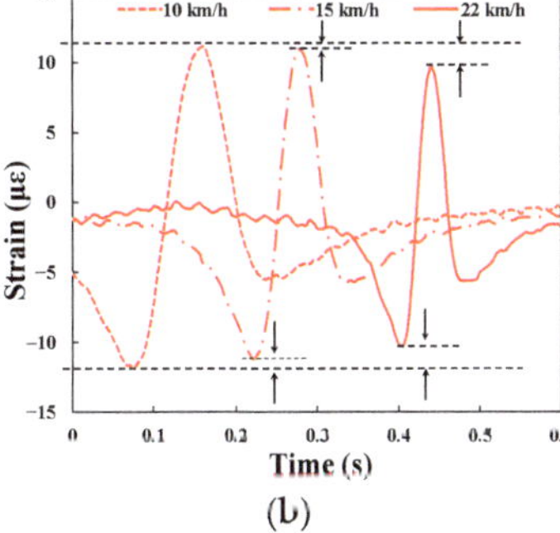
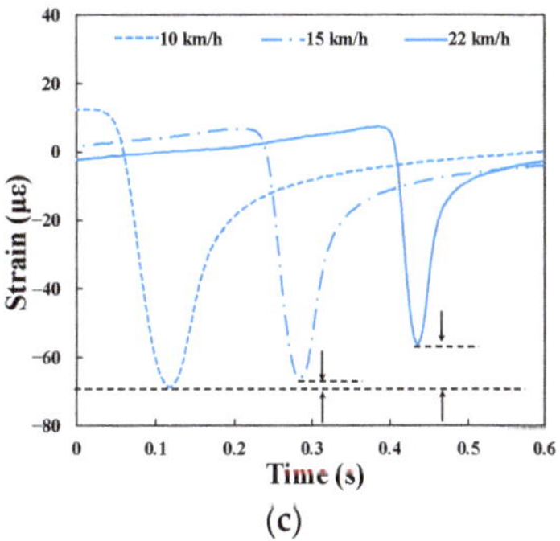

Figure 11. Strain–time-history curves at the bottom of the middle asphalt layer: (**a**) transverse, (**b**) longitudinal and (**c**) vertical strain.

It can be seen that the three-way strain amplitudes were all at the maximum level when v = 10 km/h. Then, the amplitude decreased as v increased. With the decrease in loading speed, the frequency of loading rotation for asphalt mixture increases correspondingly, and the interaction time between the wheels and the road surface lengthens. As a result, both the action time and the amplitude for the time-course curve of strain in the pavement increased, which agrees with the results published in Refs. [55,56]. Therefore, the lower the speed in the pavement service process, the more extensive the three-way strain amplitude and high-strain area over time, resulting in fatigue cracking prematurely at the asphalt layer bottom.

3.3. Dynamic Response Prediction

Following the above analysis from Sections 3.2.1–3.2.3, it was found that the measured strain amplitudes were positively correlated with temperatures and loading weights but negatively correlated with loading speeds. Combining Equation (6) with the measured data of the tested road, the prediction equations of three-way strain amplitude were developed under multiple factors, as shown in Equations (19) and (20), respectively.

$$|\Delta_\varepsilon| = \begin{cases} Longitudinal: 10.180e^{-0.039v+0.088T+0.001W}, R^2 = 0.99 \\ Transverse: 2.594e^{-0.025v+0.105T+0.001W}, R^2 = 0.98 \\ Vertical: 47.927e^{-0.024v+0.075T+0.001W}, R^2 = 0.99 \end{cases}, \quad (19)$$

$$|\Delta_\varepsilon| = \begin{cases} Longitudinal: 3.361e^{-0.020v+0.082T+0.003W}, R^2 = 0.97 \\ Transverse: 1.076e^{-0.032v+0.094T+0.009W}, R^2 = 0.97 \\ Vertical: 28.934e^{-0.028v+0.076T+0.003W}, R^2 = 0.96 \end{cases} \quad (20)$$

Figure 12 further describes the relationship between the monitoring and predicted data of three-way strain amplitude. It exhibits a good correlation, with all correlation coefficients being above 0.96, which is better than the values obtained by most previous studies [14,57].

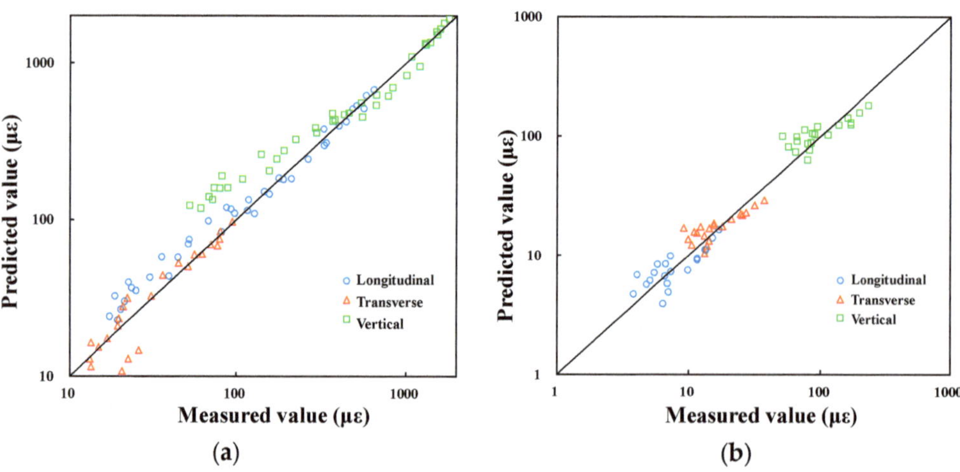

Figure 12. The measured and the predicted strain amplitudes: (**a**) the bottom of the middle layer; (**b**) the bottom of the lower layer.

3.4. Finite Element Simulation

In this section, the FE simulation results and their applications are discussed, including the following points.

3.4.1. Results of Material Parameter Inversion

Table 3 summarizes the calculated ε_M of the tested road according to Equations (19) and (20) under different working conditions, including the three-way strain at the bottom of the two measured layers.

Table 3. Calculated results of ε_M (µε) under different working conditions.

Structure Layer	Direction	Loading Speed (km/h)	Temperature (°C)			
			20	30	40	50
Bottom of the middle layer at asphalt surface	Vertical	10	177.8	376.3	796.6	1686.2
		15	157.7	333.8	706.5	1495.6
		22	133.3	282.1	597.3	1264.3
	Transverse	10	17.3	49.5	141.6	404.5
		15	15.3	43.7	124.9	356.9
		22	12.8	36.7	104.9	299.7
	Longitudinal	10	42.1	101.5	244.7	589.9
		15	34.7	83.5	201.4	485.4
		22	26.4	63.6	153.2	369.5
Bottom of the lower layer at asphalt surface	Vertical	10	116.1	248.3	530.9	1135.3
		15	100.9	215.9	461.6	986.9
		22	83.0	177.5	379.5	811.3
	Transverse	10	8.0	20.6	52.6	134.7
		15	6.8	17.5	44.8	114.7
		22	5.5	13.9	35.8	91.7
	Longitudinal	10	17.8	40.4	91.8	208.3
		15	16.1	36.6	83	188.5
		22	14	31.8	72.2	163.9

Then, the modulus inversion was performed based on Figure 9. The calculated error (S) at the final iteration round and the modulus is summarized in Table 4.

Table 4. Modulus optimization results of ε_M (µε) under different working conditions.

T (°C)	v (km/h)		Inversion Results and Error							
20	10	E	7220	7240	7260	**7280** [1]	7300	7320	7340	7360
		S	6.3	5.5	5.1	**4.6**	5.2	5.9	6.7	7.9
	15	E	7640	7660	7680	7700	**7720**	7740	7760	7780
		S	7.9	6.7	5.7	5.1	**4.8**	5.2	5.7	6.6
	22	E	8080	8100	8120	**8140**	8160	8180	8200	8220
		S	7.5	6.1	5.3	**4.9**	5.3	5.8	6.7	8.1
30	10	E	2630	2640	**2650**	2660	2670	2680	2690	2700
		S	5.7	5.2	**4.8**	5.1	5.8	6.7	8.1	9.5
	15	E	2820	2830	2840	2850	2860	**2870**	2880	2890
		S	8.9	7.6	6.5	5.7	5.1	**4.8**	5.2	5.9
	22	E	3310	3320	**3330**	3340	3350	3360	3370	3380
		S	5.8	5.0	**4.6**	4.9	5.5	6.4	7.6	8.9

Table 4. *Cont.*

T (°C)	v (km/h)		Inversion Results and Error							
40	10	E	940	945	950	**955**	960	965	970	975
		S	6.3	5.5	5.0	**4.7**	4.9	5.3	5.9	6.7
	15	E	1735	1740	**1745**	1750	1755	1760	1765	1770
		S	5.7	5.2	**4.8**	5.1	5.5	6.1	6.8	7.7
	22	E	2200	2205	2210	**2215**	2220	2225	2230	2235
		S	7.2	6.1	5.2	**4.9**	5.1	5.5	6.1	6.9
50	10	E	216	217	**218**	219	220	221	222	223
		S	5.2	5.0	**4.8**	4.9	5.1	5.4	5.7	6.1
	15	E	247	248	249	250	**251**	252	253	254
		S	5.1	5.1	5.0	4.9	**4.8**	4.9	5.0	5.1
	22	E	264	265	**266**	267	268	269	270	271
		S	5.2	4.9	**4.7**	4.8	4.9	5.1	5.3	5.6

[1] The bolded value is the optimal solution.

The corresponding modulus of the minimum S ($\leq$5%) was selected to obtain the optimal modulus of the asphalt layer at various working conditions (red font in Table 3). Therefore, its back-calculated modulus combining measured data and FE simulation was realized, which provided reliable parameter input for the subsequent simulation analysis of the key mechanical indexes.

3.4.2. Prediction Results of Fatigue Failure and Critical Conditions

Based on the analysis results from Sections 3.2 and 3.3, the mechanical indexes of the structural layers in the tested road increased nonlinearly with increasing T and W, and decreasing v. Combined with loading test conditions, it can be determined that the most unfavorable combination was 75 kN and 10 km/h at 5# temperature field. The most favorable combination was 50 kN and 22 km/h at 1# temperature field. According to Equation (20), the calculated tension and compression strains at the bottom of the asphalt layer were as follows: the maximum tension strain under the most unfavorable and favorable working conditions was 164.39 $\mu\varepsilon$ and 5.44 $\mu\varepsilon$, respectively. The maximum compression strain under the most unfavorable and favorable working conditions was 1210.36 $\mu\varepsilon$ and 82.73 $\mu\varepsilon$, respectively.

Combined with these results and Equation (16), it can be calculated that the N_{f1} of the asphalt layer is between 2.24×10^7 and 1.69×10^{13} equivalent-axle times. In the light of a previous study [41], the maximum shear strain of a typical structure occurred at the top of the middle surface layer (6 cm below the pavement), and the change rule of shear strain with time under the most unfavorable and favorable conditions was further simulated. Figure 13 suggests that the maximum shear strain of the asphalt layer of a semi-rigid pavement ranged from 5.53 $\mu\varepsilon$ to 133.48 $\mu\varepsilon$.

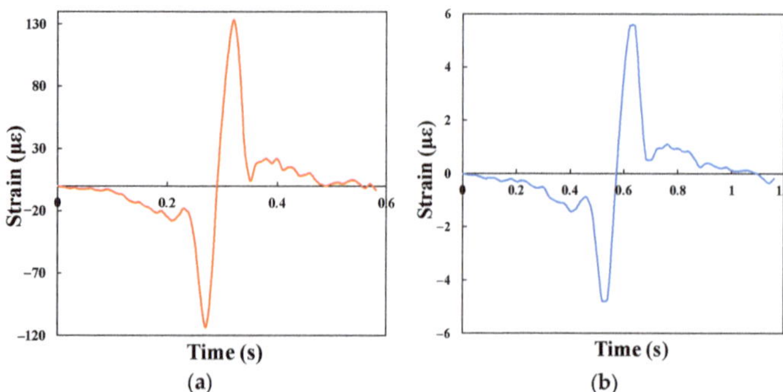

Figure 13. Maximum shear strain of asphalt layer under (**a**) the most adverse and (**b**) most favorable conditions.

The FE calculation results in Figure 14 demonstrate that the tension stress at the bottom of the base layer was between 0.101 MPa and 0.237 MPa under different working conditions. Combined with Equation (17), it could be calculated that the N_{f2} of the base layer is between 5.33×10^8 and 4.60×10^{10} equivalent-axle times.

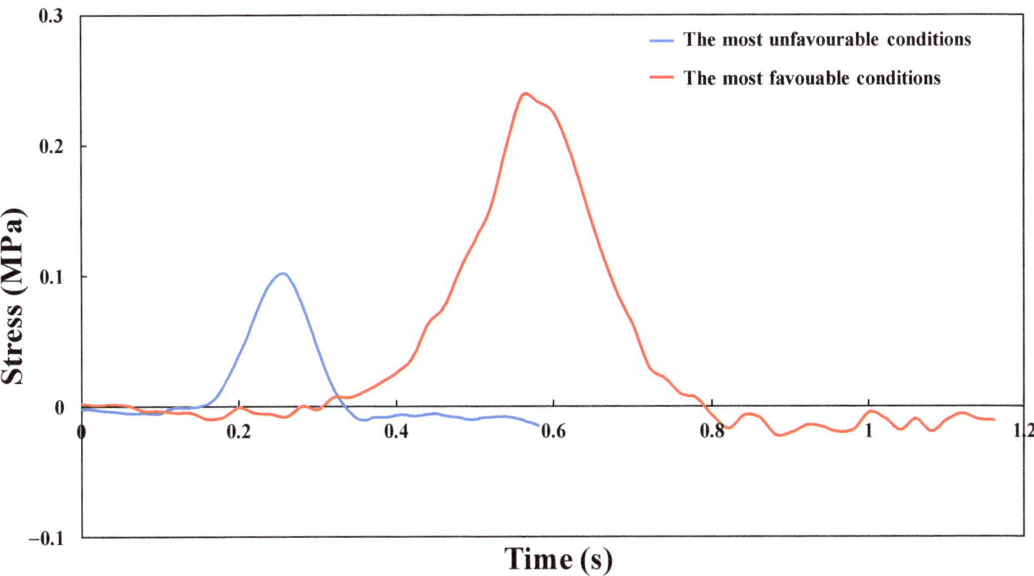

Figure 14. Maximum tension stress at the bottom of base layer.

3.5. Relationship between Pavement Structural Conditions and Remaining Life

After GPR detection of several highways in Jiangsu Province, Figure 15 illustrates the relationship between the IPCI and RLR of the asphalt surface and CSM base layers under different traffic levels. The following points summarize our findings:

- Linear fitting was performed for the data with IPCI less than 10, IPCI greater than 80 and the intermediate segment, respectively. IPCI and RLR showed a roughly negative relationship: when the IPCI was less than 10, the slope k of the line was small (about 0.2). It was significant larger when the IPCI was greater than 10 (between 1.1 to 1.3); however, it became small again after the IPCI was greater than 80.
- With the increase in traffic grade, the IPCI of different structural layers decreased to different degrees, and the IPCI of the base layer decreased more obviously due to load accumulation. On the other hand, the fitting relationship between IPCI and RLR was slightly weakened, which may be because the thickness of pavement structural layers changes under repeated load, thus affecting the calculation result of IPCI.
- Under the same IPCI value, the RLR of the base layer was lower than that of the asphalt surface layer, and this difference was more evident with the increase in traffic grade. This may be due to the increase in the distress ratio, as the performance of the CSM base material decreases significantly, and the modulus attenuation is greater.

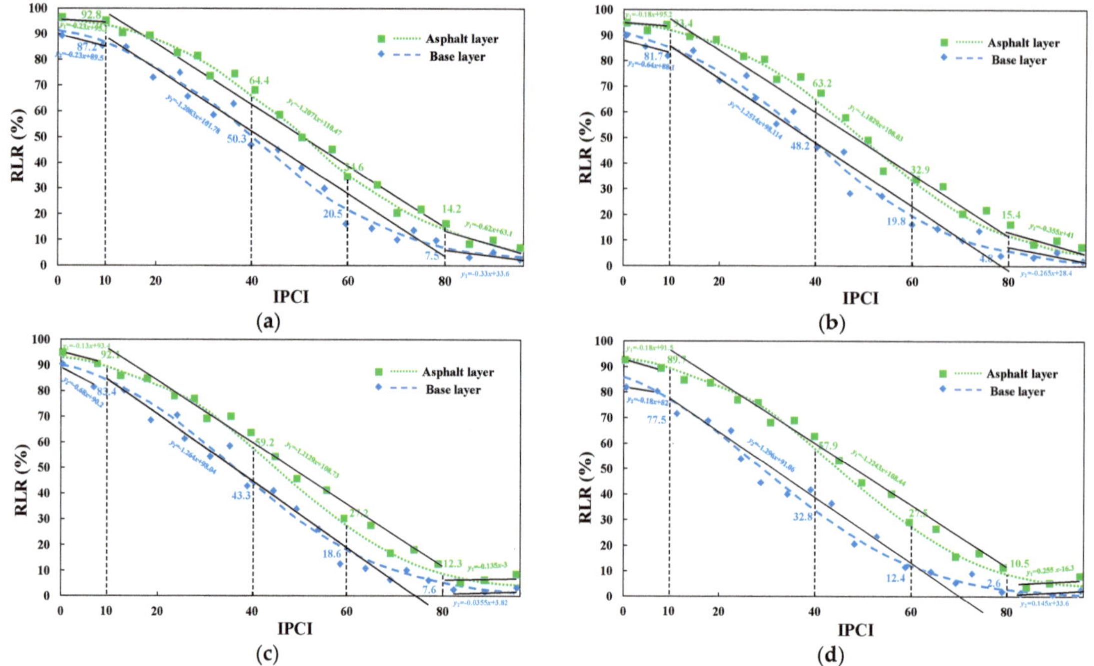

Figure 15. Relationship between the RLR and IPCI of pavement structures: (**a**) light, (**b**) moderate, (**c**) slightly heavy and (**d**) heavy traffic.

4. Conclusions

This study assesses structural conditions and remaining life by combining APT and GPR for semi-rigid base pavement. The main conclusions are as follows:

(1) Temperatures were predicted using a dual sinusoidal model for pavement structures based on the measured atmospheric temperature and structural temperatures. The good linear correlation (coefficient > 0.95) indicates that this model is reliable.

(2) The asphalt surface layer showed a three-way strain increase with increasing temperature and load weight, but a decrease with increasing loading speed. It was excellently correlated with the measured values to predict dynamic responses under multivariate factors.

(3) The material parameter inversion of the asphalt surface layer was proposed by controlling the average error of six strains between the FE-simulated and APT-measured values. Based on the established FE model, key mechanical index values can be analyzed under different conditions, along with the fatigue life of pavement structural layers.

(4) There is a good negative correlation between the IPCI and the RLR of the pavement structure. Therefore, the RLR of the pavement structure can be predicted via GPR detection and quantitative assessment of structure conditions.

This work solves the problems of assessing the condition of pavement structures quantitatively that could guide asphalt pavement maintenance with a semi-rigid base. It is worth noting that the asphalt surface layer is only studied as a whole because the material parameters are obtained via indoor testing of the specimens with on-site coring in the tested road. In addition, fluctuating loads are also an important factor. Future research could establish a finite element model of pavement structure considering load fluctuation and modify it with APT-measured data.

Author Contributions: The authors confirm contributions to the paper as follows: study conception and design: Z.L.; data collection: Z.L. and Q.Y.; analysis and interpretation of results: Z.L. and Q.Y.; draft manuscript preparation: Z.L., Q.Y. and X.G. All authors have read and agreed to the published version of the manuscript.

Funding: This research received no external funding.

Data Availability Statement: Restrictions apply to the availability of these data. Data can be obtained from the corresponding author upon request.

Conflicts of Interest: The authors declare no conflict of interest.

References

1. Wang, D.; Liu, Z.; Gu, X.; Wu, W.; Chen, Y.; Wang, L. Automatic Detection of Pothole Distress in Asphalt Pavement Using Improved Convolutional Neural Networks. *Remote Sens.* **2022**, *14*, 3892. [CrossRef]
2. Ren, D.Y.; Song, H.; Huang, C.; Xiao, C.; Ai, C.F. Comparative Evaluation of Asphalt Pavement Dynamic Response with Different Bases under Moving Vehicular Loading. *J. Test. Eval.* **2020**, *48*, 1823–1836. [CrossRef]
3. Liu, Z.; Gu, X.Y.; Ren, H.; Wang, X.; Dong, Q. Three-dimensional finite element analysis for structural parameters of asphalt pavement: A combined laboratory and field accelerated testing approach. *Case Stud. Constr. Mat.* **2022**, *17*, e01221. [CrossRef]
4. Liu, Z.; Gu, X.; Ren, H. Rutting prediction of asphalt pavement with semi-rigid base: Numerical modeling on laboratory to accelerated pavement testing. *Constr. Build. Mater.* **2023**, *375*, 130903. [CrossRef]
5. Ingrassia, L.P.; Virgili, A.; Canestrari, F. Effect of geocomposite reinforcement on the performance of thin asphalt pavements: Accelerated pavement testing and laboratory analysis. *Case Stud. Constr. Mat.* **2020**, *12*, e00342. [CrossRef]
6. Al-Qadi, I.L.; Loulizi, A.; Elseifi, M.; Lahouar, S. The Virginia Smart Road: The impact of pavement instrumentation on understanding pavement performance. *J. Assoc. Asph. Pavement* **2004**, *73*, 427–465.
7. Dessouky, S.H.; Alqadi, I.L.; Yoo, P.J. Full-Depth Flexible Pavement Response to Different Truck Tire Loadings. In Proceedings of the Transportation Research Board Meeting, Washington, DC, USA, 21–25 January 2007.
8. Bhattacharjee, S.; Mallick, R.B. Effect of temperature on fatigue performance of hot mix asphalt tested under model mobile load simulator. *Int. J. Pavement Eng.* **2012**, *13*, 166–180. [CrossRef]
9. Han, Z.; Sha, A.; Hu, L.; Jiao, L. Modeling to simulate inverted asphalt pavement testing: An emphasis on cracks in the semirigid subbase. *Constr. Build. Mater.* **2021**, *306*, 124790. [CrossRef]
10. Zhu, X.Y.; Zhang, Q.F.; Chen, L.; Du, Z. Mechanical response of hydronic asphalt pavement under temperature-vehicle coupled load: A finite element simulation and accelerated pavement testing study. *Constr. Build. Mater.* **2021**, *272*, 121884. [CrossRef]
11. Greene, J.; Choubane, B.; Upshaw, P. Evaluation of a Heavy Polymer Modified Asphalt Binder Using Accelerated Pavement Testing. In Proceedings of the 4th International Conference on Accelerated Pavement Testing, Davis, CA, USA, 19–21 September 2012.
12. Blab, R.; Kluger-Eigl, W.; Füssl, J.; Arraigada, M.; Hofko, B. Accelerated Pavement Testing on Slab and Block Pavements using the New Mobile Load Simulator MLS10. In Proceedings of the 4th International Conference on Accelerated Pavement Testing, Davis, CA, USA, 19–21 September 2012.
13. Jansen, D.; Wacker, B.; Pinkofsky, L. Full-scale accelerated pavement testing with the MLS30 on innovative testing infrastructures. *Int. J. Pavement Eng.* **2018**, *19*, 456–465. [CrossRef]
14. Zhang, H.; Ren, J.; Ji, L.; Wang, L. Mechanical response measurement and simulation of full scale asphalt pavement. *J. Harbin Inst. Technol.* **2016**, *48*, 41–48.
15. Zaumanis, M.; Haritonovs, V. Long term monitoring of full scale pavement test section with eight different asphalt wearing courses. *Mater. Struct.* **2016**, *49*, 1817–1828. [CrossRef]
16. Liu, Z.; Gu, X. Performance Evaluation of Full-scale Accelerated Pavement using NDT and Laboratory Tests: A case study in Jiangsu, China. *Case Stud. Constr. Mat.* **2023**, *18*, e02083. [CrossRef]
17. Ling, J.M.; Wei, F.L.; Zhao, H.D.; Tian, Y.; Han, B.Y.; Chen, Z.A. Analysis of airfield composite pavement responses using full-scale accelerated pavement testing and finite element method. *Constr. Build. Mater.* **2019**, *212*, 596–606. [CrossRef]
18. Yang, H.L.; Wang, S.J.; Miao, Y.H.; Wang, L.B.; Sun, F.Y. Effects of accelerated loading on the stress response and rutting of pavements. *J. Zhejiang Univ. Sci. A* **2021**, *22*, 514–527. [CrossRef]
19. Ai, C.F.; Rahman, A.; Xiao, C.; Yang, E.H.; Qiu, Y.J. Analysis of measured strain response of asphalt pavements and relevant prediction models. *Int. J. Pavement Eng.* **2017**, *18*, 1089–1097. [CrossRef]
20. Chun, S.; Kim, K.; Greene, J.; Choubane, B. Evaluation of interlayer bonding condition on structural response characteristics of asphalt pavement using finite element analysis and full-scale field tests. *Constr. Build. Mater.* **2015**, *96*, 307–318. [CrossRef]
21. Ma, X.; Dong, Z.; Chen, F.; Xiang, H.; Cao, C.; Sun, J. Airport asphalt pavement health monitoring system for mechanical model updating and distress evaluation under realistic random aircraft loads. *Constr. Build. Mater.* **2019**, *226*, 227–237. [CrossRef]
22. Li, N.; Zhan, H.; Yu, X.; Tang, W.; Yu, H.; Dong, F. Research on the high temperature performance of asphalt pavement based on field cores with different rutting development levels. *Mater. Struct.* **2021**, *54*, 70. [CrossRef]
23. Gouveia, B.C.S.; Preti, F.; Cattani, L.; Bozzoli, F.; Roberto, A.; Romeo, E.; Tebaldi, G. Numerical and Experimental Analysis of the Raise-Temperature Effect of Quicklime in Cold Recycled Mixtures. *J. Mater. Civil. Eng.* **2022**, *34*, 04022283. [CrossRef]

24. Liu, Z.; Gu, X.; Dong, X.; Cui, B.; Hu, D. Mechanism and Performance of Graphene Modified Asphalt: An Experimental Approach Combined with Molecular Dynamic Simulations. *Case Stud. Constr. Mat.* **2022**, *18*, e01749. [CrossRef]

25. Liu, Z.; Sun, L.; Gu, X.; Wang, X.; Dong, Q.; Zhou, Z.; Tang, J. Characteristics, mechanisms, and environmental LCA of WMA containing sasobit: An analysis perspective combing viscosity-temperature regression and interface bonding strength. *J. Clean. Prod.* **2023**, *391*, 136255. [CrossRef]

26. Yang, J.; Fang, Y. Rationality and Applicability of High-Temperature Performance Indexes for Polymer-Modified Asphalts. *J. Mater. Civ. Eng.* **2021**, *33*, 04021238. [CrossRef]

27. Barber, E.S. Calculation of Maximum Pavement Temperatures from Weather Reports. *Highw. Res. Board Bull.* **1957**, *168*. Available online: http://onlinepubs.trb.org/Onlinepubs/hrbbulletin/168/168-001.pdf (accessed on 29 June 2023).

28. Anupam, K.; Srirangam, S.K.; Scarpas, A.; Kasbergen, C. Influence of Temperature on Tire-Pavement Friction: Analyses. *Transp. Res. Rec.* **2013**, *2369*, 114–124. [CrossRef]

29. Si, W.; Ma, B.; Ren, J.P.; Hu, Y.P.; Zhou, X.Y.; Tian, Y.X.; Li, Y. Temperature responses of asphalt pavement structure constructed with phase change material by applying finite element method. *Constr. Build. Mater.* **2020**, *244*, 118088. [CrossRef]

30. Zhao, X.Y.; Shen, A.Q.; Ma, B.F. Temperature response of asphalt pavement to low temperatures and large temperature differences. *Int. J. Pavement Eng.* **2020**, *21*, 49–62. [CrossRef]

31. Luo, X.; Gu, F.; Lytton, R.L. Kinetics-based aging prediction of asphalt mixtures using field deflection data. *Int. J. Pavement Eng.* **2019**, *20*, 287–297. [CrossRef]

32. Liu, Z.; Gu, X.; Yang, H.; Wang, L.; Chen, Y.; Wang, D. Novel YOLOv3 Model With Structure and Hyperparameter Optimization for Detection of Pavement Concealed Cracks in GPR Images. *IEEE Trans. Intell. Transp.* **2022**, *23*, 22258–22268. [CrossRef]

33. Liu, Z.; Gu, X.; Chen, J.; Wang, D.; Chen, Y.; Wang, L. Automatic recognition of pavement cracks from combined GPR B-scan and C-scan images using multiscale feature fusion deep neural networks. *Autom. Constr.* **2023**, *146*, 104698. [CrossRef]

34. Liu, Z.; Yeoh, J.K.W.; Gu, X.; Dong, Q.; Chen, Y.; Wu, W.; Wang, L.; Wang, D. Automatic pixel-level detection of vertical cracks in asphalt pavement based on GPR investigation and improved mask R-CNN. *Autom. Constr.* **2023**, *146*, 104689. [CrossRef]

35. Liu, Z.; Gu, X.Y.; Wu, W.X.; Zou, X.Y.; Dong, Q.; Wang, L.T. GPR-based detection of internal cracks in asphalt pavement: A combination method of DeepAugment data and object detection. *Measurement* **2022**, *197*, 111281. [CrossRef]

36. Long, J.; Luo, Q.; Liu, Z.; Zhu, Z. Road distress detection and maintenance evaluation based on ground penetrating radar. In *Advances in Civil Function Structure and Industrial Architecture*; CRC Press: Boca Raton, FL, USA, 2022; pp. 474–481.

37. Wang, L.T.; Gu, X.Y.; Liu, Z.; Wu, W.X.; Wang, D.Y. Automatic detection of asphalt pavement thickness: A method combining GPR images and improved Canny algorithm. *Measurement* **2022**, *196*, 111248. [CrossRef]

38. Dai, Q.; Lee, Y.H.; Sun, H.H.; Qian, J.; Ow, G.; Yusof, M.L.M.; Yucel, A.C. A Deep Learning-Based GPR Forward Solver for Predicting B-Scans of Subsurface Objects. *IEEE Geosci. Remote Sens.* **2022**, *19*, 4025805. [CrossRef]

39. Rasol, M.; Pais, J.C.; Pérez-Gracia, V.; Solla, M.; Fernandes, F.M.; Fontul, S.; Ayala-Cabrera, D.; Schmidt, F.; Assadollahi, H. GPR monitoring for road transport infrastructure: A systematic review and machine learning insights. *Constr. Build. Mater.* **2022**, *324*, 126686. [CrossRef]

40. *JTG E20-2011*; Standard Test Methods of Bitumen and Bituminous Mixtures for Highway Engineering. China Communications Press: Beijing, China, 2011; Volume JTG E20-2011.

41. Liu, Z.; Zhou, Z.; Gu, X.; Sun, L.; Wang, C. Laboratory evaluation of the performance of reclaimed asphalt mixed with composite crumb rubber-modified asphalt: Reconciling relatively high content of RAP and virgin asphalt. *Int. J. Pavement Eng.* **2023**, *24*, 2217320. [CrossRef]

42. Liu, Z.; Gu, X.Y.; Ren, H.; Zhou, Z.; Wang, X.; Tang, S. Analysis of the dynamic responses of asphalt pavement based on full-scale accelerated testing and finite element simulation. *Constr. Build. Mater.* **2022**, *325*, 126429. [CrossRef]

43. Liu, Z.; Gu, X.; Ren, H.; Li, S.; Dong, Q. Permanent Deformation Monitoring and Remaining Life Prediction of Asphalt Pavement Combining Full-Scale Accelerated Pavement Testing and FEM. *Struct. Control Health Monit.* **2023**, *2023*, 6932621. [CrossRef]

44. Liu, Z.; Gu, X.Y.; Wu, C.Y.; Ren, H.; Zhou, Z.; Tang, S. Studies on the validity of strain sensors for pavement monitoring: A case study for a fiber Bragg grating sensor and resistive sensor. *Constr. Build. Mater.* **2022**, *321*, 126085. [CrossRef]

45. Ren, H.; Gu, X.Y.; Liu, Z. Analysis of Mechanical Responses for Semi-Rigid Base Asphalt Pavement Based on MLS66 Accelerated Loading Test. In Proceedings of the 20th and 21st Joint COTA International Conference of Transportation Professionals—Advanced Transportation, Enhanced Connection, Xi'an, China, 16–20 December 2021; pp. 732–742.

46. Loría-Salazar, L.G.; Aguiar-Moya, J.P.; Vargas-Nordcbeck, A.; Leiva-Villacorta, F. *The Roles of Accelerated Pavement Testing in Pavement Sustainability*; Springer: Berlin/Heidelberg, Germany, 2016.

47. Liu, Z.; Gu, X.; Ren, H. Dynamic response analysis of asphalt pavement with semi-rigid base based on MLS66 accelerated loading test. *J. Southeast Univ.* **2023**, *53*, 114–122. [CrossRef]

48. Liang, X.; Yu, X.; Chen, C.; Jin, Y.; Huang, J. Automatic Classification of Pavement Distress Using 3D Ground-Penetrating Radar and Deep Convolutional Neural Network. *IEEE Trans. Intell. Transp.* **2022**, *23*, 22269–22277. [CrossRef]

49. Hong, X.; Tan, W.; Xiong, C.; Qiu, Z.; Yu, J.; Wang, D.; Wei, X.; Li, W.; Wang, Z. A Fast and Non-Destructive Prediction Model for Remaining Life of Rigid Pavement with or without Asphalt Overlay. *Buildings* **2022**, *12*, 868. [CrossRef]

50. Lin, X.; Li, Q. Study on esidual Life Evaluation Method of Semi-rigid Base Structure of Expressway Asphalt Pavement. *J. Highw. Transp. Res. Dev.* **2021**, *38*, 1–8. [CrossRef]

51. Ge, N.; Li, H.; Yang, B.; Fu, K.; Yu, B.; Zhu, Y. Mechanical responses analysis and modulus inverse calculation of permeable asphalt pavement under dynamic load. *Int. J. Transp. Sci. Technol.* **2021**, *11*, 243–254. [CrossRef]
52. *JTG F50-2011*; Technical Specification for Construction of Highway Bridge and Culverts. China Communications Press: Beijing, China, 2011; Volume JTG F50-2011.
53. Guan, Z.; Zhuang, C.; Lin, M. Accelerated loading dynamic response of full-scale asphalt concrete pavement. *J. Traffic Transp. Eng.* **2012**, *12*, 24–31.
54. Dong, Z.; Xu, Q.; Lu, P. Dynamic Response of Semi-rigid Base Asphalt Pavement Based on Accelerated Pavement Test. *China J. Highw. Transp.* **2011**, *24*, 1.
55. Wang, H.; Zhao, J.; Hu, X.; Zhang, X. Flexible Pavement Response Analysis under Dynamic Loading at Different Vehicle Speeds and Pavement Surface Roughness Conditions. *J. Transp. Eng. B Pavements* **2020**, *146*, 04020040. [CrossRef]
56. Li, J.; Li, Y.; Xin, C.; Zuo, H.; An, P.; Zuo, S.; Liu, P. Dynamic Strain Response of Hot-Recycled Asphalt Pavement under Dual-Axle Accelerated Loading Conditions. *Coatings* **2022**, *12*, 843. [CrossRef]
57. Hu, X.-D.; Walubita, L.F. Modeling mechanistic responses in asphalt pavements under three-dimensional tire-pavement contact pressure. *J. Cent. South Univ. Technol.* **2011**, *18*, 250–258. [CrossRef]

Communication

LDANet: A Lightweight Dynamic Addition Network for Rural Road Extraction from Remote Sensing Images

Bohua Liu [1], Jianli Ding [1,*], Jie Zou [1], Jinjie Wang [1] and Shuai Huang [2]

[1] College of Geography and Remote Sensing Sciences, Xinjiang University, Urumqi 830046, China
[2] College of Geography and Environment, Liaocheng University, Liaocheng 252000, China
*   Correspondence: watarid@xju.edu.cn; Tel.: +86-13579265967

Abstract: Automatic road extraction from remote sensing images has an important impact on road maintenance and land management. While significant deep-learning-based approaches have been developed in recent years, achieving a suitable trade-off between extraction accuracy, inference speed and model size remains a fundamental and challenging issue for real-time road extraction applications, especially for rural roads. For this purpose, we developed a lightweight dynamic addition network (LDANet) to exploit rural road extraction. Specifically, considering the narrow, complex and diverse nature of rural roads, we introduce an improved Asymmetric Convolution Block (ACB)-based Inception structure to extend the low-level features in the feature extraction layer. In the deep feature association module, the depth-wise separable convolution (DSC) is introduced to reduce the computational complexity of the model, and an adaptation-weighted overlay is designed to capture the salient features. Moreover, we utilize a dynamic weighted combined loss, which can better solve the sample imbalance and boosts segmentation accuracy. In addition, we constructed a typical remote sensing dataset of rural roads based on the Deep Globe Land Cover Classification Challenge dataset. Our experiments demonstrate that LDANet performs well in road extraction with fewer model parameters (<1 MB) and that the accuracy and the mean Intersection over Union reach 98.74% and 76.21% on the test dataset, respectively. Therefore, LDANet has potential to rapidly extract and monitor rural roads from remote sensing images.

Keywords: remote sensing image; rural roads; lightweight neural networks; Inception; dynamic weight

Citation: Liu, B.; Ding, J.; Zou, J.; Wang, J.; Huang, S. LDANet: A Lightweight Dynamic Addition Network for Rural Road Extraction from Remote Sensing Images. *Remote Sens.* **2023**, *15*, 1829. https://doi.org/10.3390/rs15071829

Academic Editors: Alessandro Mei, Xianfeng Zhang and Valerio Baiocchi

Received: 22 February 2023
Revised: 22 March 2023
Accepted: 27 March 2023
Published: 29 March 2023

1. Introduction

As basic geographic information, rural roads are an important part of the transport system and play a significant role in urban planning, traffic navigation and digital map updating. With the advancement of science and technology, efficient and high-accuracy road extraction must be carried out in geographic mapping. Traditional methods for road area labeling are mostly manual and GPS-based methods, but the former is burdensome and the latter usually loses road details such as width, edges, etc. [1]

Remote sensing images have the advantages of being large-scale, fast-updating, easy to access and rich in information. With increasing resolution, the value and universality of remote sensing images have been greatly expanded. This introduces the possibility and potential of various Earth observation tasks by providing powerful data support [2]. Therefore, using remote sensing images for rapid and efficient road extraction has been a key research topic for many scholars in recent years. There are two main areas of study in remote-sensing-based road network extraction technology. The first is the shallow feature observation method, in which early scholars used the inherent geometric, textural and spectral features of images for road network extraction. However, these features are so simple that the method resulted in low accuracy [3–6], so some researchers combined it with multi-source data fusion, template matching and model orientation application methods to improve accuracy and efficiency. Zhao et al. [7] used the Extended Kalman Filter (EKF) and

Particle Filter (PF) models for road extraction, achieving acceptable performance in moderately and highly noisy backgrounds. Perciano et al. [8] used a two-layer Markov random field (MRF) to analyze multi-source fused data and to improve accuracy. Zang et al. [9] proposed a non-periodic directional structure measure (ADSM) method by introducing the representation of road-like features to enhance its effectiveness. Chinnathev et al. [10] referenced morphological features to develop an automatic road-centerline-extraction field-programmable gate array (FPGA) architecture to meet the demand for real-time road extraction. Although these studies have been effective, there are still problems with low stability and adaptability to complex information, gradient loss and overfitting of results.

The second method is the deep feature mining method, which uses multilayer networks to expand the non-linear mapping of image features and extract them. Hinton et al. [11] introduced a neural network that made the deep learning approach gain widespread attention. Compared with traditional machine learning, deep learning focuses on automatic feature learning from huge datasets with multilayer neuron organization. With the development of artificial intelligence, many scholars have been interested in the intelligent extraction of road networks. Zhong et al. [12] used full convolutional networks (FCNs) for road extraction from remote sensing images and obtained acceptable performance. Varia et al. [13] validated the effectiveness of deep learning methods by using conditional generative adversarial networks (GANs) and FCNs to extract roads from the data of unmanned aerial vehicles (UAVs). Doshi et al. [14] combined the ResNet network and the Inception network to propose a Residual Inception Skip Network and greatly improved the performance of road network extraction. Zhou et al. [15] constructed a D-LinkNet model by improving the extended convolutional layers in LinkNet, which expanded the perceptual field without reducing the resolution of the feature map. Li et al. [16] developed the hybrid convolutional network (HCN) by referring to FCN, Unet and VGG. Boonpook et al. [17] proposed a deep residual deconvolutional network with SegNet, which improves model extraction accuracy by enhancing feature relationships to overcome interference with complex scenes. Lu et al. [18] constructed a multi-scale and multi-task deep learning framework based on Unet, which concerns both road detection and centerline extraction operations, and it outperformed in deep-learning-based road extraction methods. However, it was found that while the higher complexity may achieve better performance, it may also result in greater requirements. Therefore, most of the deep mining models in road extraction are seeking to overcome the problems of training and application.

For meeting the requirements of practical applications, many lightweight networks have been proposed, with the primary intention of lightening the network in size and speed, while maintaining as much accuracy as possible. MobileNets [19–21] replaces the standard convolutions with depth-wise separable convolutions (DSC), which effectively reduces the number of parameters by decomposing the convolution operations. ShuffleNets [22,23] came up with a split–shuffle structure to speed up the model training process and enhance inter-channel correlation. ENet [24] was proposed to solve the problem of the large number of floating point operations in the network by compressing the channel, but with the resulting loss of spatial information, the accuracy does not perform well. In recent years, multiple networks have been proposed that employ optimized convolution, which can balance the network's number of parameters, inference speed and segmentation accuracy. LiteSeg [25] applies the Atrous Spatial Pyramid Pooling module to MobileNetV2 and has shown strong performance. ERFNet [26] uses residual connections and factorized convolutions to remain efficient while retaining remarkable accuracy. EDANet [27] proposes a bottleneck structure that combines asymmetric convolution with depth-wise convolution. ESPNet [28] and ESPNetv2 [29] introduce aurous convolution into their networks because they can obtain more features over larger areas using the same parameters. ELANet [30] uses an attention mechanism to strengthen different levels of information. MADNet [31] proposes a dense lightweight network by combining the residual multiscale module with an attention mechanism with the dual residual path block, which effectively reduces the model parameters and complexity. Despite these achievements, we believe that the balance between the accuracy and efficiency of these design strategies for the extraction of single vulnerable targets still need to be improved. Designing a reasonable network structure that achieves greater accuracy with few computational resources by making full use of remote sensing image features is important to facilitate the application of deep learning to road extraction.

To achieve a lightweight, efficient and high-accuracy real-world application of rural road extraction, we designed a lightweight dynamic addition network (LDANet) based on the encoder–decoder framework. Firstly, we constructed an improved ACB-based Inception structure to extend the low-level features in the feature extraction layer. Then, we developed a deep feature association module by introducing DSC and an adaptation-weighted overlay to reduce the computational complexity. Finally, we built a typical rural road dataset to evaluate the performance and utilize a dynamically weighted combined loss function to solve the sample imbalance. The main contributions of this article are as follows:

1. A lightweight rural road extraction model is proposed and shows significant performance on two datasets, enhancing the applicability of remote sensing techniques.
2. We extended shallow features using ACB-based Inception and designed a lightweight deep correlation module by referring to DSC and an adaptation-weighted overlay.
3. We designed a dynamic hybrid loss function to improve the accuracy of unbalanced samples.

2. Data

In this paper, we evaluate the performance of the LDANet on two datasets: (1) the typical rural road dataset and (2) the Massachusetts roads dataset [32].

2.1. The Typical Rural Roads Dataset

This dataset is constructed based on the DeepGlobe Land Cover Classification Challenge dataset [33], which is a publicly available dataset of high-resolution remote sensing images that can be obtained from the website http://Deepglobe.org/challenge.html (accessed on 10 March 2022). The dataset's spatial resolution is 50 cm/pixel and includes 1000 images, each 1024×1024 in size. The training set, validation set and test set were divided from the typical rural road dataset in a ratio of 75%, 15% and 10%, respectively, and the types of roads contained in the dataset include unpaved roads, paved roads and dirt roads. In the experiment, we used the data augmentation method which includes geometric transformations (scaling and rotation) and pixel transformations (noise addition and gamma transform) [34] to expand the training set and the validation set to a size of 7500 and 1500 images, respectively, with image sizes of 256×256. In the test set, 100 images of size 1024×1024 were used. Sample images are shown in Figure 1.

2.2. The Massachusetts Roads Dataset

This dataset was downloaded from https://www.cs.toronto.edu/~vmnih/data/ (accessed on 15 December 2021). The dataset's spatial resolution is 1.2 m/pixel and the image size is 1500×1500. The Massachusetts roads dataset covers both urban and suburban areas. In this paper, we split the images from the original dataset into multiple 256×256 images and expanded the training set and validation set to 6000 and 1200 images, respectively, for the experiment. In the test set, we used 100 images of size 1500×1500. Sample images are shown in Figure 2.

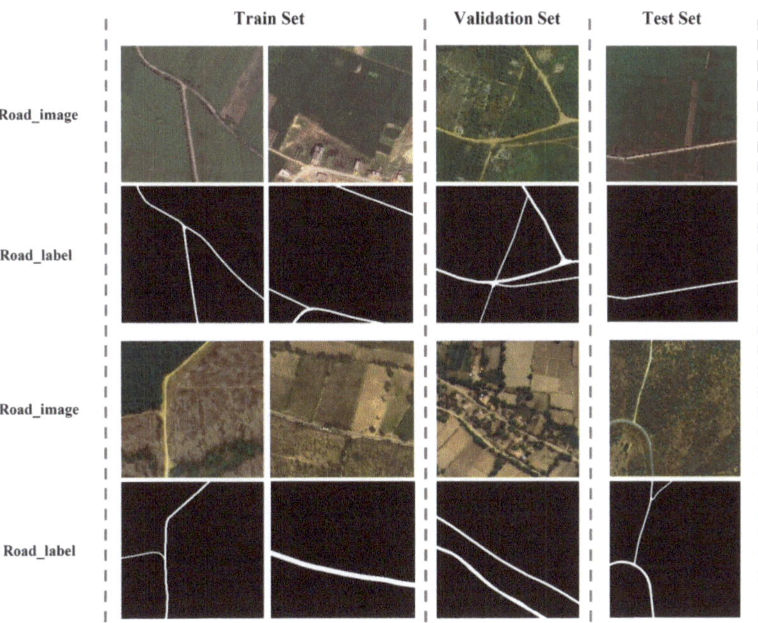

Figure 1. Sample images of the training set, validation set and test set of the rural road dataset.

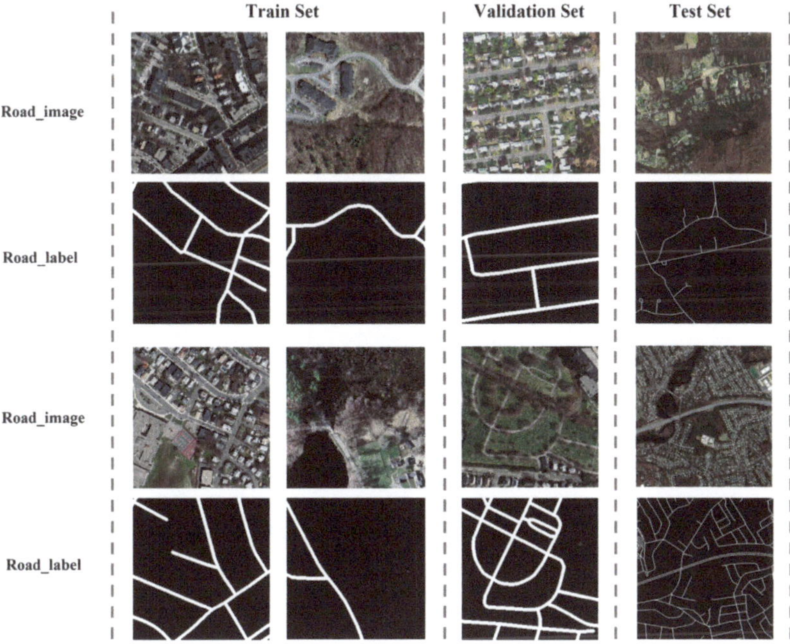

Figure 2. Massachusetts roads dataset training set, validation set and test set sample images.

3. Methodology

In this work, we aimed to construct a lightweight dynamic addition network (LDANet) for rural road extraction. LDANet consists of two modules: the feature expansion module based on the improved Inception framework and the deep feature association module based on the DSC and dynamic feature superposition. The model framework is shown in Figure 3.

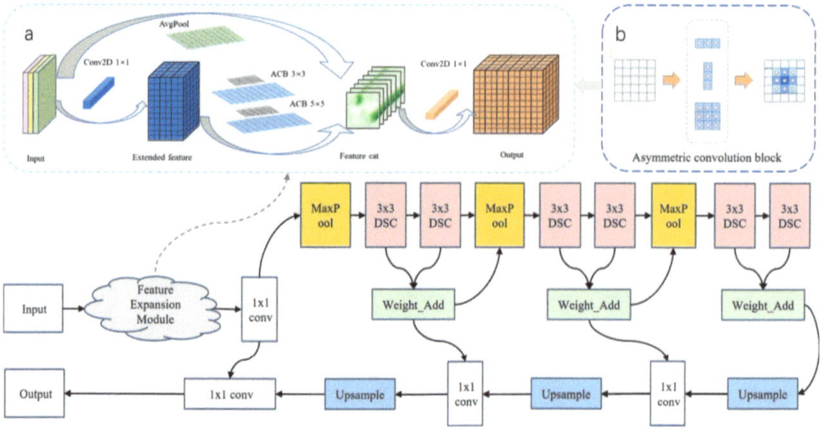

Figure 3. Structure of the lightweight dynamic addition network. (**a**) Feature expansion module. (**b**) Asymmetric convolution block (ACB), depth-wise separable convolution (DSC).

In the feature expansion module, the model obtains global and multi-scale features from the input image by averaging pooling and convolving at different scales, and uses 1×1 convolution for the fusion and dimensionality enhancement of feature layers. The deep feature association module is divided into two stages: the encoder and the decoder. In the encoder stage, six convolutional layers are constructed to extract the deep features of the expanded feature layers, where every two layers are pooled once to compress the image size, and the two feature layers are dynamically weighted and superimposed. In the decoder stage, the image size of the deep feature layers is expanded by upsampling, then the expanded layers are fused with the upper superimposed layers, and finally, a result map the same size as the input data is produced.

3.1. Feature Expansion Module

Based on previous research [35], we have found that road texture, color and geometric features have a positive effect on road extraction. Rural roads are narrow, complex and diverse, so enriching shallow information is essential and important. To fully reflect the performance of shallow features, we built a feature expansion module based on the Inception network module (Figure 3a). First, the input layer is up-dimensioned using a 1×1 convolution kernel, which not only reduces the convolution parameters but also integrates features across channels to ensure the validity of the features while simplifying the model. Then, we use the asymmetric convolution block (ACB) [36] to expand the feature layers. The ACB (Figure 3b) contains three branches—a 3×1 horizontal kernel, a 1×3 vertical kernel and a 3×3 global kernel—and each branch extracts the layer features separately and then fuses the results. Compared with double-layer convolution, ACB not only reduces the model parameters but also improves the saliency of node features while ensuring feature globalization. As roads have certain geospatial correlations, expanding the perceptual field can further enhance the road feature information, so we set the convolutional kernel size to 3×3 versus 5×5. Finally, the features are up-dimensioned and output by 1×1 convolution.

3.2. Deep Feature Association Module

As shown in Figure 4b, we use a multi-layer convolutional network for deep feature extraction and to enhance feature association with dynamic weighted superposition. This module is based on the Unet [37] network framework (Figure 4a). In the encoder stage, we expand the feature space dimension by 3×3 convolution to enhance the deep information and then compress the feature layer size by MaxPool to reduce the model's computational complexity and to improve the computational efficiency. In the decoder stage, we recover and output the image by jump connection and upsampling. In this module, we introduce an adaptation-weighted overlay layer. In the encoder stage, the layer has an overlap and dynamically adjusts the weights of the convolutional output feature layer, which increases the expressiveness of salient features and improves the efficiency of feature extraction. In the decoder stage, dynamically weighted overlay layers and 1×1 convolution realize the jump connection in the network. Compared with the complex jump connection in Unet++ [38] and Unet+++ [39], this method can not only strengthen the association of salient features but can also better preserve the structure of the feature space and achieve model simplification based on ensuring the accuracy of extraction. As shown in Figure 4b, this module takes fewer feature dimensions than Unet because, on the one hand, the road is a two-class model, and too rich a feature space will cause greater calculation pressure, and on the other hand, this module uses DSC instead of traditional convolution calculation to reduce model parameters. Because one convolution kernel is responsible for one channel in DSC, and one channel is only convolved by one convolution kernel, the number of features mapped is the same as the number of channels in the input layer [40]. So, we added an intermediate feature layer to the DSC, first by up-dimensioning the input features by 1×1 convolution, then using DSC to generate the intermediate feature layer, and finally using the 1×1 convolution output; the convolution process is shown in Figure 5.

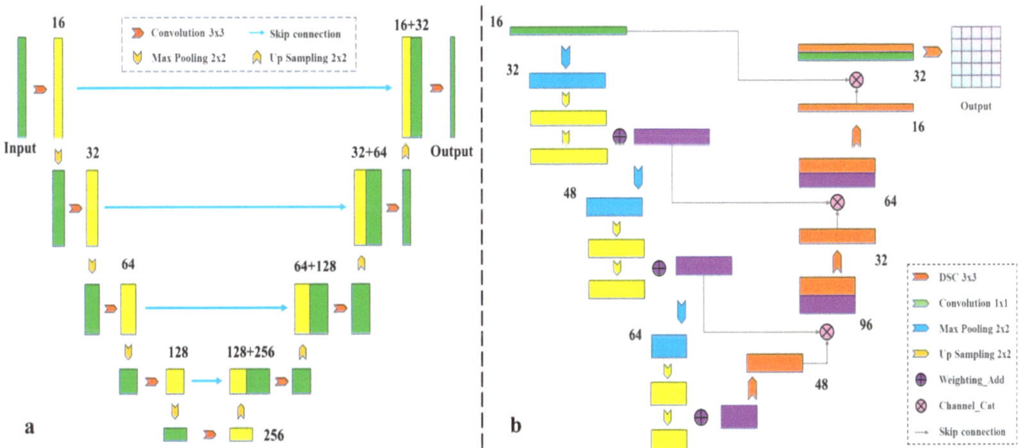

Figure 4. Module structure of (**a**) Unet, and (**b**) deep feature association module.

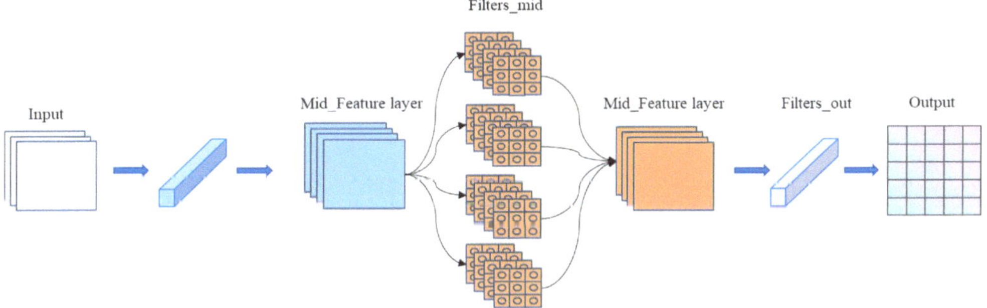

Figure 5. Depth-wise separable convolution (DSC) process.

3.3. Loss Function

Loss functions are often used to measure the extent to which a model's predictions differ from the actual data, and selecting the correct loss function can guide model training in the right direction.

Cross-Entropy Loss is a loss function that evaluates the difference between the probability distribution and the true distribution of the current training set to judge the training effect on the model. The model in this paper is dichotomous, so Binary Cross-Entropy Loss (BCE Loss) was chosen as the loss function.

$$\text{BCE Loss} = -\frac{1}{N} \sum_i [y_i \cdot \log(p_i) + (1 - y_i) \cdot \log(1 - p_i)] \tag{1}$$

where y_i is the sample value, the positive class is 1 and the negative class is 0. p_i is the predicted value, taking values within $(0, 1)$.

Roads behave narrowly in the dataset, covering a small area, and can suffer from severe sample imbalance during training. Dice Loss solves the problem of having too small a proportion of foregrounds by measuring the overlap of two samples, and it has acceptable performance in binary classification problems.

$$\text{Dice Loss} = 1 - \sum_1^N \frac{2y_i p_i + \varepsilon}{y_i + p_i + \varepsilon} \tag{2}$$

where y_i is the sample value, the positive class is 1 and the negative class is 0. p_i is the predicted value, taking values within $(0, 1)$.

Dice Loss has acceptable performance for scenarios with a severe imbalance between positive and negative samples, but the loss tends to be unstable when training small targets, which leads to drastic gradient changes. Therefore, this paper proposes a Combined Weighted Loss (CWL), which sets the weights according to the ratio of BCE Loss to Dice Loss, giving higher weights to functions with larger loss values and lower weights to functions with smaller loss values, to increase the proportion of high-value loss functions while using low-value loss functions to maintain the stability of the model, thereby accelerating the convergence of the model and improving its accuracy.

$$\text{CWL} = W_{CB} \cdot \text{BCE Loss} + W_{CD} \cdot \text{Dice Loss} \tag{3}$$

$$W_{CB} = \frac{|\text{BCE Loss}_n|}{|\text{BCE Loss}_n| + |\text{Dice Loss}_n|} \tag{4}$$

$$W_{CD} = 1 - W_{CB} \tag{5}$$

where W_{CB} is the proportion of BCE Loss, and W_{CD} is the proportion of Dice Loss.

4. Experimental Study

4.1. Model Evaluation Criteria

In this paper, four precision metrics, Precision, Recall, F1 Score and IoU (Intersection over Union), were chosen to evaluate the road extraction results, and their expressions are as follows.

Precision is the ratio of the area of the real road area in the resulting image to the area of the road in the labeled image.

$$\text{Precision} = \frac{TP}{TP + FP} \tag{6}$$

Recall is the proportion of correctly identified roads to the total roads in the tagged image.

$$\text{Recall} = \frac{TP}{TP + FN} \tag{7}$$

The F1 Score, a statistical measure of the accuracy of a dichotomous model, is often used to determine the overall performance of a dichotomous model.

$$\text{F1 Score} = 2 * \frac{\text{Precision} * \text{Recall}}{\text{Precision} + \text{Recall}} \tag{8}$$

IoU is the degree of overlap between the pre-target measurement results and the target label.

$$\text{IoU} = \frac{\text{Result(Road)} \cap \text{Label(Road)}}{\text{Result(Road)} \cup \text{Label(Road)}} \tag{9}$$

where TP (true positive) represents a positive label and a positive prediction, FP (false positive) represents a negative label and a positive prediction and FN (false negative) represents a positive label and a negative prediction.

4.2. Loss Function Selection

In this section, we use BCE Loss, Dice Loss, BCE Loss + Dice Loss and CWL as model loss functions and test them on the rural road dataset. The experimental results were evaluated using both Precision and IoU, and the results are shown in Table 1. From the comparison of the single-loss-function applications, it can be seen that BCE Loss performs better in terms of extraction accuracy compared to Dice Loss, but Dice Loss has better performance in terms of IoU, which is because Dice Loss focuses more on the overlapping area of image samples, while BCE Loss focuses more on the global performance. The combined loss results show that the Precision and IoU of BCE Loss + Dice Loss and CWL are higher than that of using a single loss function, indicating that the combination of loss functions can effectively improve the performance of the model. The results show that the Precision and IoU of CWL are the highest, which shows that the dynamic weighting method can give full play to the advantages of the combination of loss functions, so that the model in this paper can fully utilize the advantages of Dice Loss in the secondary classification while ensuring stability and improving the applicability of the model to situations of serious imbalance between positive and negative samples. Therefore, CWL was selected as the loss function for road extraction in this paper.

Table 1. Effect of different loss functions on road extraction results.

	Precision	IoU
BCE Loss	0.9862	0.7322
Dice Loss	0.9850	0.7561
BCE Loss + Dice Loss	0.9866	0.7605
CWL	0.9874	0.7621

4.3. Results and Discussion

To estimate the performance of our model, we compared the model with five other models: Unet, Unet++, Unet+++, MACUnet [41] and MobileNet. Unet++, Unet+++ and MACUnet are all Unet-based networks, with Unet++ and Unet+++ refining the jump layer and enhancing global feature linkage, and MACUnet improving upon Unet with the ACB module and multi-scale jumps to enhance network feature acquisition. MobileNet was used as a benchmark for the comparison of lightweight models. All experiments were implemented on an NVIDIA Tesla P100 GPU with Adam as the optimizer and a learning rate of 0.001. LDANet uses CWL as the loss function and the rest of the models use a Cross-Entropy Loss function.

4.3.1. Results of the Typical Rural Roads Dataset

In this section, we compare LDANet with five other image segmentation models on the rural road dataset. Table 2 shows the results of four accuracy evaluation metrics, Precision, Recall, F1 Score and IoU, for the six models on the rural road test set. Since we aimed to build fast and efficient lightweight models for rural road extraction, it was necessary to evaluate the model complexity, so we conducted statistics on the training time and model parameters for each of the six models mentioned above to measure the model performance.

Table 2. Comparison of six models evaluated on the rural roads dataset.

	Precision	Recall	F1 Score	IoU	Parameters (M)	Train Time/Epoch (S)
Unet	0.9754	0.9728	0.9741	0.7482	9.85	580
Unet++	0.9831	0.9822	0.9826	0.7593	11.80	1318
Unet+++	0.9881	0.9875	0.9878	0.7644	6.75	1530
MACUnet	0.9840	0.9817	0.9829	0.7617	5.15	725
MobileNet	0.9683	0.9632	0.9657	0.7431	0.17	178
LDANet	0.9874	0.9870	0.9872	0.7621	0.20	183

From Table 2, Unet++ has the highest Precision and IoU, scoring 0.9881 and 0.7644, respectively. This is because it deepens the network context and can fully learn the characteristics of the target. However, the complex network structure degradation increases its operation difficulty, making it not perform well in terms of model parameters and training time. In terms of training efficiency, MobileNet uses DSC to replace the traditional convolution process, which reduces the cost of convolution and makes it perform best in terms of model parameters and training time. However, due to image restoration only occurring through the upsampling method, the global feature association of the target is insufficient in the modeling process, resulting in low extraction accuracy.

In terms of model accuracy, LDANET has a Precision of 0.9874 and an IoU of 0.7621, which are 1.98% and 2.13% lower than Une+++, respectively, but LDANET greatly improves training efficiency with a model parameter count of 0.20 M and a training speed of 183 s/epoch, thus having 3375% fewer parameters and being 836% faster than Unet+++. Meanwhile, LDANET performed comparably to MobileNet in terms of training efficiency, but with 1.91% and 1.9% improvements in accuracy and IoU, respectively.

The results show that compared with Unet++, LDANet is relatively lightweight because it introduces ACB and DSC at the cost of slightly reducing model accuracy, which greatly enhances the applicability of the model. Compared with MobileNet, although LDANet's calculation parameters have increased, it can significantly improve the accuracy of the model. This is because LDANet uses the encoder and decoder structure to strengthen the feature association and to significantly improve the expression of dominant features by introducing dynamically weighted overlay layers.

Figure 6 shows the extraction results of the six models on the rural road test set. From the figure, it can be seen that all six models have some defects in the road extraction effects, but compared with other models, Unet+++ and LDANet have an obvious improvement in terms of road extraction. In summary, the use of LDANet for rural road extraction has a high application value.

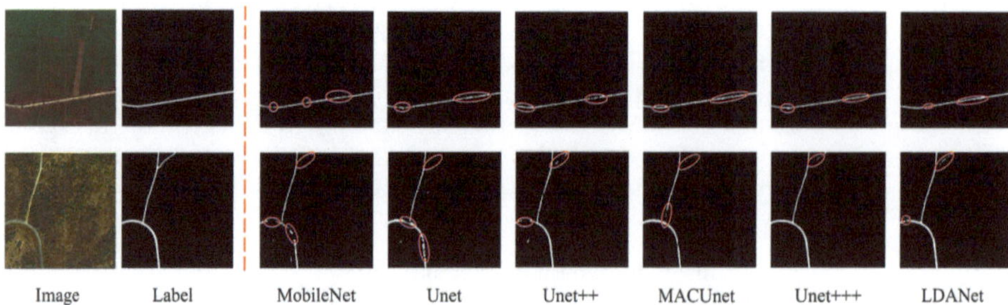

Figure 6. Extraction results of six models on the rural roads dataset.

4.3.2. Results of the Massachusetts Roads Dataset

To further validate the superiority and robustness of our LDANet, we also compared the LDANet with other methods on the Massachusetts roads dataset, using Precision, Recall, F1 Score and IoU metrics for evaluation.

The comparison results are shown in Table 3. The table shows that LDANet also performs well on the Massachusetts roads dataset, with Precision, Recall, F1 Score and IoU reaching 0.9755, 0.9707, 0.9731 and 0.6834, respectively. Figure 7 shows the extraction results of LDANet on the Massachusetts roads dataset compared to the ground truth map comparison, and it can be seen that our method achieves adequate visual results of road extraction.

Table 3. Comparison of the six models evaluated on the Massachusetts roads dataset.

	Precision (%)	Recall	F1 Score	IoU	Parameters (M)	Train Time/Epoch (S)
Unet	0.9612	0.9584	0.9598	0.6513	9.85	480
Unet++	0.9710	0.9667	0.9688	0.6769	11.80	1130
Unet+++	0.9768	0.9716	0.9742	0.6957	6.75	1280
MACUnet	0.9721	0.9683	0.9702	0.6774	5.15	605
MobileNet	0.9533	0.9412	0.9472	0.6455	0.17	152
LDANet	0.9755	0.9707	0.9731	0.6834	0.20	163

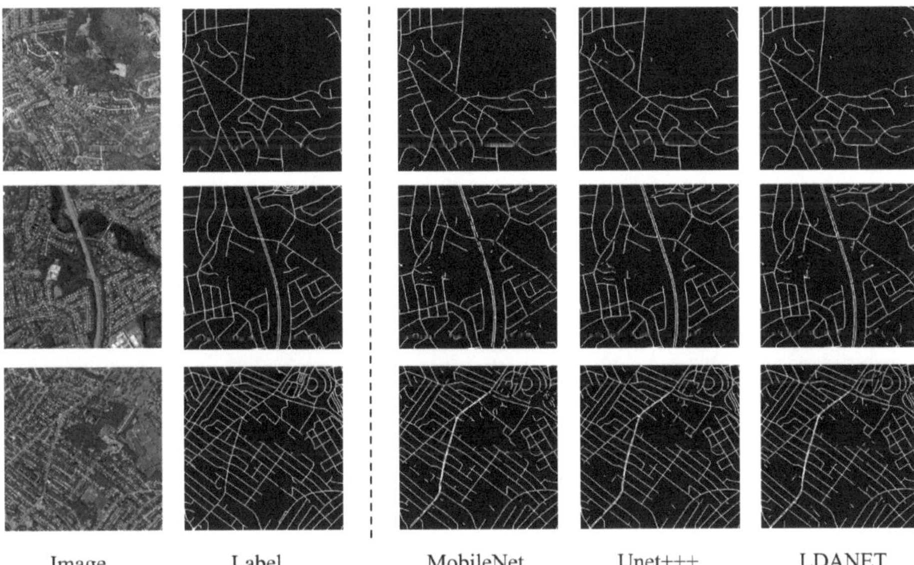

Figure 7. LDANet extraction results on the Massachusetts roads dataset.

4.3.3. Discussion

Deep learning methods to achieve road extraction from remote sensing images are important to promote the application of remote sensing images and the development of cities. In this work, a lightweight model was constructed based on the Unet network structure. The model enhances the shallow feature information by introducing a feature expansion module and uses dynamic weighted superposition to improve the feature representation. Compared with Boonpook and Lu et al., this method can significantly reduce the modeling parameters by using DSC, making the model lighter and faster. Compared with MobileNets and ShuffleNets, this model can obtain high-accuracy road extraction results by upsampling and hopping connections of the Unet network structure to fully learn contextual features.

5. Conclusions

We have proposed a lightweight extraction model based on rural road data. The model is composed of a feature expansion module and a deep feature association module. In addition, we used a dynamically weighted loss function according to the small proportion of rural roads. Compared with complex methods which are expensive to calculate, our method focuses on enriching shallow features and strengthening the correlation of deep salient features, which can effectively balance reliability and speed. In the typical rural roads dataset, our model's accuracy was 0.9874, and its IoU was 0.7621. In the Massachusetts roads dataset, our model also performed well. Our model has the characteristics of a small number of parameters and a fast training speed, which can greatly reduce the requirements for hardware while still ensuring extraction accuracy in practical applications. Therefore, it is of great significance to promote the portable and rapid application of remote sensing technology.

Future work will involve optimization strategies based on combined model applications to achieve multi-objective learning applications by enhancing global information interactivity based on constrained model complexity.

Author Contributions: Conceptualization, J.D. and B.L.; methodology, B.L. and J.Z.; software, B.L.; validation, B.L.; formal analysis, B.L. and S.H.; resources, B.L.; data curation, B.L.; writing—original draft preparation, B.L.; writing—review and editing, B.L.; visualization, B.L.; supervision, J.D.; project administration, J.D. and J.W.; funding acquisition, J.D. All authors have read and agreed to the published version of the manuscript.

Funding: This research was funded by National Natural Science Foundation of China (42171269); Key Project of Natural Science Foundation of Xinjiang Uygur Autonomous Region (2021D01D06); National Natural Science Foundation of China (41961059).

Data Availability Statement: Not applicable.

Conflicts of Interest: The authors declare no conflict of interest.

References

1. Chen, Z.; Wang, C.; Li, J.; Fan, W.; Du, J.; Zhong, B. Adaboost-like End-to-End Multiple Lightweight U-Nets for Road Extraction from Optical Remote Sensing Images. *Int. J. Appl. Earth Obs. Geoinf.* **2021**, *100*, 102341. [CrossRef]
2. Jiang, X.; Li, Y.; Jiang, T.; Xie, J.; Wu, Y.; Cai, Q.; Jiang, J.; Xu, J.; Zhang, H. RoadFormer: Pyramidal Deformable Vision Transformers for Road Network Extraction with Remote Sensing Images. *Int. J. Appl. Earth Obs. Geoinf.* **2022**, *113*, 102987. [CrossRef]
3. Li, C.; Zeng, Q.; Fang, J.; Wu, N.; Wu, K. Road Extraction in Rural Areas from High Resolution Remote Sensing Image Using a Improved Full Convolution Network. *Natl. Remote Sens. Bull.* **2021**, *25*, 1978–1988. [CrossRef]
4. Herumurti, D.; Uchimura, K.; Koutaki, G.; Uemura, T. Urban Road Extraction Based on Hough Transform and Region Growing. In Proceedings of the FCV 2013—19th Korea-Japan Joint Workshop on Frontiers of Computer Vision, Incheon, Republic of Korea, 30 January–1 February 2013.
5. Shi, W.; Miao, Z.; Debayle, J. An Integrated Method for Urban Main-Road Centerline Extraction from Optical Remotely Sensed Imagery. *IEEE Trans. Geosci. Remote Sens.* **2014**, *52*, 3359–3372. [CrossRef]
6. Lian, R.; Wang, W.; Mustafa, N.; Huang, L. Road Extraction Methods in High-Resolution Remote Sensing Images: A Comprehensive Review. *IEEE J. Sel. Top. Appl. Earth Obs. Remote Sens.* **2020**, *13*, 5489–5507. [CrossRef]
7. Zhao, J.Q.; Yang, J.; Li, P.X.; Lu, J.M. Semi-Automatic Road Extraction from SAR Images Using EKF and PF. In Proceedings of the International Archives of the Photogrammetry, Remote Sensing and Spatial Information Sciences—ISPRS Archives, Kona, HI, USA, 21–23 July 2015; Volume 40.
8. Perciano, T.; Tupin, F.; Hirata, R.; Cesar, R.M. A Two-Level Markov Random Field for Road Network Extraction and Its Application with Optical, SAR, and Multitemporal Data. *Int. J. Remote Sens.* **2016**, *37*, 3584–3610. [CrossRef]
9. Zang, Y.; Wang, C.; Cao, L.; Yu, Y.; Li, J. Road Network Extraction via Aperiodic Directional Structure Measurement. *IEEE Trans. Geosci. Remote Sens.* **2016**, *54*, 3322–3335. [CrossRef]
10. Sujatha, C.; Selvathi, D. FPGA Implementation of Road Network Extraction Using Morphological Operator. *Image Anal. Stereol.* **2016**, *35*, 93–103. [CrossRef]
11. Hinton, G.E.; Salakhutdinov, R.R. Reducing the Dimensionality of Data with Neural Networks. *Science* **2006**, *313*, 504–507. [CrossRef]

12. Zhong, Z.; Li, J.; Cui, W.; Jiang, H. Fully Convolutional Networks for Building and Road Extraction: Preliminary Results. In Proceedings of the International Geoscience and Remote Sensing Symposium (IGARSS), Beijing, China, 10–15 July 2016; Volume 2016.

13. Varia, N.; Dokania, A.; Senthilnath, J. DeepExt: A Convolution Neural Network for Road Extraction Using RGB Images Captured by UAV. In Proceedings of the 2018 IEEE Symposium Series on Computational Intelligence, SSCI, Bangalore, India, 18–21 November 2018.

14. Doshi, J. Residual Inception Skip Network for Binary Segmentation. In Proceedings of the IEEE Computer Society Conference on Computer Vision and Pattern Recognition Workshops, Salt Lake City, UT, USA, 18–22 June 2018; Volume 2018.

15. Zhou, L.; Zhang, C.; Wu, M. D-Linknet: Linknet with Pretrained Encoder and Dilated Convolution for High Resolution Satellite Imagery Road Extraction. In Proceedings of the IEEE Computer Society Conference on Computer Vision and Pattern Recognition Workshops, Salt Lake City, UT, USA, 18–22 June 2018; Volume 2018.

16. Li, Y.; Guo, L.; Rao, J.; Xu, L.; Jin, S. Road Segmentation Based on Hybrid Convolutional Network for High-Resolution Visible Remote Sensing Image. *IEEE Geosci. Remote Sens. Lett.* **2019**, *16*, 613–617.

17. Boonpook, W.; Tan, Y.; Bai, B.; Xu, B. Road Extraction from UAV Images Using a Deep ResDCLnet Architecture. *Can. J. Remote Sens.* **2021**, *47*, 450–464. [CrossRef]

18. Lu, X.; Zhong, Y.; Zheng, Z.; Liu, Y.; Zhao, J.; Ma, A.; Yang, J. Multi-scale and multi-task deep learning framework for automatic road extraction. *IEEE Trans. Geosci. Remote Sens.* **2019**, *57*, 9362–9377. [CrossRef]

19. Howard, A.G.; Zhu, M.; Chen, B.; Kalenichenko, D.; Wang, W.; Weyand, T.; Andreetto, M.; Adam, H. MobileNets: Efficient Convolutional Neural Networks for Mobile Vision Applications. *arXiv* **2017**, arXiv:1704.04861.

20. Sandler, M.; Howard, A.; Zhu, M.; Zhmoginov, A.; Chen, L.C. MobileNetV2: Inverted Residuals and Linear Bottlenecks. In Proceedings of the IEEE Computer Society Conference on Computer Vision and Pattern Recognition, Salt Lake City, UT, USA, 18–22 June 2018.

21. Howard, A.; Sandler, M.; Chen, B.; Wang, W.; Chen, L.C.; Tan, M.; Chu, G.; Vasudevan, V.; Zhu, Y.; Pang, R.; et al. Searching for MobileNetV3. In Proceedings of the IEEE International Conference on Computer Vision, Seoul, Republic of Korea, 27 October–2 November 2019; Volume 2019.

22. Ma, N.; Zhang, X.; Zheng, H.T.; Sun, J. *Shufflenet V2: Practical Guidelines for Efficient Cnn Architecture Design*; Lecture Notes in Computer Science (Including Subseries Lecture Notes in Artificial Intelligence and Lecture Notes in Bioinformatics); In Proceedings of the European Conference on Computer Vision (ECCV), Munich, Germany, 8–14 September 2018; Springer: Berlin/Heidelberg, Germany, 2018; Volume 11218 LNCS.

23. Zhang, X.; Zhou, X.; Lin, M.; Sun, J. ShuffleNet: An Extremely Efficient Convolutional Neural Network for Mobile Devices. In Proceedings of the IEEE Computer Society Conference on Computer Vision and Pattern Recognition, Salt Lake City, UT, USA, 18–22 June 2018.

24. Paszke, A.; Chaurasia, A.; Kim, S.; Culurciello, E. ENet: A Deep Neural Network Architecture for Real-Time Semantic Segmentation. *arXiv* **2016**, arXiv:1606.02147.

25. Emara, T.; Abd El Munim, H.E.; Abbas, H.M. Liteseg: A novel lightweight convnet for semantic segmentation. In Proceedings of the 2019 Digital Image Computing: Techniques and Applications (DICTA), Perth, WA, Australia, 2–4 December 2019; IEEE: Piscataway, NJ, USA, 2019; pp. 1–7.

26. Romera, E.; Alvarez, J.M.; Bergasa, L.M.; Arroyo, R. ERFNet: Efficient Residual Factorized ConvNet for Real-Time Semantic Segmentation. *IEEE Trans. Intell. Transp. Syst.* **2018**, *19*, 263–272. [CrossRef]

27. Chen, L.-C.; Zhu, Y.; Papandreou, G.; Schroff, F.; Adam, H. Encoder-Decoder with Atrous Separable Convolution for Semantic Image Segmentation. In Proceedings of the European Conference on Computer Vision (ECCV), Munich, Germany, 8–14 September 2018.

28. Mehta, S.; Rastegari, M.; Caspi, A.; Shapiro, L.; Hajishirzi, H. *ESPNet: Efficient Spatial Pyramid of Dilated Convolutions for Semantic Segmentation*; Lecture Notes in Computer Science (Including Subseries Lecture Notes in Artificial Intelligence and Lecture Notes in Bioinformatics); In Proceedings of the European Conference on Computer Vision (ECCV), Glasgow, UK, 8–14 September 2018; Springer: Berlin/Heidelberg, Germany, 2018; Volume 11214 LNCS.

29. Mehta, S.; Rastegari, M.; Shapiro, L.; Hajishirzi, H. ESPNetv2: A Light-Weight, Power Efficient, and General Purpose Convolutional Neural Network. In Proceedings of the IEEE Computer Society Conference on Computer Vision and Pattern Recognition, Long Beach, CA, USA, 15–20 June 2019; Volume 2019.

30. Yi, Q.; Dai, G.; Shi, M.; Huang, Z.; Luo, A. ELANet: Effective Lightweight Attention-Guided Network for Real-Time Semantic Segmentation. *Neural Process. Lett.* **2023**, *55*, 1–18. [CrossRef]

31. Lan, R.; Sun, L.; Liu, Z.; Lu, H.; Pang, C.; Luo, X. MADNet: A fast and lightweight network for single-image super resolution. *IEEE Trans. Cybern.* **2020**, *51*, 1443–1453. [CrossRef]

32. Mnih, V. *Machine Learning for Aerial Image Labeling*, University of Toronto: Toronto, ON, Canada, 2013.

33. Demir, I.; Koperski, K.; Lindenbaum, D.; Pang, G.; Huang, J.; Basu, S.; Hughes, F.; Tuia, D.; Raska, R. DeepGlobe 2018: A Challenge to Parse the Earth through Satellite Images. In Proceedings of the IEEE Computer Society Conference on Computer Vision and Pattern Recognition Workshops, Salt Lake City, UT, USA, 18–22 June 2018; Volume 2018.

34. Ran, S.; Ding, J.; Liu, B.; Ge, X.; Ma, G. Multi-U-Net: Residual module under multisensory field and attention mechanism based optimized U-Net for VHR image semantic segmentation. *Sensors* **2021**, *21*, 1794. [CrossRef] [PubMed]

35. Wang, W.; Yang, N.; Zhang, Y.; Wang, F.; Cao, T.; Eklund, P. A Review of Road Extraction from Remote Sensing Images. *J. Traffic Transp. Eng. Engl. Ed.* **2016**, *3*, 271–282. [CrossRef]
36. DIng, X.; Guo, Y.; DIng, G.; Han, J. ACNet: Strengthening the Kernel Skeletons for Powerful CNN via Asymmetric Convolution Blocks. In Proceedings of the IEEE International Conference on Computer Vision, Seoul, Republic of Korea, 27 October–2 November 2019; Volume 2019.
37. Ronneberger, O.; Fischer, P.; Brox, T. U-Net: Convolutional Networks for Biomedical Image Segmentation. In Proceedings of the Medical Image Computing and Computer-Assisted Intervention—MICCAI 2015: 18th International Conference, Munich, Germany, 5–9 October 2015; Springer: Berlin/Heidelberg, Germany, 2015; pp. 234–241.
38. Zhou, Z.; Siddiquee, M.M.R.; Tajbakhsh, N.; Liang, J. Unet++: Redesigning Skip Connections to Exploit Multiscale Features in Image Segmentation. *IEEE Trans. Med. Imaging* **2019**, *39*, 1856–1867. [CrossRef] [PubMed]
39. Huang, H.; Lin, L.; Tong, R.; Hu, H.; Zhang, Q.; Iwamoto, Y.; Han, X.; Chen, Y.W.; Wu, J. UNet 3+: A Full-Scale Connected UNet for Medical Image Segmentation. In Proceedings of the ICASSP, IEEE International Conference on Acoustics, Speech and Signal Processing, Barcelona, Spain, 4–8 May 2020; Volume 2020.
40. Ren, Y.; Zhang, X.; Ma, Y.; Yang, Q.; Wang, C.; Liu, H.; Qi, Q. Full Convolutional Neural Network Based on Multi-Scale Feature Fusion for the Class Imbalance Remote Sensing Image Classification. *Remote Sens.* **2020**, *12*, 3547. [CrossRef]
41. Li, R.; Duan, C.; Zheng, S.; Zhang, C.; Atkinson, P.M. MACU-Net for Semantic Segmentation of Fine-Resolution Remotely Sensed Images. *IEEE Geosci. Remote Sens. Lett.* **2022**, *19*, 8007205. [CrossRef]

remote sensing

MDPI

Article

ISTD-PDS7: A Benchmark Dataset for Multi-Type Pavement Distress Segmentation from CCD Images in Complex Scenarios

Weidong Song [1], Zaiyan Zhang [1,2,*], Bing Zhang [1], Guohui Jia [3], Hongbo Zhu [1] and Jinhe Zhang [1]

[1]   School of Geomatics, Liaoning Technical University, Fuxin 123000, China
[2]   College of Mining Engineering, Heilongjiang University of Science and Technology, Harbin 150022, China
[3]   School of Resources and Civil Engineering, Liaoning Institute of Science and Technology, Benxi 117004, China
*   Correspondence: 2014800205@usth.edu.cn; Tel.: +86-1894-600-7768

Abstract: The lack of large-scale, multi-scene, and multi-type pavement distress training data reduces the generalization ability of deep learning models in complex scenes, and limits the development of pavement distress extraction algorithms. Thus, we built the first large-scale dichotomous image segmentation (DIS) dataset for multi-type pavement distress segmentation, called ISTD-PDS7, aimed to segment highly accurate pavement distress types from natural charge-coupled device (CCD) images. The new dataset covers seven types of pavement distress in nine types of scenarios, along with negative samples with texture similarity noise. The final dataset contains 18,527 images, which is many more than the previously released benchmarks. All the images are annotated with fine-grained labels. In addition, we conducted a large benchmark test, evaluating seven state-of-the-art segmentation models, providing a detailed discussion of the factors that influence segmentation performance, and making cross-dataset evaluations for the best-performing model. Finally, we investigated the effectiveness of negative samples in reducing false positive prediction in complex scenes and developed two potential data augmentation methods for improving the segmentation accuracy. We hope that these efforts will create promising developments for both academics and the industry.

Keywords: pavement CCD images; deep learning; distress semantic segmentation; ISTD-PDS7

Citation: Song, W.; Zhang, Z.; Zhang, B.; Jia, G.; Zhu, H.; Zhang, J. ISTD-PDS7: A Benchmark Dataset for Multi-Type Pavement Distress Segmentation from CCD Images in Complex Scenarios. *Remote Sens.* **2023**, *15*, 1750. https://doi.org/10.3390/rs15071750

Academic Editor: John Trinder

Received: 15 February 2023
Revised: 20 March 2023
Accepted: 22 March 2023
Published: 24 March 2023

1. Introduction

The automatic detection of road pavement distress is necessary to realize the maintenance and monitoring of complex traffic networks, and is an effective way to improve the quality of road service [1]. For both cement concrete and asphalt pavements, in the actual operation process, unidirectional cracks, alligator cracks, broken slabs, potholes, and other types of distress can appear under the comprehensive influence of the traffic volume, load, temperature, moisture, and weathering, and are collectively called pavement distress [2]. Pavement distress accelerates highway aging, greatly reduces driving comfort, increases vehicle wear, and can increase avoidance actions that may lead to collisions, posing potential threats to highway and driving safety [3]. According to a survey conducted by the Ministry of Transport of China, the road maintenance mileage in China has reached 5.28 million kilometers, which is about 99.4% of the total road mileage in 2021 [4]. Thus, with the increasing demand for road maintenance, computer vision-based road condition assessments have become a research hotspot in the industry.

The traditional manual inspection approach is time-consuming, laborious, and bi-ased. To solve this problem, optical imaging with onboard charge-coupled device (CCD) sensors combined with digital image processing technologies has attracted much attention because it can automatically monitor pavement conditions [5]. For several decades, researchers have been working on the application of computer vision technologies for pavement distress assessment. Early studies based on digital image processing technologies [6–10] and machine learning technologies [11–14] are extensive [15], but their flaws are obvious as these are affected by factors, such as the environment, traffic load, and maintenance conditions. There are three obvious characteristics in the CCD distress images of highway pavements in natural scenes: (1) the image quality is greatly affected by the light intensity, highway dryness, shadow, and other interference; (2) the background is complex and changeable, there is strong speckle noise, and a low target signal-to-noise ratio (SNR); and (3) there are many distress classes, complex topological structures, and the gray feature difference is very small. For example, as shown in Figure 1, the impulse noise brought by the grain-like pavement texture breaks the crack and undermines its continuity, and shadows reduce the contrast between the crack and the background. As a result, these traditional methods struggle to achieve complete multi-type pavement distress segmentation from complex backgrounds [16,17].

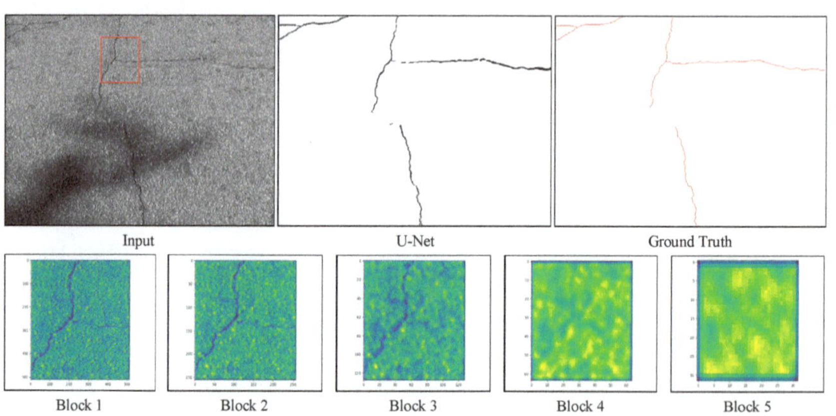

Figure 1. A real example of distress segmentation using U-Net. The second row shows the feature maps of blocks with different depths in U-Net (for the image patch denoted by the rectangle in the input image).

Recent studies have shown that, with the introduction of deep learning (DL) models in the fields of photogrammetry, remote sensing, and computer vision, DL based methods are now performing a dominant role in the detection of road surface distress with CCD images [18–20]. Studies, such as [16,21–23], have shown that deep convolutional neural networks (DCNNs) can be used for the automatic segmentation of pavement cracks. These deep architectures build high-level features from low-level primitives by hierarchical convolution of the input. As can be seen in the second row of Figure 1. These DCNN architectures can build high-level features from low-level primitives by hierarchically convolving the sensory inputs. Typically, DL based crack detection methods can be classified into three categories: (1) image classification based methods, (2) object detection based methods, and (3) semantic segmentation based methods [20].

However, it must be noted that, compared with classification [24–26] and detection [27–29], semantic segmentation can provide more accurate geometric target description for a wide range of applications, such as geometric feature measurement of distress [2,30], severity division [21], and quantitative assessment of pavement conditions [16], which is also the research focus of this paper. In addition, most studies have focused on only one or two types of distress. Ouma's research focused on linear crack detection [31] and Siriborvornratanakul's re-search focus was on crater detection [32].

The tasks and datasets for image segmentation are closely related in the deep learning era. Some of the segmentation tasks, such as those considered in [33–36], are even directly built upon the datasets. However, most of the pavement data collection systems are complex and complicated, and it is costly to label this amount of data [23]. As shown in Table 1, the datasets published so far do often consist of less than 500 pavement distress images [37–40].

Moreover, most of the available pavement distress datasets have been collected from highways, which are regularly maintained. As a result, the collected pavement datasets contain images with less variety of pavement distress [23], fewer scenes, and almost all the datasets are derived from images of asphalt road. Therefore, a bench-mark dataset is urgently required to address these challenging problems. Previous studies have found that the multiple distress types have complex topological structures, different sizes, and similar gray-level values, and labeling experts commonly disagree on the accurate classification of the categories [22]. Therefore, the construction of a multi-type pavement distress dataset according to the conventional semantic segmentation task is not conducive to accurate segmentation. Recent research has found that a category agnostic dichotomous image segmentation (DIS) task defined on non-conflicting annotations can be used to accurately segment objects with different structural complexities regardless of their characteristics [40].

Table 1. Comparison of the established publicly available datasets for pavement distress segmentation. AP and CCP are the abbreviations for asphalt pavement and cement concrete pavement, respectively.

Task	Dataset	Illumination	Proportion of Pavement Type/%	Equipment
	CrackLS315 [30]	Laser	AP: 100.0%	Area-array camera
	CRKWH100 [30]	Visible light	AP: 100.0%	Linear-array camera
	CrackTree260 [30]	Visible light	AP: 100.0%	Area-array camera
Crack segmentation	AigleRN [39]	Visible light	AP: 100.0%	Area-array camera
	CFD [37]	Visible light	CCP: 1.7%; AP: 98.3%	Smartphone
	CRACK500 [38]	Visible light	CCP: 2.4%; AP: 97.6%	Smartphone
	GAPs384 [38]	Laser	AP: 100.0%	Linear-array camera
Multi-type distress DIS	ISTD-PDS7	HID lamp	CCP: 29.4%; AP: 70.6%	Area-array camera

Inspired by these observations, in this study, we built a highly detailed DIS dataset, named ISTD-PDS7, for multi-type pavement distress segmentation. The dataset has a sufficient scale, labeling precision, and scene diversity. We hope that the ISTD-PDS7 dataset can contribute to improving the robustness and reliability of automatic pavement distress extraction algorithms in complex scenarios, and further promote the research progress into automatic assessment of large-scale highway pavement conditions. The ISTD-PDS7 dataset and benchmarks will be made publicly available at: https://ciigis.lntu.edu.cn/. The main contributions of this paper can be summarized as the following three aspects:

1. A large-scale extendable DIS dataset—ISTD-PDS7—containing 18,527 CCD images and 7 types of pavement distress, was built and annotated manually by 4 experts in the field of pavement distress detection. Finally, highly detailed binary segmentation masks were generated. The dataset was analyzed in detail from three aspects: image dimension, image complexity, and annotation complexity.
2. Based on the new ISTD-PDS7 dataset, we compared the cutting-edge segmentation models with different network structures and made a comprehensive evaluation and analysis of their pavement distress segmentation performance. These results will serve as the baseline results for future works.
3. We briefly review the numerous previously published datasets. We also describe the detailed evaluation and comparative experiments conducted between these datasets and ISTD-PDS7, and propose their comparison results in crack segmentation.

The rest of this paper is organized as follows: A brief review of the related work is provided in Section 2. Section 3 presents the details of the dataset collection and labeling, data analysis, and splitting of the dataset. Subsequently, we provide a description of the baseline algorithms for the benchmark evaluation in Section 4. Section 5 first describes the implementation details and the dataset setup of this study, then a performance analysis of the baseline algorithms on ISTD-PDS7 is given. We also evaluate and compare the crack segmentation performance obtained with the different public datasets and the new ISTD-PDS7 dataset, in addition to the impact of negative samples. Finally, our conclusions are given in Section 6.

2. Related Work

2.1. Automated Crack Detection

Crack detection has always been a research hotspot in the field of automatic distress detection. In 2016, Zhang et al. [41] used smartphones to collect highway images to build a dataset, and LeNet-5 [42] was used for crack detection for the first time. Subsequently, a pre-trained VGG-16 model based on ImageNet was transferred to the crack detection task for asphalt and cement pavements by Gopalakrishnan et al. [43] and achieved a superior crack detection performance. However, the above two methods need to be converted into a fully convolutional network to obtain the image segmentation results. Therefore, crack segmentation based on an encoder–decoder network structure is becoming more common. In view of the excellent performance of the U-Net [44] architecture in the field of biomedical image segmentation, Jenkins et al. [45] developed a semantic segmentation algorithm for pavement cracks based on U-Net. However, due to the lack of data (80 training images and 20 validation images), the generalization ability of the model was insufficient. Researchers have since made a variety of improvements to the network structure based on the U-Net architecture. For example, Lau et al. [46] replaced the encoder phase of U-Net with a pre-trained ResNet-34 model, and the method achieved an F1-score of 96% and 73% on the CrackForest dataset (CFD) [37] and Crack500 dataset [38], respectively. Escalona et al. [47] implemented three different U-Net models for crack segmentation, and performed segmentation performance tests on the CFD [37] and AigleRN [39] datasets. Furthermore, in order to balance the segmentation efficiency and accuracy, Polovnikov et al. [48] proposed a lightweight U-Net-based network architecture called DAUNet, which was tested on the publicly available datasets, and was found to be able to effectively detect cracks in complex scenes.

In addition, Zou et al. [30] improved the SegNet [49] architecture and proposed an end-to-end trainable DeepCrack network to detect cracks where the pavement image pixels are distinguished into crack and non-crack background forms. In addition, Fan et al. [50] proposed a novel highway crack detection algorithm based on DL and adaptive image segmentation. Xu et al. [51] proposed a new network architecture, an enhanced high-resolution semantic network (EHRS-Net), which was suitable for tiny cracks and noised pavement cracks. In order to compensate the transformer for the deficiency of local features, Xu et al. [20] proposed a new network for pavement crack detection from CCD images, called LETNet, which has strong robustness.

2.2. Multi-Type Distress Segmentation Approaches

However, cracks are not the only type of distress in pavements and some researchers have recently turned their attention to multi-type distress segmentation. For example, Lõuk et al. [52] applied a U-Net-like network architecture with different context resolution levels to integrate more contextual information. The pavement distress detection system (PDDS) proposed by Lõuk et al. [52] can output multi-class pavement distress regions, and they have stated that their future research work will focus on the 11 defect categories defined by the Estonian Road Administration. Majidifard et al. [16] developed a method based on the combination of U-Net and YOLO, which can effectively distinguish cracks, dense cracks, and potholes. In addition, Zhang et al. [23] collected urban highway pavement images in Montreal, Canada, and produced semantic segmentation datasets for potholes, patches, lane lines, unidirectional cracks, and network cracks. The authors also proposed and evaluated a method for the automatic detection and classification of pavement distress classes using a convolutional neural network (CNN) and low-cost video data, where the detection rate and classification accuracy of the model both reached 83.8%.

It must be noted that DL is a data-driven technology, and massive labeled data can effectively reduce the risk of overfitting and improve the generalization performance of a model. However, none of the above studies proposed a model based on a comprehensive dataset which covers all the highway pavement distress classes under different conditions in natural scenes [16]. In addition, due to the lack of standardized images as a test set, it is difficult to measure the geometric features and evaluate the accuracy of pavement distress segmentation results.

2.3. Existing Datasets for Pavement Distress Segmentation

Although many different distress detection methods have been proposed to date, there is still a lack of public datasets that are both large enough and annotated in a standardized manner. The datasets released to date typically contain fewer than 500 images, e.g., CrackTree260, CrackLS315, and CrackWH100 were annotated manually at a single-pixel width [30]. CFD [37], CRACK500 [38], and AigleRN [39] were annotated manually according to the actual crack width. See Table 1 for details.

- **CrackTree260** [30]: The CrackTree260 dataset consists of 260 pavement crack images with a size of 800 × 600 pixels, for which the pavement images were captured by an area-array camera under visible light illumination.
- **CrackLS315** [30]: The CrackLS315 dataset contains 315 road pavement images captured under laser illumination. These images were captured by a linear-array camera, at the same ground sampling distance.
- **CRKWH100** [30]: The CRKWH100 dataset contains 100 road pavement images captured by a linear-array camera under visible light illumination. The linear-array camera captures the pavement at a ground sampling distance of 1 mm.
- **CFD** [37]: The CFD dataset consists of 118 iPhone 5 images of cracks in the urban pavement of Beijing in China, where each image is manually labeled with the ground-truth contour and the size is adjusted to 480 × 320 pixels. The dataset also includes a few images that are contaminated by small oil spots and water stain noise.
- **Crack500** [38]: The Crack500 dataset consists of 500 2000 × 1500 pixel crack images, with a few containing oil spots and shadow noise.
- **AigleRN** [39]: The AigleRN dataset contains 38 preprocessed grayscale images of pavements in France, with the size of one half of the AigleRN dataset being 991 × 462 pixels, and the size of the other half being 311 × 462 pixels.
- **GAPs** [53]: In 2017, a freely available large pavement distress detection dataset called the German Asphalt Pavement (GAPs) dataset was released by Eisenbach et al. [51], which has since received considerable attention from several research groups (e.g., [54–56]). The GAPs dataset was the first attempt at creating a standard benchmark pavement distress image dataset for DL applications. It includes 1969 grayscale pavement images (1418 for training, 51 for validation, and 500 for testing) with various distress types, including cracks (longitudinal/transverse, alligator, sealed/filled), potholes, patches, open joints, and bleeding [53]. Unfortunately, the method of bounding box annotation is not very friendly for semantic segmentation tasks. To solve this problem, Yang et al. [38] selected and annotated 384 crack images from the GAPs dataset at the pixel level and built a new segmented dataset called GAPs384. It is worth noting that all the images in the GAPs dataset were collected from the pavements of three different German federal highways. The shooting conditions were dry and warm, so the GAPs dataset is suitable for studying the segmentation and extraction problems of pavement distress in urban highways and expressways with good highway conditions. More recently, Stricker et al. [22] released the publicly available GAPs-10 m dataset for semantic segmentation. This dataset contains 20 high-resolution images (5030 × 11505 pixels, each corresponding to 10 m of highway pavement) that cover 200 m of asphalt roads with different asphalt surface types and a wide variety of distress classes [22]. The corresponding multi class distress labels were annotated by experts; this dataset is currently the only publicly available dataset with high-resolution images.
- **Others:** Although some of the larger datasets recently published, such as the dataset made up of 700K Google Street View images [57] or the Global Road Damage Detection Challenge (GRDDC) 2020 dataset [58], are mostly used for object detection tasks in

severely damaged images as they do not have the resolution level required for highway damage condition assessment.

In summary, although automatic pavement distress detection algorithms based on DL have made good progress in recent years, the existing publicly available datasets for pavement distress segmentation still have the problems of a small scale, few scenes, single pavement type, low resolution, and unmeasurable segmentation results. Therefore, a benchmark dataset is urgently required to address these challenging problems.

3. ISTD-PDS7 Dataset

3.1. Data Collection and Annotation

Data Collection: To solve the data problem (see Section 2) based on complex scenes and high-resolution CCD images, we built a detailed DIS dataset for multi-type pavement distress segmentation named ISTD-PDS7. The original images of the ISTD-PDS7 dataset were acquired using a mobile acquisition vehicle (see Figure 2).

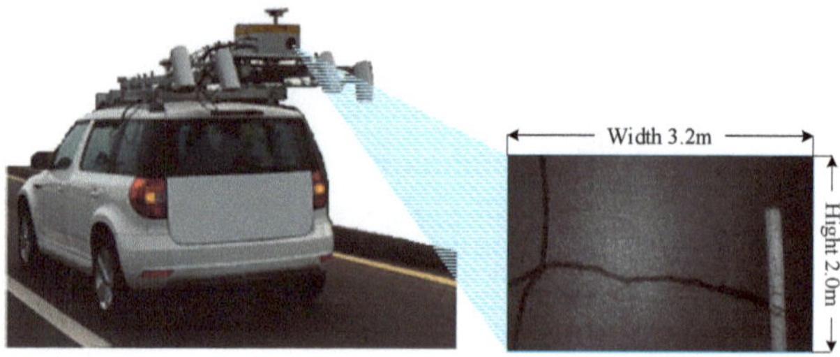

Figure 2. Mobile pavement distress detection equipment. The vehicle was equipped with an acA4096 array camera and two hernia lamps to reduce the influence of light intensity. The image resolution was 3517 × 2193 pixels, and the single pixel size was 0.91 × 0.91 mm.

In this study, we first collected the original images from 504 roads in different regions of China and manually screened 30,000 original CCD images based on four pre-designed keywords according to the extent of pavement damage: high-quality asphalt pavement, low-quality asphalt pavement, high-quality cement concrete pavement, and low-quality cement concrete pavement. Then, we developed a lossless cropping tool, and it took four experts six months to crop the seven kinds of distress (transverse crack, longitudinal crack, cement concrete crack, alligator network crack, broken slabs, patch, and pothole) and negative sample areas in the earlier screened images. Finally, 18,527 sample images were obtained according to the nine kinds of complex scene (clear pavement, fuzzy pavement, bright light, weak light, dry pavement, slippery pavement, shadow, stain, and sundries) for each distress type, covering 6553 distress sample images and 11,974 negative sample images with interference noise (Figure 3), which can effectively reduce the false extraction of complex interference noise (we illustrate this point in the experiments). Note that the selection and tailoring strategy was similar to the approach of Everingham et al. [59] and Zou et al. [30]. The distressed area typically occupies a small proportion of the whole image, and many background areas have no practical significance for the training process. Therefore, most of the selected distress-containing images contain only a single target to provide rich and highly detailed structures. Meanwhile, the segmentation and labeling confusion caused by the co-occurrence of multiple distress types from different categories is avoided as much as possible. Specifically, the selection criteria for the distress images can be summarized as follows:

- We covered more categories while reducing the number of "redundant" samples with simple structures that are already included in the other existing datasets. The focus was on increasing the scene richness of each type of distress sample, which is crucial for improving the reasoning ability of the network model. As shown in Figure 3, we screened seven types of distress sample images, where each type of distress covered nine types of complex scenes (Figure 4).

- We enlarged the intra-type dissimilarities of the selected distress types by adding more diversified intra-type images (see Figure 4). First, the same distress type may appear with different lengths, widths, and topologies, due to the diversity of the distress formation mechanisms in rural pavements. Second, the appearance of road pavement distress is greatly affected by dust accumulation and humidity. For example, the black appearance of the crack and the white appearance of the crack shown in the first row of Figure 4. Finally, the vibration during the shooting also affects the clarity of the distress in the imagery.
- We included more images that are highly similar to the road surface distress in terms of gray-level and texture characteristics, which are called negative samples, such as shadows, water or oil stains, dropped objects, pavement appendages, etc. (Figure 5). These are common in actual distress detection tasks, but they are ignored by the other datasets due to their complex types or collection difficulties.

The purpose of the DIS task is to obtain accurate pixel regions of the different distress types to analyze the condition of the pavement, which seems to be contradictory with the image collection based on pre-designed keywords/types in this paper. The main reasons for this include: (1) To facilitate image retrieval and organization in the construction of the large-scale pavement distress dataset. (2) Collecting samples according to the distress types is a reasonable way to ensure the characteristics (such as texture, topological structure, contrast, background complexity, etc.) of the distress sample diversity, which can improve the robustness and generalization of type-agnostic segmentation. (3) Prior to the development of the different pavement distress inspection systems, the existing datasets needed to be reorganized and extended according to the task requirements. The type information provided in this paper will help developers to quickly track down the required samples. Therefore, the type-based collection approach is intrinsically consistent with the objective of the pavement distress DIS task.

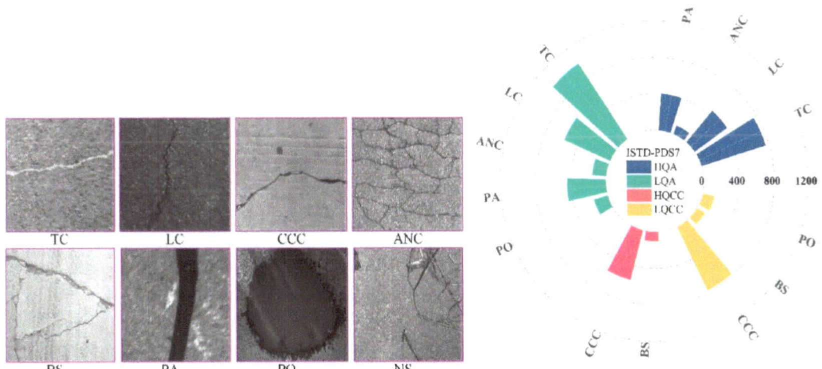

Figure 3. (**Left**) example images of ISTD-PDS7, where "TC", "LC", "CCC", "ANC", "BS", "PA", "PO", and "NS" are the abbreviations for "transverse crack", "longitudinal crack", "cement concrete crack", "alligator network crack", "broken slab", "patch", "pothole", and "negative sample". (**Right**) distress types and groups of the ISTD-PDS7 dataset, where "HQA", "LQA", "HQCC", and "LQCC" are the abbreviations for high quality asphalt pavement, low-quality asphalt pavement, high-quality cement concrete pavement, and low-quality cement concrete pavement.

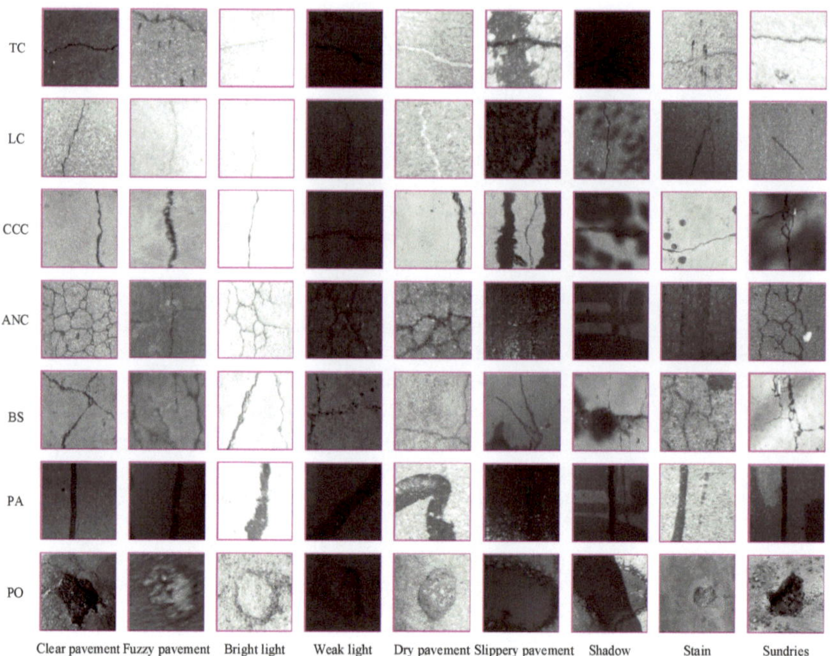

Figure 4. Samples from the ISTD-PDS7 dataset: nine scenes of each distress class are shown. There are 6533 images within seven classes.

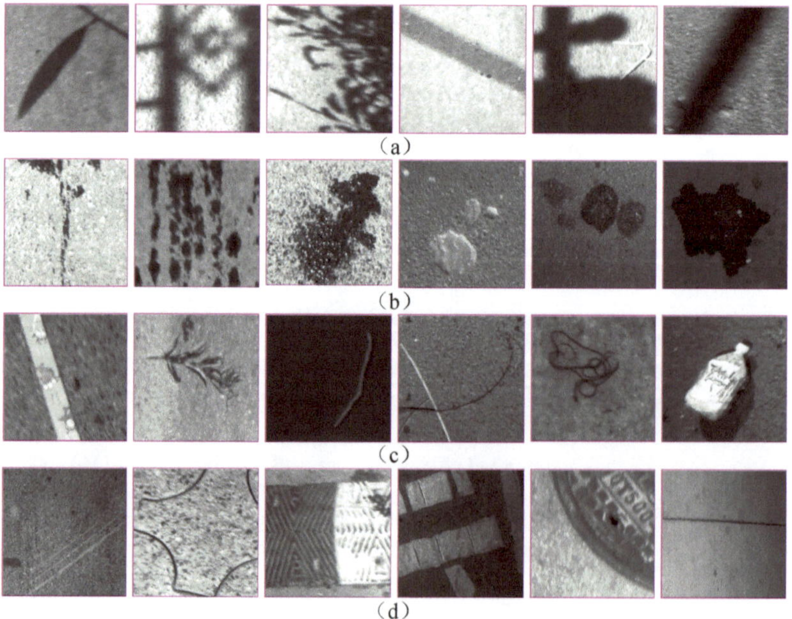

Figure 5. Negative samples from the ISTD-PDS7 dataset. (**a**) Under the influence of interference, such as trees, light poles, and acquisition equipment, shadows appear in pavement images, leading to

deviations in the overall gray-level distribution of the image. Some shadows have similar structural characteristics to the patch, but the roughness of the patch is greater than that of strip shadows; (**b**) Water stains, oil stains, mud stains, and other interference noise, some of which are similar to the linear characteristics of cracks, and some are similar to the appearance of potholes; (**c**) Litter, such as branches, weeds, and garbage, has obvious edge characteristics; (**d**) Other pavement interference noise: scratches, cutting marks, speed bumps, zebra crossings, manhole covers, and cement pavement expansion joints. The existence of negative samples further increases the challenge of the dataset and helps to evaluate the robustness and generalization of different distress detection models.

Dataset Annotation: Each image was manually annotated with pixel-wise precision by four pavement distress detection field experts using LabelMe (Figure 6). The average labeling time for each image was about 20 min and some alligator network crack images took up to 1 h. Figure 6a–g shows the different levels of details in annotation and the differences in detail labeling during the ISTD-PDS7 dataset and the existing public datasets in terms of crack labeling. Figure 6h shows annotated samples of seven types of pavement distress and negative samples in complex scenes. Figure 6i demonstrates the greater diversity of the intra-type structure complexities of the ISTD-PDS7 dataset.

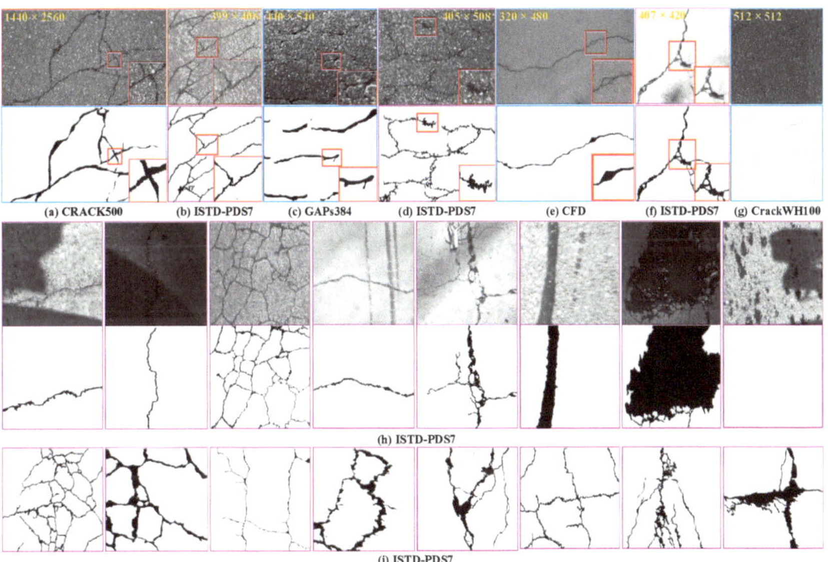

Figure 6. Qualitative comparison of the different datasets: (**a**–**d**) indicate that ISTD-PDS7 provides more detailed crack labeling; (**e**,**f**) show the differences in detail labeling between the CFD [37] dataset and the ISTD-PDS7 dataset; (**g**) is a sample of the CRACKWH100 dataset [30], which was annotated manually with a single-pixel width; (**h**) shows labeling cases of seven types of distress and negative samples in complex scenarios; and (**i**) demonstrates the structural complexity and diversity of the alligator network cracks within the ISTD-PDS7 dataset.

3.2. Data Analysis

- For a deeper insight into the ISTD-PDS7 dataset, we compared the dataset with seven other related datasets: three crack segmentation datasets annotated by a single-pixel width, i.e., CrackLS315, CrackWH100, and CrackTree260 [30], and four datasets annotated manually by the actual width of the crack, i.e., AigleRN [39], CFD [37], CRACK500 [38], and GAPs384 [38]. The comparison was made mainly from the four metrics of image number, image dimension, image complexity, and annotation complexity, and they are described as follows: **Image dimension** is crucial to the segmentation task, as it directly affects the accuracy, efficiency, and computational cost of the segmentation [40].

- **Image Complexity** is described by the image information entropy (IE), which can quantitatively represent the difficulty of object recognition or extraction in complex scenes [60]. The IE ($IE = \sum_{i=0}^{i=255} P_i \log_2 P_i$), where P_i represents the proportion of pixels whose gray value is i in the image measures the information contained in the aggregation features of the gray-level distribution in an image from the perspective of information theory. The greater the IE in an image, the more information the image contains [61,62].

- **Annotation Complexity** is described by the three metrics of the isoperimetric inequality quotient (IPQ) [63–65], the number of distress contours ($Cnum$), and the number of points marked ($Pnum$). Among these metrics, $IPQ = L^2/4\pi A^2$, where L and A denote the distress perimeter and the region area, respectively. The IPQ represents the structural complexity of the labeled distress types. The $Cnum$ is the number of closed contours involved in the labeling, which can quantitatively reflect the complexity of the topological structure of the distress. The $Pnum$ metric is the number of labeling points needed to delineate the outline of the distress example [66], which can quantitatively reflect the fineness of the labeling and the labor cost.

Table 2 lists the statistical findings for four indicators across various data sets. Note that red denotes the best outcomes, green the second-best outcomes, blue the third-best outcomes, and negative samples are excluded from labeling complexity statistics. In addition, these four metrics are complementary and can provide a comprehensive analysis of the complexity of the original imagery and annotated objects, see Figure 7 for details.

The mean values (H, W, D) and standard deviations ($\sigma H, \sigma W, \sigma D$) of the image height, width, and diagonal length of each dataset are listed in Table 2. The CRACK500 dataset has the largest average image dimensions, but it only contains 500 images. In view of the distress area typically occupying a small proportion of the whole CCD image, and many background areas having no practical significance for the training process, targeted cropping was carried out on the distress area in the ISTD-PDS7 dataset, so that the average image dimension of ISTD-PDS7 is relatively small. In addition, the targets of the seven open-source datasets are primarily cracks, which limits their application in diversified tasks. From the mean value and standard deviation of the IE in Table 2 and Figure 7a, it is apparent that, compared with the other public datasets, the ISTD-PDS7 dataset has the highest image complexity. Figure 6h also intuitively indicates that ISTD-PDS7 is closer to the actual application scenario.

Table 2. Data analysis of the existing datasets.

Task	Dataset	Number	Image Dimension			Image Complexity	Annotation Complexity			Annotation Method
		I num	H $\pm\sigma H$	W $\pm\sigma W$	D $\pm\sigma D$	IE $\pm\sigma IE$	IPQ $\pm\sigma IPQ$	$Cnum$ $\pm\sigma Cnum$	$Pnum$ $\pm\sigma Pnum$	
Crack	CrackLS315 [6]	315	512.00 ± 0.00	512.00 ± 0.00	724.00 ± 0.00	50.12 ± 11.46	297.70 ± 199.48	3.40 ± 2.60	618.29 ± 415.57	Single-pixel width
	CrackWH100 [6]	100	512.00 ± 0.00	512.00 ± 0.00	724.00 ± 0.00	36.72 ± 9.15	432.68 ± 452.24	2.90 ± 4.27	855.45 ± 887.29	
	CrackTree260 [6]	260	624.92 ± 48.77	833.23 ± 65.03	1041.54 ± 81.29	53.45 ± 12.15	1122.18 ± 996.63	8.64 ± 16.07	2551.08 ± 2195.26	
	AigleRN [38]	38	522.35 ± 151.02	692.05 ± 280.26	890.23 ± 244.33	32.18 ± 9.27	437.02 ± 416.94	17.40 ± 14.62	1374.35 ± 1014.11	Actual width
	CFD [36]	118	320.00 ± 0.00	480.00 ± 0.00	577.00 ± 0.00	36.42 ± 9.36	106.60 ± 56.30	3.69 ± 4.19	661.78 ± 457.74	
	CRACK500 [37]	500	1568.38 ± 313.60	2594.61 ± 240.45	3042.13 ± 303.57	59.66 ± 13.21	91.24 ± 95.42	18.11 ± 25.92	3603.03 ± 2065.43	
	GAPs384 [37]	384	551.65 ± 99.43	540.00 ± 0.00	775.15 ± 69.60	55.29 ± 12.19	48.44 ± 40.89	5.26 ± 5.57	452.22 ± 286.27	
DIS	ISTD-PDS7	18527	375.47 ± 99.50	371.00 ± 114.44	529.45 ± 145.73	89.28 ± 13.63	134.82 ± 265.07	9.10 ± 16.95	1083.38 ± 1054.07	

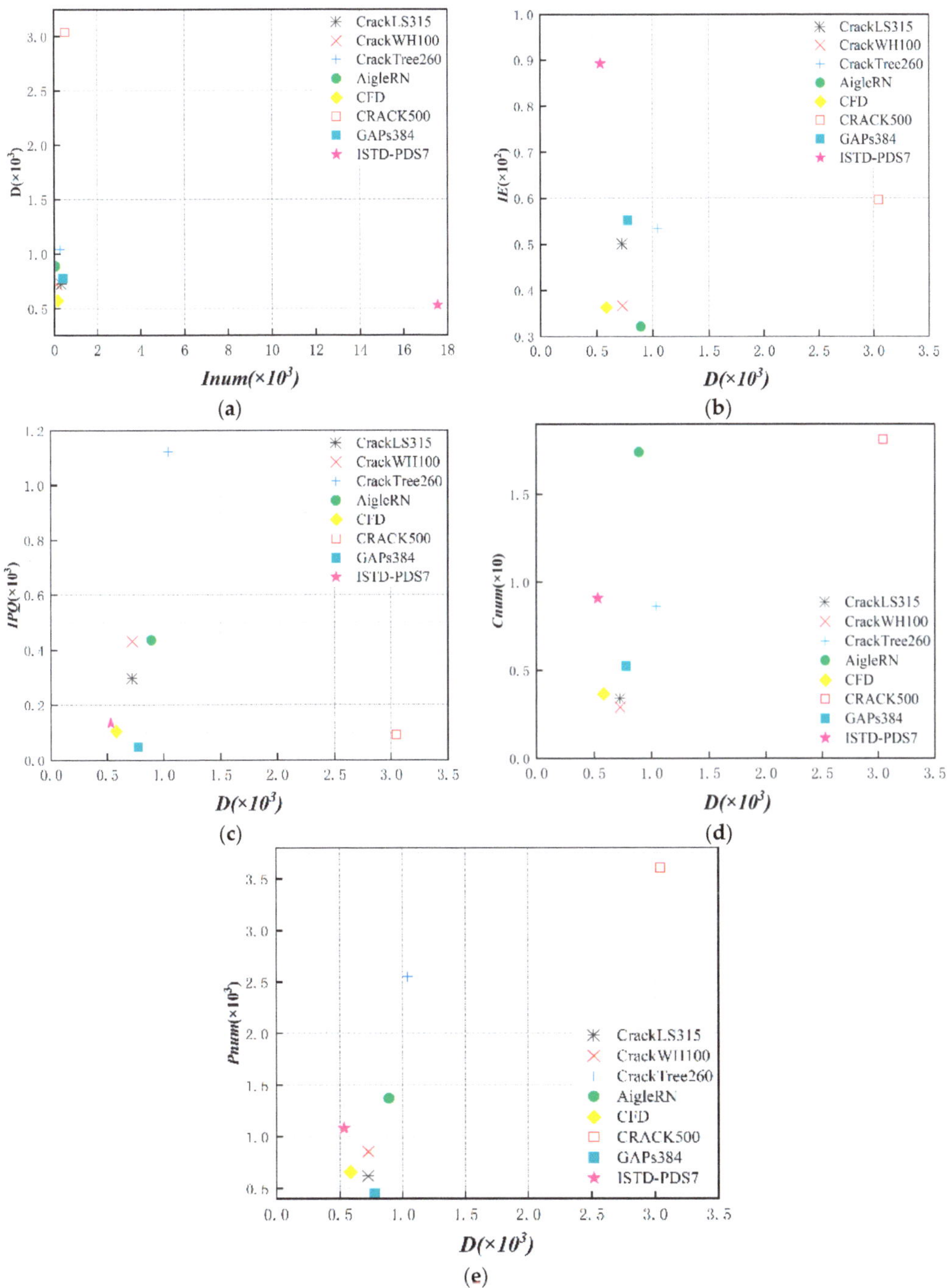

Figure 7. Correlations between the different complexity metrics. These four complexity measurements are complementary and can provide a comprehensive analysis of the complexity of the original imagery and annotated objects (**a**–**e**).

As shown in Table 2 and Figure 7, the CrackTree260 dataset achieves the highest complexity in terms of the average structural complexity *IPQ*, but its *Inum* and *IE* values are far lower than for the ISTD-PDS7 dataset. In addition, although the ISTD-PDS7 dataset contains simply shaped patches (single contour), potholes (single contour), and negative sample labels (no contour), the mean and standard deviation of *Cnum* still achieve high scores, suggesting that the cracks in the new dataset are a more refined structure composed of multiple contours. The average *Pnum* of the ISTD-PDS7 dataset is over 1000, which is around two times more complicated than the GAPs384 dataset. This shows that the ISTD-PDS7 dataset provides more detailed annotation on images with smaller dimensions than the other datasets.

3.3. Dataset Splitting

As shown in Table 3, we split the 18,527 images in the ISTD-PDS7 dataset into three subsets: ISTD-TR (15,774), ISTD-VD (1753), and ISTD-TE (1000) for training, validation, and testing, respectively. The distress types in ISTD-TR and those in ISTD-VD and ISTD-TE are mainly consistent. In addition, using the complexity of the scene and topology structure as the screening index, the 1000 images of ISTD-TE are further split into a subset, containing 532 crack images and 18 negative samples with texture similarity noise, which is called ISTD-CRTE, to evaluate the segmentation performance for pavement cracks. Overall, the ISTD-PDS7 dataset is designed to meet the challenges of model training and performance evaluation in complex scenarios.

Table 3. Dataset splitting of ISTD-PDS7.

Class	ISTD-TR	ISTD-TR Percentage/%	ISTD-VD	ISTD-VD Percentage/%	ISTD-TE	ISTD-TE Percentage/%
Transverse crack	1581	10.0%	176	10.0%	150	15.0%
Longitudinal crack	921	5.8%	102	5.8%	126	12.6%
Alligator network crack	236	1.5%	26	1.5%	249	24.9%
Patch	763	4.8%	85	4.8%	88	8.8%
Pothole	275	1.7%	31	1.7%	15	1.5%
Cement concrete crack	1206	7.6%	134	7.6%	119	11.9%
Broken slab	189	1.2%	21	1.2%	60	6.0%
Negative sample	10,603	67.2%	1178	67.2%	193	19.3%
Total	15,774	100.0%	1753	100.0%	1000	100.0%

4. Baseline Methods

4.1. Methods

We evaluated the performance of seven state-of-the-art convolution-based and transformer-based semantic segmentation models on the ISTD-PDS7 dataset.

- **SegNet [49]:** The SegNet network achieves end-to-end learning and segmentation by sequentially using an encoder network and a decoder network. It can process input images of any size. VGG16 [67] without the fully connected layer is used as the encoding phase to achieve feature extraction. The sizes of the input images and the network parameters are reduced step by step through maximum pooling, and the pooling index position in the image is recorded at the same time. The decoding phase restores the resolution of the image through multiple upsampling. Finally, the semantic segmentation results are output by the SoftMax classifier.
- **PSPNet [68]:** The PSPNet network uses a pyramid pooling module (PPM) to aggregate contextual information from different regions to improve the ability to obtain global information. This network came first in the ImageNet Scene Parsing Challenge 2016, the PASCAL VOC Challenge 2012, and the Cityscapes test (2016).
- **DeepLabv3+ [69]:** At the decoder stage, atrous convolution is introduced to increase the receptive field, and atrous spatial pyramid pooling (ASPP) is used to extract multi-scale information. DeepLabv3+ achieved a test set performance of 89% and 82.1%

without any post-processing on the PASCAL VOC Challenge 2012 and Cityscapes test (2018).

- **U-Net** [44]: The U-Net network performs skip-layer fusion for end-to-end boundary segmentation and formulates the training target with a single loss function. This model is the most commonly used model in medical image segmentation. We modified the padding mode to be the "same" so that the input and output image sizes remained the same.

- **HRNet** [70]: Differing from the above four methods, HRNet is a network model that breaks away from the traditional encoder-decoder architecture. HRNet maintains high-resolution representation by connecting high-resolution to low-resolution convolution in parallel and enhancing the high-resolution representations by repeatedly performing multi-resolution fusion across parallel convolution. In this way, it can learn high-resolution representations that are more sensitive to location.

- **Swin-Unet** [71]: Swin-Unet is a UNet-like pure transformer for image segmentation, using a transformer-based U-shaped encoder-decoder architecture with skip connections for local-global semantic feature learning.

- **SegFormer** [72]: An efficient encoder-decoder architecture for image segmentation, using multiple layers Transformer-Encoder to get multiscale features. At the same time, a lightweight multilayer perceptron (MLP) is used to aggregate semantic information of different layers. SegFormer has shown a state-of-the-art performance on the ADE20K dataset, performing better than the Segmentation Transformer (SETR) model [73], Auto-DeepLabv3+ [69], and OCRNet [74].

4.2. Loss Function Selection

Differing from semantic segmentation on the Pascal VOC2012 dataset [41], there are only two classes in the new dataset. Segmentation of the distress types in this paper can be seen as a DIS problem. Generally speaking, the ground-truth distress pixels stand as a minority class in the distress image (i.e., the proportion of background pixels in the new dataset is 97.42%, and the proportion of distress pixels is 2.58%), which makes it an imbalanced segmentation task. Some works [75,76] have dealt with this problem by adding larger weights to the minority class. However, in crack detection [30], it has been found that adding larger weights to the cracks results in more false positives. In order to tackle both types of imbalance during training and inference, we introduce a hybrid loss function consisting of contributions from both dice loss [77] and cross-entropy loss [78]. Specifically, the dice loss (Equation (1)) learns the class distribution, alleviating the imbalance problem, while the cross-entropy loss (Equation (2)) is used to penalize false positives/negatives while performing curve smoothing at the same time. The two loss terms are combined as shown in Equation (3), and more weight is given to the dice loss term because it can better handle the category imbalance problem. Thus, we define the pixel-wise prediction loss as shown in the following equations:

$$L(\mathbf{W})_{Dice} = 1 - \frac{2\sum_{n=1}^{N} y_n \hat{y}_n(x_n, \mathbf{W}) + \varepsilon}{\sum_{n=1}^{N} y_n + \sum_{n=1}^{N} \hat{y}_n(x_n, \mathbf{W}) + \varepsilon} \tag{1}$$

$$L(\mathbf{W})_{CE} = -\frac{1}{N}\sum_{n=1}^{N} y_n \log \hat{y}_n(x_n, \mathbf{W}) + (1 - y_n)\log(1 - \hat{y}_n(x_n, \mathbf{W})) \tag{2}$$

$$L(\mathbf{W}) = 0.9 \cdot L(\mathbf{W})_{Dice} + 0.1 \cdot L(\mathbf{W})_{CE} \tag{3}$$

Given one set of batch-size training data with M input images, $S = \{(X^m, Y^m), m = 1, \cdots, M\}$, N denotes the total number of pixels and is equal to the number of pixels in a single image multiplied by $x_n \in \{[1, 255], n = 1, \cdots, N\}$, where $x_n \in \{[1, 255], n = 1, \cdots, N\}$ denotes the pixel of the input image; $y_n \in \{0, 1\}$ denotes the ground-truth distress label map corresponding to x_n; $\hat{y}_n \in \{0, 1\}$ is the predicted probability for x_n being distress or background; $\mathbf{W}$ is the set of standard parameters in the network layers; $L(\mathbf{W})_{Dice}$ denotes the dice loss, ε is used for smoothing purposes; $L(\mathbf{W})_{CE}$ denotes the cross-entropy loss; and $L(\mathbf{W})$ denotes the total loss.

4.3. Evaluation Metrics

In this paper, to provide a comprehensive evaluation, the *Precision*, *Recall*, *F1*, and *mIoU* are used to quantitatively evaluate the performance of the different segmentation models. For each image, the *Precision* and *Recall* can be calculated by comparing the detected distress with the human-annotated ground truth. The *F1* ($2 \times \frac{Precision \times Recall}{(Precision + Recall)}$) is the *Precision* and *Recall* harmonic average. The intersection over union (*IoU*) reflects the overlap degree between the recognized samples of the same class and the real samples, and the *mIoU* metric is the average value of the *IoU*. The *mIoU* can be calculated as follows:

$$mIoU = \frac{1}{N}\sum_{k=1}^{N} \frac{TP_K}{TP_K + FP_K + FN_K} \tag{4}$$

where TP_K, FP_K, and FN_K represent the true positives, false positives, and false negatives, respectively, $N = 2$.

5. Experiments and Results

In this section, we first introduce the experimental settings. The distress segmentation performance on the test dataset of the seven representative end-to-end semantic segmentation models (see Section 4.1) selected for training on the new dataset is then compared and discussed in detail. In addition, we describe the comprehensive comparative analysis conducted between ISTD-PDS7 and the open-source datasets on the optimal baseline model. Finally, the influence of negative samples and the effect of data augmentation methods on the segmentation accuracy for road pavement distress in complex scenes are analyzed. Figure 8 shows the specific experimental flowchart of this paper.

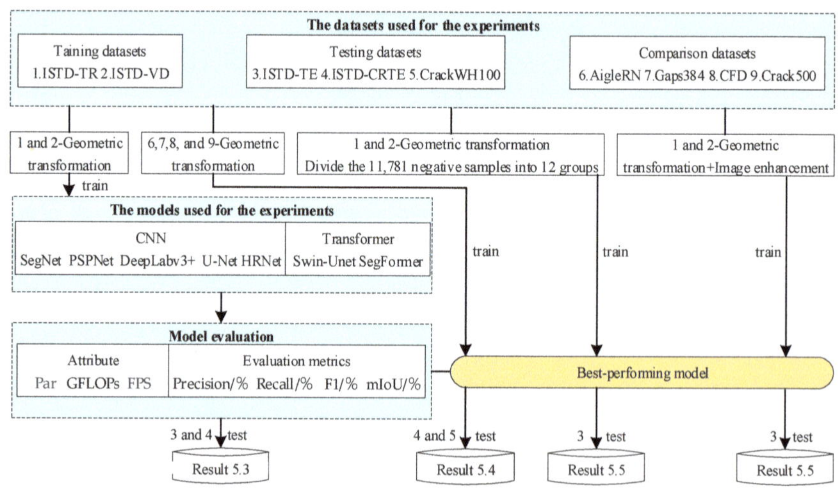

Figure 8. Experimental research flowchart.

5.1. Implementation Details

We implemented the different networks using the publicly available PyTorch, which is well-known in this community. To improve the learning performance, transfer learning [79] was adopted to train all the experimental models. The pre-trained models selected the optimal weight of the backbone network trained on the Cityspaces dataset. The output category is set to 2. The models were then trained in a freezing phase (50EP) and a thawing phase (100EP). The batch processing size of the two phases was set to 4 and 8, respectively, and the learning rates were set to 10^{-4} and 10^{-5}, respectively. The momentum and weight decay were set to 0.9 and 0, respectively. The model training process was optimized using the AdamW optimizer with a learning rate of 1×10^{-4}. All the experiments described in this paper were performed using a single GeForce RTX 3090 GPU.

5.2. Dataset Setup

As described in Section 3.1, the ISTD-PDS7 dataset was built from the ground up and covers a highly diverse set of geometric structures and image scenarios for pavement distress. Thus, its diversity (i.e., resolution, image features, object complexity, and markup accuracy) and distribution differ from the existing datasets.

- **Baseline evaluation of datasets:** As described in Section 3.3, ISTD-TR and ISTD-VD were used for the training and validation of the baseline evaluation networks. Data augmentation was performed to enlarge the number of distress samples in the training set, including vertical flip, horizontal flip, and flip and transpose, to balance the ratio of positive and negative samples. After the data augmentation, we obtained a training set of 30,475 images in total, containing 14,620 distress samples and 11,974 negative samples. It is worth noting that the data augmentation was not applied to the ISTD-TE or ISTD-CRTE. All the baseline models used ISTD-TE and ISTD-CRTE as the test sets to evaluate their performance in the multi-type of distress DIS task and crack segmentation task in complex scenarios.

- **Dataset comparison:** In this study, the best-performing model from the baseline assessment was used as the evaluation tool to make cross-dataset evaluations [80] of CFD [37], CRACK500 [38], AigleRN [39], and GAPs384 [38], which are labeled with the actual crack width. In order to ensure the fairness of the evaluation, we used the same data enhancement methods described above, and increased the training data size to about 14,600 pieces. CrackWH100 [30] and ISTD-CRTE were selected as the test sets. It must be noted that the images in the CrackWH100 dataset were acquired by a linear CCD camera, and we re-annotated their ground truth by the actual pixel width of the crack using LabelMe.

- **Influence of negative samples and different data augmentation methods:** We took the best-performing model in the baseline assessment as the experimental tool, randomly divided the 11,781 negative samples into 12 groups with roughly 1000 images per group and added one group of negative samples each time to participate in the model training. Then, we use two data augmentation methods, geometric transformation (vertical flip, horizontal flip, and flip and transpose) and image enhancement (shift scale rotate, random contrast, random brightness, blur, and CLAHE), on the training set to explore the influence of different data augmentation methods on the performance of the distress segmentation. Image geometric transformation simulates the change of direction and angle when an image is taken. Blur change takes into account the possible instability of the imaging camera when the light is low and the lens is unfocused. Brightness transition mainly enhances or weakens the illumination, considering the unstable situations that may occur in the case of insufficient light or severe exposure during shooting. The ISTD-TE dataset was taken as the test set in this part.

5.3. ISTD-PDS7 Benchmark

5.3.1. Quantitative Evaluation

Table 4 lists the overall performance of all the comparative models on the ISTD-PDS7 validation and test sets. It can be observed that the performance of the different models in the crack segmentation task is lower than that in the multi-type of distress segmentation task, because the crack structure is more complex and fine, so it requires the models to keep as much spatial information as possible, which is challenging to most models. Compared with the other models, SegFormer [72] based on the transformer achieves the most competitive performance in the four evaluation indicators. In contrast, the mIoU values of Swin-Unet [71] on ISTD-TE and ISTD-CRTE decrease by 7.85 and 9.00, respectively. It is worth noting that although Swin-Unet has a similar number of network parameters as SegFormer, it has a low computational complexity (GFLOPs = 52.56) and high computational speed (frames per second (FPS) = 110.95). In addition, the HRNet [70] and U-Net [44] models based on convolution achieve an average performance. Compared with the other two models, HRNet [70] performs better in the distress DIS task, while U-Net [44] performs better in the crack segmentation task. The two models achieve the lowest FPS values and show a slow reasoning speed. SegNet [49], with the simple encoder-decoder architecture, performs the worst on the test sets. These baseline results will provide scientific data support for subsequent model research.

Table 4. Quantitative evaluation on the ISTD-PDS7 validation and test sets. V-16 = VGG16 [67], MV2 = MobileNetV2 [81], R-50 = ResNet50 [18], H-V2 = HRNetV2 [70], X = Xception [82], and ST = swin transformer [71].

Dataset	Metric	SegNet [49]	PSPNet [68]		DeepLabv3+ [69]		U-Net [44]	HRNet [70]	Swin-Unet [71]	SegFormer [72]
	Backbone	V16	MV2	R-50	MV2	X	V-16	H-V2	ST	MiT-B2
	Input size	512×512	473×473	473×473	512×512	512×512	512×512	480×480	224×224	512×512
Attribute	Par/M	16.32	2.38	46.71	5.813	54.709	24.89	29.538	27.18	27.348
	CC/GFLOPs	601.78	5.28	118.43	52.87	166.841	450.602	79.915	52.56	113.427
	Speed/FPS	57.91	131.38	75.53	100.32	47.05	26.99	23.32	110.95	34.75
ISTD-VD	Precision/%	74.84	86.31	82.69	83.41	86.87	87.05	87.22	85.02	88.39
	Recall/%	57.12	82.83	80.2	80.05	80.43	87.81	83.66	77.33	89.69
	F1/%	64.79	84.53	81.43	81.70	83.53	87.43	85.40	80.99	89.04
	mIoU/%	54.15	75.72	72.08	72.64	74.33	79.47	76.80	71.3	81.67
ISTD-TE	Precision/%	81.14	92.14	85.67	89.44	96.31	92.77	92.80	89.56	93.64
	Recall/%	63.98	87.27	80.56	87.94	83.49	91.30	92.65	88.66	94.82
	F1/%	71.55	89.64	83.04	88.68	89.44	92.03	92.72	89.11	94.23
	mIoU/%	60.40	82.27	73.55	81.03	81.13	85.96	87.07	81.64	89.49
ISTD-CRTE	Precision/%	87.31	79.16	76.08	84.46	90.62	85.17	84.15	82.79	87.22
	Recall/%	60.45	68.78	73.11	79.64	70.42	86.67	86.02	76.93	87.06
	F1/%	71.44	73.61	74.57	81.98	79.25	85.91	85.07	79.75	87.14
	mIoU/%	57.05	63.50	65.00	72.66	67.50	77.50	76.43	70.12	79.12

5.3.2. Qualitative Evaluation

Figure 9 presents a qualitative comparison between the seven baseline methods. As shown in the first column of Figure 9, the seven kinds of pavement distress and negative sample images were randomly selected, some of which are affected by interference noise, such as shadows, oil stains, or zebra crossings. A visual inspection shows that SegFormer based on the hierarchical attention mechanism module outperforms the other six methods in the multi-type of pavement distress DIS task, especially for the distress types with different sizes, illumination, and interference noise, and the false positive predictions are reduced. For the semantic segmentation model based on convolutional operations, SegNet can only achieve a rough segmentation of the distress examples, and shows poor processing of details, such as poor continuity of the crack extraction results, and it has difficulty in extracting the planar area of pothole. In addition, Table 3 and columns 4–7 in Figure 9 show that the PPM and ASPP modules are, respectively, used in PSPNet and DeepLabv3+ for contextual information. Due to the multi-scale pooling and atrous convolution operations, a large amount of detailed information is lost, along with detailed boundaries of the pavement cracks, thereby reducing the continuity of the fine cracks. In contrast, U-Net (column 8 in Figure 9) and HRNet (column 9 in Figure 9), which take into account the fusion of multi-scale feature information, are more suitable for the task of pavement distress prediction and obtain better prediction results, but the prediction speed is the slowest (see Table 2). Columns 10 and 11 of Figure 9 demonstrate the potential of a pure transformer backbone for dense prediction tasks, compared to a CNN. In terms of operation speed, Swin-Unet obtains the second-fastest prediction speed, with 110.95 FPS, due to the low computational complexity and few parameters, but the pure transformer operation in this model results in a loss of detailed information and insufficient edge information for the crack prediction.

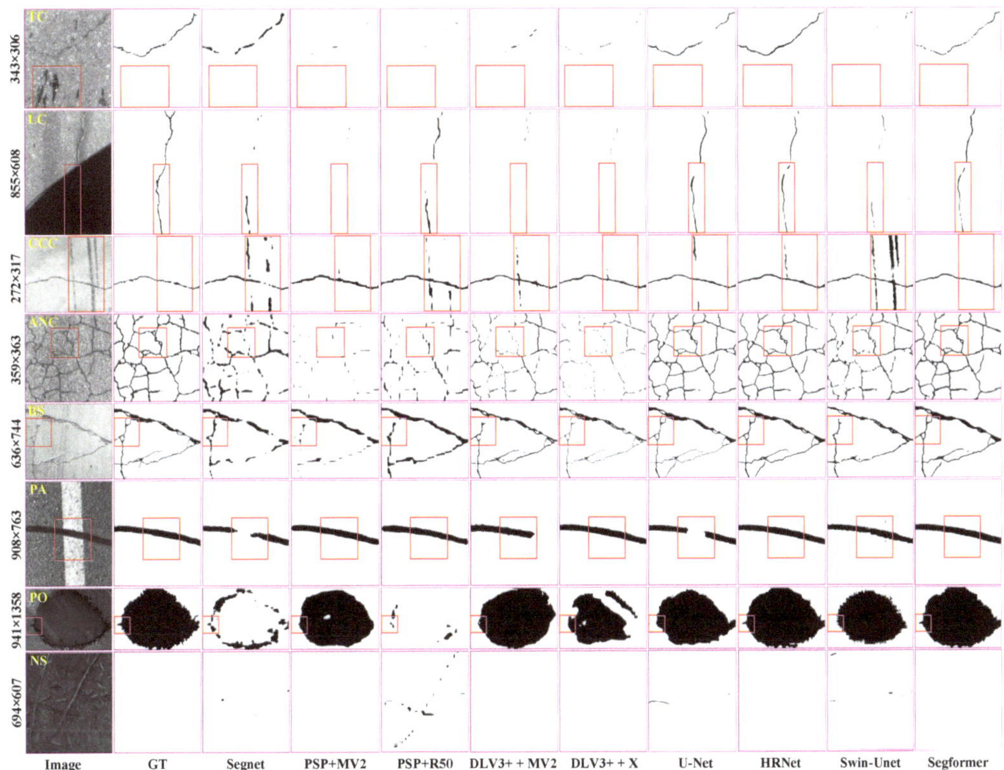

Figure 9. Segmentation results comparison for the seven classes of distress and distractors in complex scenes by the seven methods on the ISTD-PDS7 dataset. The red boxes in each image in the first row are the focus areas.

In contrast, a transformer encoder with hierarchical structure is used in SegFormer, which outputs multi-scale features instead of single low-resolution features, as in the vision transformer (ViT) model [83]. In addition, positional encoding is removed, which can avoid the problem of intensive prediction performance degradation caused by the interpolation of positional codes when the testing resolution differs from the training resolution [72]. The proposed MLP decoder aggregates information from different layers and combines local attention and global attention to render powerful feature representation capabilities, thus effectively reducing false positive predictions. The above findings and baseline results will provide certain reference information for the design of subsequent intelligent distress segmentation networks.

5.4. Comparison with the Public Datasets

Table 5 provides the quantitative comparison results of SegFormer trained by the AigleRN, CFD, CRACK500, GAPs384, and ISTD-PDS7 datasets on the two test sets. Obviously, the model trained based on ISTD-PDS7 shows the best crack extraction performance in the two test sets. Specifically, compared to the AigleRN and CRACK500 datasets. On the CrackWH100 test set, the SegFormer achieves an increase of 5.07% to 10.09 % in the F1 values and 6.17% to 11.62% in the mIoU values after being trained using the ISTD-PDS7 dataset. On the ISTD-CRTE test set, the SegFormer achieves an increase of 3.36% to 8.39% in the F1 values, and 25.65% to 5.35% in the mIoU values after being trained using the ISTD-PDS7 dataset. In addition, the results of Table 5 show that the SegFormer trained using the four public datasets, F1, and mIoU scores on the CrackWH100 test set were better than those on the ISTD-CRTE test set. According to a comparison of the two test sets, it was found that the crack images in the ISTD-CRTE contain more texture similarity noises, such as shadows, branches, stains, etc. (Figure 5). The new test set also includes dramatically changed illumination conditions and various crack width sizes (Figure 4), which were more complex than the existing public datasets. Consequently, when evaluating the crack detection performance of different crack detection models in complex scenarios, it can be much more challenging to use it as the test set.

Table 5. Performance comparison for SegFormer trained on the different datasets.

Dataset	Metric	AigleRN [39]	Gaps384 [38]	CFD [37]	Crack500 [38]	ISTD-PDS7
CrackWH100 [6]	Precision/%	78.38	76.75	75.48	84.29	85.73
	Recall/%	80.36	87.28	93.83	84.47	93.50
	F1/%	79.36	81.68	83.66	84.38	89.45
	mIoU/%	70.58	72.32	73.39	76.03	82.20
ISTD-CRTE	Precision/%	89.86	71.65	75.72	78.16	87.22
	Recall/%	55.67	84.58	88.81	90.26	87.06
	F1/%	68.75	77.58	81.74	83.78	87.14
	mIoU/%	53.47	66.40	71.12	73.77	79.12

Figure 10 shows the crack detection results obtained on the two test sets. A visual inspection shows that SegFormer trained with the five different datasets shows a different performance on the two test sets, and the model trained on ISTD-PDS7 performs better in the crack detection tasks than the others, especially for cracks with different appearances, contrast, and interference conditions. Specifically, as shown in Figure 10a,f, although the cracks in the original images are affected by water stains and shadows, the model trained with the new dataset can still effectively depict cracks under the low contrast condition, while the models trained with the other datasets can extract cracks in the rectangular box area with a low degree of confidence. In addition, as shown in Figure 10b,c, the four public dataset trained models misclassify grass roots/branches (which are highly similar to the crack texture) as cracks to generate false positive predictions. It is worth noting that, as shown in Figure 10d, when an image with a "white crack" exists, the model trained with the new dataset can correctly and completely delineate the crack, while the models trained based on the other datasets fail since they are lacking in "white crack" samples. It can also be observed that, as shown in Figure 10e, for the alligator network crack with a clear appearance, the continuity of the cracks delineated by the model trained on the AigleRN dataset is the worst, which is also consistent with the result for the lowest image complexity of this dataset (see Table 2). The models trained with the other datasets produce more complete predictions, but the model trained with the new dataset shows more detailed crack extraction results. Thus, it can be concluded that training the model on the ISTD-PDS7 dataset can effectively improve the precision of crack extraction and the robustness of background noise suppression.

5.5. Influence of Negative Samples and Data Augmentation Methods

Influence of Negative Samples: The ISTD-PDS7 dataset includes more negative samples that are highly similar to pavement distress in terms of gray-level and texture characteristics. In this experiment, SegFormer was used as the experimental model (see Section 5.3 for details) to analyze the influence of negative samples on the segmentation accuracy of pavement distress in complex scenes. Figure 11 shows the variation trend of F1 and mIoU on the ISTD-TE dataset with the increase in the number of negative samples involved in the training.

As can be seen from Figure 11, with the increase in the negative samples, the segmentation performance of the trained model is gradually improved. When all the negative samples participate in the model training, the F1 and mIoU are increased by 4.25% and 7.00%, respectively. When the number of negative samples participating in the training reaches 5000, the difference between the negative sample images participating in the training and the overall negative samples in the ISTD-PDS7 dataset is significantly reduced, resulting in the growth trend of the F1-score and mIoU of the subsequent model slowing down, but the model does not reach a saturation state.

Figure 12 visually demonstrates the effectiveness of negative samples in reducing the false extraction of road interference noise in complex scenes. When the original CCD imagery contains complex interference noise, such as shadows, oil stains, water stains, branches, scratches, and tire indentation (Figure 12, line 1), if negative samples are not used in the model training, SegFormer can overcome some of the noise interference, but in the face of interference noise that is very similar to the pavement distress characteristics, there are still many false positive predictions (Figure 12, line 2). In contrast, when the number of negative samples involved in the model training reaches 5000, the false positive predictions decrease significantly (Figure 12, line 3). When all the negative samples participate in the model training, most of the false positive predictions disappear (Figure 12, line 4). Therefore, it can be concluded that only using normal samples in the training cannot achieve a satisfactory precision. However, after the introduction of the targeted negative samples in the model training, the DCNN can extract richer semantic information from the imagery. By judging the gap between features, the network focuses on the distress areas, which makes the model have a higher feature expression ability and effectively reduces the false extraction of complex interference noise. In order to meet the practical engineering requirements, in our future dataset maintenance and algorithm research, we will expand the negative samples through the developed manual inspection software.

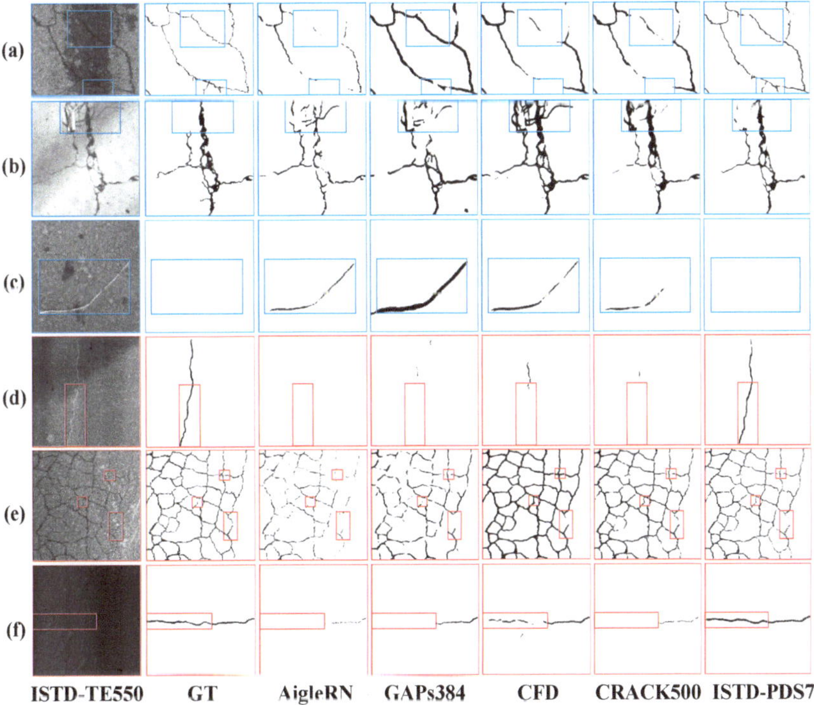

Figure 10. Comparison of the segmentation results of SegFormer after training on different datasets, (**a–c**) come from ISTD-PDS7; (**d–f**) are come from CrackWH100. The boxes in each image are the focus areas.

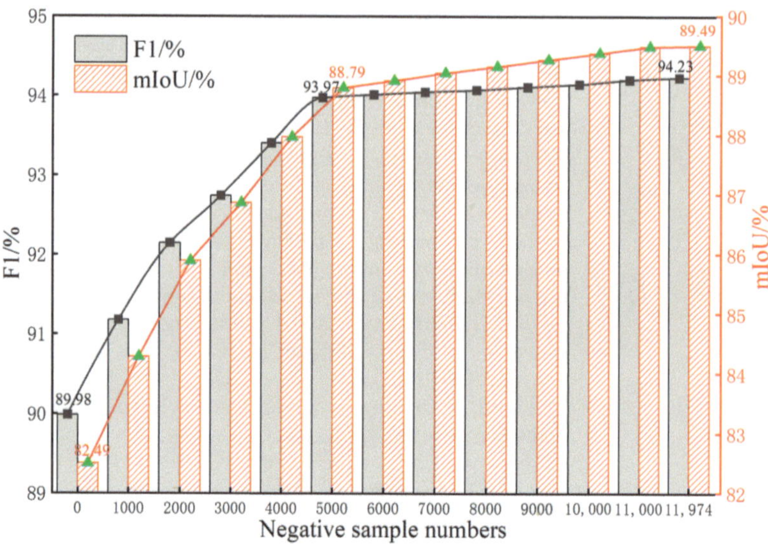

Figure 11. Effects of different numbers of negative samples on the distress segmentation performance.

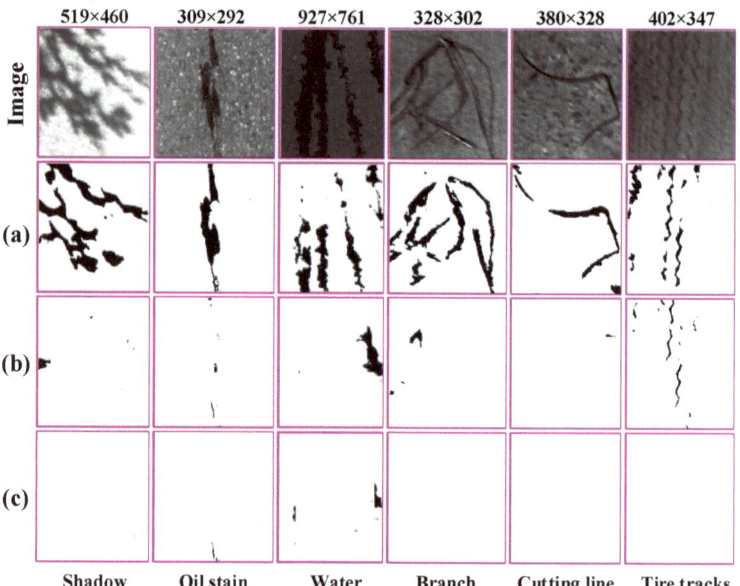

Figure 12. The influence of negative samples on fracture extraction. (**a**) SegFormer prediction results when negative samples are not used in the training. (**b**) SegFormer prediction results when 5000 negative samples are used in the training. (**c**) SegFormer prediction results when 11,781 negative samples are used in the training.

Influence of Data Augmentation Methods: In computer vision, image augmentation has become a common implicit regularization technique to solve the overfitting problem in DL models and is widely used to improve performance. In this experiment, SegFormer was used as a test model to verify the impact of two data augmentation techniques.

According to the quantitative comparison in Table 6 overall, the two data augmentation techniques adopted on the ISTD-PDS7 dataset can effectively improve the performance of SegFormer. While it also can be seen from Table 6 that the most direct geometric transformation improves the mIoU by 5.60%, the image enhancement results in a relatively small improvement in the distress segmentation performance, and the mIoU is improved by only 1.90%. ISTD-PDS7 retains the two types of enhancement methods and can improve the mIoU by 6.79%. The geometric transformation technique increases the variation of the distress, such as the position, viewpoint, scale, etc. These operations can simulate multiple variations of objects in natural scenes, thus improving the within-class richness. Although the image enhancement method also increases the richness of the imagery, the reason for its small performance improvement may be that, although the boundary features of the distress in the original image are changed, the corresponding label is not changed.

Table 6. Results of different data augmentation techniques on the test set.

Method	Training Set	Precision/%	Recall/%	F1/%	mIoU/%
None	17,147	92.04	89.3	90.65	83.89
Image enhancement	47,601	92.21	91.63	91.92	85.79
Geometric transformation	30,475	93.64	94.82	94.23	89.49
All	60,929	93.92	95.74	94.82	90.68

Furthermore, Figure 13 shows the influence of the two types of data augmentation techniques adopted in this study on the details of the distress segmentation. In comparison, it is found that both types of data enhancement technique improve the accuracy and completeness of the distress segmentation.

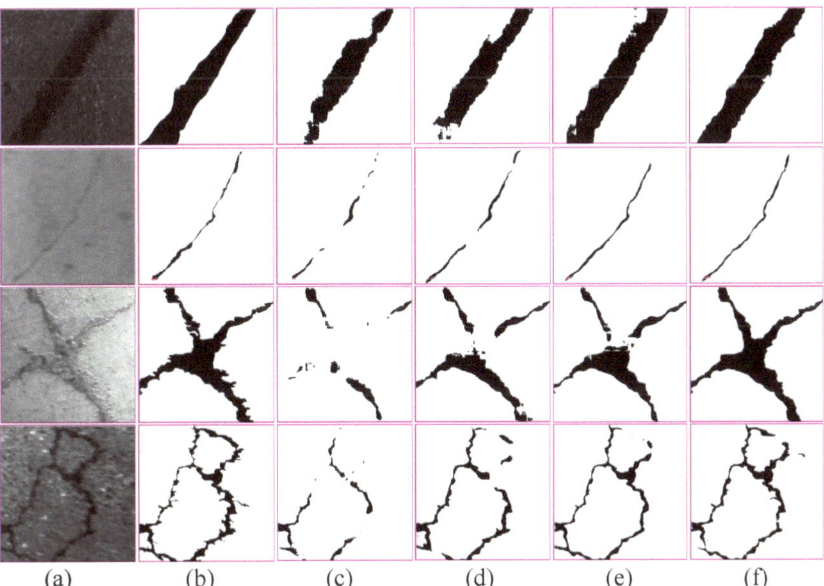

(a)   (b)   (c)   (d)   (e)   (f)

Figure 13. The segmentation performance with different data augmentation techniques. (**a**) The original image of road pavement distress. (**b**) The ground truth. (**c**) The segmentation results without data augmentation. (**d**) The segmentation results using only image enhancement. (**e**) The segmentation results using only geometric enhancement. (**f**) The segmentation results of using the two data augmentation techniques.

6. Conclusions

In this paper, we first reviewed the related research on the automatic detection of highway pavement distress based on DL techniques. We found that the current status of the public datasets and test results is insufficient for engineering-level application research, which seriously limits the progress of automatic pavement distress segmentation in complex scenes. Therefore, to solve the problem, we built a new large-scale DIS dataset for multi-type pavement distress segmentation, i.e., the ISTD-PDS7 dataset, based on measured highway pavement CCD images, which is currently the largest and most challenging dataset for the semantic segmentation of pavement distress. Additionally, we evaluated a set of representative semantic segmentation methods on the new dataset, which can serve as baseline results for future works. Quantitative assessments and qualitative inspections demonstrate that the Segformer model with multiple layers Transformer-Encoder is more suitable for the segmentation of multiple types of pavement diseases in complex scenes. The ISTD-PDS7 dataset can effectively improve the precision of crack extraction and the robustness of background noise suppression when compared to other publicly accessible datasets. The inclusion of negative samples in model training can effectively avoid false positive detection of models. Additionally, the necessary data augmentation methods can considerably improve the pavement distress segmentation performance of the model. Furthermore, based on the new dataset, we intend to carry out additional research on data augmentation methods, training strategies, and model improvement tactics to promote the development of applications in this field.

Author Contributions: Conceptualization, W.S.; data curation, Z.Z., H.Z. and J.Z.; formal analysis, Z.Z., B.Z. and G.J.; funding acquisition, W.S.; investigation, G.J.; methodology, W.S. and Z.Z.; project administration, W.S.; software, H.Z.; visualization, J.Z.; writing—original draft, W.S. and Z.Z.; writing—review and editing, B.Z., G.J., H.Z. and J.Z. All authors have read and agreed to the published version of the manuscript.

Funding: This work was funded in part by The National Natural Science Foundation of China under Grant 42071343 and 42204031, and in part by the Basic Scientific Research Expenses of Heilongjiang Provincial Universities, China, under Grant 2020-KYYWF-0690.

Data Availability Statement: The ISTD-PDS7 dataset can be obtained on request by e-mail from https://ciigis.lntu.edu.cn/, accessed on 1 October 2023.

Conflicts of Interest: The authors declare no conflict of interest.

References

1. Xiao, C.; Zhang, X.F.; Li, B.; Ren, M.; Zhou, B. A new method for automated monitoring of road pavement aging conditions based on recurrent neural network. *IEEE Trans. Intell. Transp. Syst.* **2022**, *23*, 24510–24523.
2. Babkov, V.F. *Road Conditions and Traffic Safety*; Mir Publishers: Portsmouth, NH, USA, 1975.
3. Elghriany, A.F. *Investigating Correlations of Pavement Conditions with Crash Rates on In-Service U.S. Highways*; University of Akron: Akron, OH, USA, 2016.
4. Hao, Y.R. 2021 Statistical Bulletin on the Development of the Transportation Industry [WWW Document]. Available online: http://www.gov.cn/shuju/2022-05/25/content_5692174.htm (accessed on 25 May 2022).
5. Mei, Q.; Gul, M. A Cost Effective solution for pavement crack inspection using cameras and deep neural networks. *Constr. Build. Mater.* **2020**, *256*, 119397. [CrossRef]
6. Cheng, H.; Shi, X.J.; Glazier, C. Real-Time image thresholding based on sample space reduction and interpolation approach. *J. Comput. Civ. Eng.* **2003**, *17*, 264–272. [CrossRef]
7. Ayenu-Prah, A.Y.; Attoh-Okine, N.O. Evaluating pavement cracks with bidimensional empirical mode decomposition. *EURASIP J. Adv. Signal Process.* **2008**, *2008*, 861701. [CrossRef]
8. He, Y.Q.; Qiu, h.x; Jian, W.; Wei, Z.; Xie, J.F. Studying of road crack image detection method based on the mathematical morphology. In Proceedings of the 2011 4th International Congress on Image and Signal Processing, Shanghai, China, 15–17 October 2011; Volume 2, pp. 967–969.
9. Amhaz, R.; Chambon, S.; Idier, J.; Baltazart, V. Automatic crack detection on Two-Dimensional pavement images: An algorithm based on minimal path selection. *IEEE Trans. Intell. Transp. Syst.* **2016**, *17*, 2718–2729. [CrossRef]
10. Zhang, D.J.; Li, Q.T.; Chen, Y.; Cao, M.; He, L.; Zhang, B.L. An efficient and reliable coarse-to-fine approach for asphalt pavement crack detection. *Image Vis. Comput.* **2017**, *57*, 130–146. [CrossRef]
11. Li, N.N.; Hou, X.D.; Yang, X.Y.; Dong, Y.F. Automation recognition of pavement surface distress based on support vector machine. In Proceedings of the 2009 Second International Conference on Intelligent Networks and Intelligent Systems, Tianjian, China, 1–3 November 2009; pp. 346–349.

12. Carvalhido, A.G.; Marques, S.; Nunes, F.D.; Correia, P.L. Automatic Road Pavement Crack Detection Using SVM. Master's Thesis, Instituto Superior Técnico, Lisbon, Portugal, 2012.

13. Ai, D.H.; Jiang, G.Y.; Siew Kei, L.; Li, C.W. Automatic Pixel-Level pavement crack detection using information of Multi-Scale neighborhoods. *IEEE Access* **2018**, *6*, 24452–24463. [CrossRef]

14. Hoang, N. An artificial intelligence method for asphalt pavement pothole detection using least squares support vector machine and neural network with steerable Filter-Based feature extraction. *Adv. Civ. Eng.* **2018**, *2018*, 7419058. [CrossRef]

15. Li, G.; Zhao, X.X.; Du, K.; Ru, F.; Zhang, Y.B. Recognition and evaluation of bridge cracks with modified active contour model and greedy search-based support vector machine. *Autom. Constr.* **2017**, *78*, 51–61. [CrossRef]

16. Majidifard, H.; Adu-Gyamfi, Y.; Buttlar, W.G. Deep machine learning approach to develop a new asphalt pavement condition index. *arXiv* **2020**, arXiv:2004.13314. [CrossRef]

17. Mei, A.; Zampetti, E.; Di Mascio, P.; Fontinovo, G.; Papa, P.; D'Andrea, A. ROADS—Rover for bituminous pavement distress survey: An unmanned ground nehicle (UGV) prototype for pavement distress evaluation. *Sensors* **2022**, *22*, 3414. [CrossRef] [PubMed]

18. He, K.M.; Zhang, X.; Ren, S.Q.; Sun, J. Deep residual learning for image recognition. In Proceedings of the 2016 IEEE Conference on Computer Vision and Pattern Recognition (CVPR), Las Vegas, NV, USA, 27–30 June 2016; pp. 770–778.

19. Cao, W.; Liu, Q.F.; He, Z.Q. Review of Pavement Defect Detection Methods. *IEEE Access* **2020**, *8*, 14531–14544. [CrossRef]

20. Xu, Z.S.; Guan, H.; Kang, J.N.; Lei, X.D.; Ma, L.F.; Yu, Y.T.; Chen, Y.P.; Li, J. Pavement crack detection from CCD images with a locally enhanced transformer network. *Int. J. Appl. Earth Obs. Geoinf.* **2022**, *110*, 102825. [CrossRef]

21. Song, W.D.; Jia, G.H.; Jia, D.; Zhu, H. Automatic pavement crack detection and classification using multiscale feature attention network. *IEEE Access* **2019**, *7*, 171001–171012. [CrossRef]

22. Stricker, R.; Aganian, D.; Sesselmann, M.; Seichter, D.; Engelhardt, M.; Spielhofer, R.; Gross, H.M. Road surface segmentation—pixel-perfect distress and object detection for road assessment. In Proceedings of the 2021 IEEE 17th International Conference on Automation Science and Engineering (CASE), Lyon, France, 23–27 August 2021; pp. 1789–1796.

23. Zhang, C.; Nateghinia, E.; Miranda-Moreno, L.; Sun, L. Pavement distress detection using convolutional neural network (CNN): A case study in Montreal, Canada. *Int. J. Transp. Sci. Technol.* **2021**, *11*, 298–309. [CrossRef]

24. Dung, C.V.; Sekiya, H.; Hirano, S.; Okatani, T.; Miki, C. A vision-based method for crack detection in gusset plate welded joints of steel bridges using deep convolutional neural networks. *Autom. Constr.* **2019**, *102*, 217–229. [CrossRef]

25. Xu, H.Y.; Su, X.; Wang, Y.; Cai, H.Y.; Cui, K.A.; Chen, X.D. Automatic bridge crack detection using a convolutional neural network. *Appl. Sci.* **2019**, *9*, 2867. [CrossRef]

26. Li, H.T.; Xu, H.Y.; Tian, X.; Wang, Y.; Cai, H.Y.; Cui, K.A.; Chen, X.D. Bridge crack detection based on SSENets. *Appl. Sci.* **2020**, *10*, 4230. [CrossRef]

27. Tran, T.S.; Tran, V.P.; Lee, H.J.; Flores, J.M.; Le, V.P. A two-step sequential automated crack detection and severity classification process for asphalt pavements. *Int. J. Pavement Eng.* **2020**, *23*, 2019–2033. [CrossRef]

28. Wu, Z.Y.; Kalfarisi, R.; Kouyoumdjian, F.; Taelman, C. Applying deep convolutional neural network with 3D reality mesh model for water tank crack detection and evaluation. *Urban Water J.* **2020**, *17*, 682–695. [CrossRef]

29. Jeong, D. Road damage detection using YOLO with smartphone images. In Proceedings of the 2020 IEEE International Conference on Big Data (Big Data), Atlanta, GA, USA, 10–13 December 2020; pp. 5559–5562.

30. Zou, Q.; Zhang, Z.; Li, Q.; Qi, X.; Wang, Q.; Wang, S. DeepCrack: Learning hierarchical convolutional features for crack detection. *IEEE Trans. Image Process.* **2019**, *28*, 1498–1512. [CrossRef] [PubMed]

31. Ouma, Y.O.; Hahn, M. Wavelet-morphology based detection of incipient linear cracks in asphalt pavements from RGB camera imagery and classification using circular Radon transform. *Adv. Eng. Inform.* **2016**, *30*, 481–499. [CrossRef]

32. Siriborvornratanakul, T. An automatic road distress visual inspection system using an onboard In-Car camera. *Adv. Multim.* **2018**, *2018*, 2561953. [CrossRef]

33. Fan, D.P.; Ji, G.P.; Cheng, M.M.; Shao, L. Concealed object detection. *IEEE Trans. Pattern Anal. Mach. Intell.* **2021**, *44*, 6024–6042. [CrossRef]

34. Liew, J.H.; Cohen, S.D.; Price, B.L.; Mai, L.; Feng, J. Deep interactive thin object selection. In Proceedings of the 2021 IEEE Winter Conference on Applications of Computer Vision (WACV), Waikoloa, HI, USA, 5–9 January 2021; pp. 305–314.

35. Lin, S.C.; Yang, L.J.; Saleemi, I.; Sengupta, S. Robust High-Resolution video matting with temporal guidance. In Proceedings of the 2022 IEEE/CVF Winter Conference on Applications of Computer Vision (WACV), Waikoloa, HI, USA, 4–8 January 2022; pp. 3132–3141.

36. Wang, L.J.; Lu, H.C.; Wang, Y.F.; Feng, M.Y.; Wang, D.; Yin, B.C.; Ruan, X. Learning to Detect Salient Objects with Image-Level Supervision. In Proceedings of the 2017 IEEE Conference on Computer Vision and Pattern Recognition (CVPR), Honolulu, HI, USA, 21–26 July 2017; pp. 3796–3805.

37. Shi, Y.; Cui, L.M.; Qi, Z.Q.; Meng, F.; Chen, Z.S. Automatic road crack detection using random structured forests. *IEEE Trans. Intell. Transp. Syst.* **2016**, *17*, 3434–3445. [CrossRef]

38. Yang, F.; Zhang, L.; Yu, S.J.; Prokhorov, D.V.; Mei, X.; Ling, H.B. Feature pyramid and hierarchical boosting network for pavement crack detection. *IEEE Trans. Intell. Transp. Syst.* **2019**, *21*, 1525–1535. [CrossRef]

39. Chambon, S.; Moliard, J. Automatic road pavement assessment with image processing: Review and Comparison. *Int. J. Geophys.* **2011**, *2011*, 989354. [CrossRef]

40. Qin, X.B.; Dai, H.; Hu, X.B.; Fan, D.P.; Shao, L.; Gool, A.L. Highly accurate dichotomous image segmentation. In Proceedings of the European Conference on Computer Vision, Tel Aviv, Israel, 23–27 October 2022.

41. Zhang, L.; Yang, F.; Zhang, Y.D.; Zhu, Y.J. Road crack detection using deep convolutional neural network. In Proceedings of the 2016 IEEE International Conference on Image Processing (ICIP), Phoenix, AZ, USA, 25–28 September 2016; pp. 3708–3712.

42. LeCun, Y.; Bottou, L.; Bengio, Y.; Haffner, P. Gradient-based learning applied to document recognition. *Proc. IEEE.* **1998**, *86*, 2278–2324. [CrossRef]

43. Gopalakrishnan, K.; Khaitan, S.K.; Choudhary, A.N.; Agrawal, A. Deep convolutional neural networks with transfer learning for computer vision-based data-driven pavement distress detection. *Constr. Build. Mater.* **2017**, *157*, 322–330. [CrossRef]

44. Ronneberger, O.; Fischer, P.; Brox, T. U-Net: Convolutional networks for biomedical image segmentation. *arXiv* **2017**, arXiv:1505.04597.

45. Jenkins, M.D.; Carr, T.A.; Iglesias, M.I.; Buggy, T.W.; Morison, G. A deep convolutional neural network for semantic Pixel-Wise segmentation of road and pavement surface cracks. In Proceedings of the 2018 26th European Signal Processing Conference (EUSIPCO), Roma, Italy, 3–7 September 2018; pp. 2120–2124.

46. Lau, S.L.; Chong, E.K.; Yang, X.; Wang, X. Automated pavement crack segmentation using U-Net-Based convolutional neural network. *IEEE Access* **2018**, *8*, 114892–114899. [CrossRef]

47. Escalona, U.; Arce, F.; Zamora, E.; Azuela, J.H. Fully convolutional networks for automatic pavement crack segmentation. *Comput. Sist.* **2019**, *23*, 451–460. [CrossRef]

48. Polovnikov, V.; Alekseev, D.; Vinogradov, I.; Lashkia, G.V. DAUNet: Deep augmented neural network for pavement crack segmentation. *IEEE Access* **2021**, *9*, 125714–125723. [CrossRef]

49. Badrinarayanan, V.; Kendall, A.; Cipolla, R. SegNet: A deep convolutional Encoder-Decoder architecture for image segmentation. *IEEE Trans. Pattern Anal. Mach. Intell.* **2015**, *39*, 2481–2495. [CrossRef] [PubMed]

50. Fan, R.; Bocus, M.J.; Zhu, Y.; Jiao, J.; Wang, L.; Ma, F.; Cheng, S.; Liu, M. Road crack detection using deep convolutional neural network and adaptive thresholding. In Proceedings of the 2019 IEEE Intelligent Vehicles Symposium (IV), Paris, France, 9–12 June 2019; pp. 474–479.

51. Xu, Z.C.; Sun, Z.Y.; Huyan, J.; Wang, F.P. Pixel-level pavement crack detection using enhanced high-resolution semantic network. *Int. J. Pavement Eng.* **2021**, *23*, 4943–4957. [CrossRef]

52. Lõuk, R.; Riid, A.; Pihlak, R.; Tepljakov, A. Pavement defect segmentation in orthoframes with a pipeline of three convolutional neural networks. *Algorithms* **2020**, *13*, 198. [CrossRef]

53. Eisenbach, M.; Stricker, R.; Seichter, D.; Amende, K.; Debes, K.; Sesselmann, M.; Ebersbach, D.; Stoeckert, U.; Groß, H. How to get pavement distress detection ready for deep learning? A systematic approach. In Proceedings of the 2017 International Joint Conference on Neural Networks (IJCNN), Anchorage, AK, USA, 14–19 May 2017; pp. 2039–2047.

54. Riid, A.; Lõuk, R.; Pihlak, R.; Tepljakov, A.; Vassiljeva, K. Pavement distress detection with deep learning using the orthoframes acquired by a mobile mapping system. *Appl. Sci.* **2019**, *9*, 4829. [CrossRef]

55. Augustauskas, R.; Lipnickas, A. Improved Pixel-Level Pavement-Defect segmentation using a deep autoencoder. *Sensors* **2020**, *20*, 2557. [CrossRef]

56. Minhas, M.S.; Zelek, J.S. Defect detection using deep learning from minimal annotations. *VISIGRAPP* **2020**, *4*, 506–513. [CrossRef]

57. Ma, K.; Hoai, M.; Samaras, D. Large-scale continual road inspection: Visual Infrastructure assessment in the wild. In Proceedings of the British Machine Vision Conference 2017, London, UK, 4–7 September 2017.

58. Maeda, H.; Kashiyama, T.; Sekimoto, Y.; Seto, T.; Omata, H. Generative adversarial network for road damage detection. *Comput. Aided Civ. Infrastruct. Eng.* **2020**, *36*, 47–60. [CrossRef]

59. Everingham, M.; Gool, L.V.; Williams, C.K.; Winn, J.M.; Zisserman, A. The pascal visual object classes (VOC) Challenge. *Int. J. Comput. Vis.* **2010**, *88*, 303–338. [CrossRef]

60. Perju, V.L.; Casasent, D.P.; Mardare, I. Image complexity matrix for pattern and target recognition based on Fourier spectrum analysis. *Def. Commer. Sens.* **2009**, *7340*, 73400.

61. Lihua, Z. Trademark retrieval based on image information entropy. *Comput. Eng.* **2000**, *26*, 86–87.

62. Yang, L.; Zhou, Y.; Yang, J.; Chen, L. Variance WIE based infrared images processing. *Electron. Lett.* **2006**, *42*, 857–859. [CrossRef]

63. Osserman, R. The isoperimetric inequality. *Bull. Am. Math. Soc.* **2008**, *84*, 1182–1238. [CrossRef]

64. Watson, A. Perimetric complexity of binary digital images. *Math. J.* **2012**, *14*, ARC-E-DAA-TN3185. [CrossRef]

65. Yang, C.L.; Wang, Y.L.; Zhang, J.M.; Zhang, H.; Lin, Z.L.; Yuille, A.L. Meticulous object segmentation. *arXiv* **2020**, arXiv:2012.07181.

66. Ramer, U. An iterative procedure for the polygonal approximation of plane curves. *Comput. Graph. Image Process* **1972**, *1*, 244–256. [CrossRef]

67. Simonyan, K.; Zisserman, A. Very deep convolutional networks for Large-Scale image recognition. *arXiv* **2014**, arXiv:1409.1556.

68. Zhao, H.S.; Shi, J.P.; Qi, X.J.; Wang, X.G.; Jia, J.Y. Pyramid scene parsing network. In Proceedings of the 2017 IEEE Conference on Computer Vision and Pattern Recognition (CVPR), Honolulu, HI, USA, 21–26 July 2017; pp. 6230–6239.

69. Chen, L.C.; Zhu, Y.K.; Papandreou, G.; Schroff, F.; Adam, H. Encoder-Decoder with atrous separable convolution for semantic image segmentation. In Proceedings of the European Conference on Computer Vision, Munich, Germany, 8–14 September 2018.

70. Wang, J.D.; Sun, K.; Cheng, T.H.; Jiang, B.R.; Deng, C.R.; Zhao, Y.; Liu, D.; Mu, Y.D.; Tan, M.K.; Wang, X.G.; et al. Deep High-Resolution representation learning for visual recognition. *IEEE Trans. Pattern Anal. Mach. Intell.* **2019**, *43*, 3349–3364. [CrossRef]

71. Cao, H.; Wang, Y.Y.; Chen, J.; Jiang, D.S.; Zhang, X.P.; Tian, Q.; Wang, M.N. Swin-Unet: Unet-like pure transformer for medical image segmentation. *arXiv* **2021**, arXiv:2105.05537.
72. Xie, E.; Wang, W.H.; Yu, Z.D.; Anandkumar, A.; Álvarez, J.M.; Luo, P. SegFormer: Simple and efficient design for semantic segmentation with transformers. *Neural Inf. Process. Syst.* **2021**, *34*, 12077–12090.
73. Zheng, S.X.; Lu, J.C.; Zhao, H.S.; Zhu, X.T.; Luo, Z.K.; Wang, Y.B.; Fu, Y.W.; Feng, J.F.; Xiang, T.; Torr, P.H.; et al. Rethinking semantic segmentation from a Sequence-to-Sequence perspective with transformers. In Proceedings of the 2021 IEEE/CVF Conference on Computer Vision and Pattern Recognition (CVPR), Nashville, TN, USA, 20–25 June 2021; pp. 6877–6886.
74. Yuan, Y.H.; Chen, X.L.; Wang, J.D. Object-Contextual representations for semantic segmentation. *arXiv* **2019**, arXiv:1909.11065.
75. Xie, S.N.; Tu, Z.W. Holistically-Nested edge detection. *Int. J. Comput. Vis.* **2015**, *125*, 3–18. [CrossRef]
76. Liu, Y.; Cheng, M.M.; Hu, X.; Wang, K.; Bai, X. Richer convolutional features for edge detection. In Proceedings of the 2017 IEEE Conference on Computer Vision and Pattern Recognition (CVPR), Honolulu, HI, USA, 21–26 July 2017; pp. 5872–5881.
77. Li, X.Y.; Sun, X.F.; Meng, Y.X.; Liang, J.J.; Wu, F.; Li, J.W. Dice loss for Data-Imbalanced NLP tasks. *arXiv* **2019**, arXiv:1911.02855.
78. Zhou, Z.Y.; Huang, H.; Fang, B.H. Application of weighted Cross-Entropy loss function in intrusion detection. *J. Comput. Commun.* **2021**, *9*, 1–21. [CrossRef]
79. Pan, S.J.; Yang, Q. A Survey on Transfer Learning. *IEEE Trans. Knowl. Data Eng.* **2010**, *22*, 1345–1359. [CrossRef]
80. Torralba, A.; Efros, A.A. Unbiased look at dataset bias. In Proceedings of the Computer Vision and Pattern Recognition (CVPR), Colorado Springs, CO, USA, 20–25 June 2011; pp. 1521–1528.
81. Sandler, M.; Howard, A.G.; Zhu, M.L.; Zhmoginov, A.; Chen, L. MobileNetV2: Inverted residuals and linear bottlenecks. In Proceedings of the 2018 IEEE/CVF Conference on Computer Vision and Pattern Recognition, Salt Lake City, UT, USA, 18–23 June 2018; pp. 4510–4520.
82. Chollet, F. Xception: Deep learning with depthwise separable convolutions. In Proceedings of the 2017 IEEE Conference on Computer Vision and Pattern Recognition (CVPR), Honolulu, HI, USA, 21–26 July 2017; pp. 1800–1807.
83. Dosovitskiy, A.; Beyer, L.; Kolesnikov, A.; Weissenborn, D.; Zhai, X.; Unterthiner, T.; Dehghani, M.; Minderer, M.; Heigold, G.; Gelly, S.; et al. An Image is Worth 16x16 Words: Transformers for image recognition at Scale. *arXiv* **2020**, arXiv:2010.11929.

 *remote sensing*

Article

A Low-Cost Deep Learning System to Characterize Asphalt Surface Deterioration

Diogo Inácio [1], Henrique Oliveira [2,3,*], Pedro Oliveira [4] and Paulo Correia [1,2]

[1] Instituto Superior Técnico, Universidade de Lisboa, 1049-001 Lisboa, Portugal;
     diogobpalma@tecnico.ulisboa.pt (D.I.); plc@lx.it.pt (P.C.)
[2] Instituto de Telecomunicações (IT-Lisboa), 1049-001 Lisboa, Portugal
[3] Instituto Politécnico de Beja, 7800-111 Beja, Portugal
[4] Tecnofisil, Av. Luís Bívar 85A, 1050-143 Lisboa, Portugal; pedro.oliveira@tecnofisil.pt
*   Correspondence: hjmo@lx.it.pt

Abstract: Every day millions of people travel on highways for work- or leisure-related purposes. Ensuring road safety is thus of paramount importance, and maintaining good-quality road pavements is essential, requiring an effective maintenance policy. The automation of some road pavement maintenance tasks can reduce the time and effort required from experts. This paper proposes a simple system to help speed up road pavement surface inspection and its analysis towards making maintenance decisions. A low-cost video camera mounted on a vehicle was used to capture pavement imagery, which was fed to an automatic crack detection and classification system based on deep neural networks. The system provided two types of output: (i) a cracking percentage per road segment, providing an alert to areas that require attention from the experts; (ii) a segmentation map highlighting which areas of the road pavement surface are affected by cracking. With this data, it became possible to select which maintenance or rehabilitation processes the road pavement required. The system achieved promising results in the analysis of highway pavements, and being automated and having a low processing time, the system is expected to be an effective aid for experts dealing with road pavement maintenance.

Keywords: crack segmentation; crack classification; deep neural networks; characterization of highways condition

Citation: Inácio, D.; Oliveira, H.; Oliveira, P.; Correia, P. A Low-Cost Deep Learning System to Characterize Asphalt Surface Deterioration. *Remote Sens.* **2023**, *15*, 1701. https://doi.org/10.3390/rs15061701

Academic Editors: Valerio Baiocchi, Alessandro Mei and Xianfeng Zhang

Received: 30 December 2022
Revised: 13 March 2023
Accepted: 20 March 2023
Published: 22 March 2023

1. Introduction

Due to their constant use, road pavement surfaces are subject to continuous degradation, with cracks being the first sign of pavement surface deterioration [1]. Therefore, crack detection is crucial for monitoring and maintaining pavement surfaces, ensuring the security of the drivers. However, if traditional human inspection procedures are used, crack detection can be extremely tedious, time-consuming, and subjective [2]. Implementing automatic pavement condition monitoring systems can overcome these limitations, allowing a more precise, faster, and safer analysis than traditional methods, minimizing experts' effort and human subjectivity.

The automatic detection and classification of cracks in road pavements are challenging due to the different characteristics of pavement materials and the wide variability of crack shapes and their inhomogeneous textures, among other factors. Moreover, depending on the sensors used to gather road pavement imagery, the resulting images may be affected by different types of noise, shadows, debris, or oil and water spots. These challenges make developing a robust and reliable automatic crack detection and classification system a difficult task.

1.1. Contributions

This paper proposes an automatic crack detection and classification system based on deep neural networks to help speed up highway pavement surface monitoring. This work was conducted in collaboration with the Portuguese company Tecnofisil [3]. The contributions of this paper are:

1. A segmentation deep neural network, based on U-Net architecture, for identifying crack regions in images that was trained using five publicly available datasets.
2. A classification deep neural network used to classify detected cracks into one of four classes: (i) alligator; (ii) longitudinal; (iii) transverse; or (iv) non-crack.
3. A system used to automatically estimate the percentage of cracking present in a road pavement segment to serve as an aid for experts in their maintenance planning.
4. An anomaly detection method based on isolated cracking classifications that relates uncertainty to the proposed system results.

The paper is organized as follows. After this introduction, Section 1.2 briefly reviews the related work. Section 2 presents the proposed methodology and provides details about the development process, while Section 3 presents the experimental results, and Section 4 discusses the experimental results. This paper ends with Section 5, in which conclusions and proposals for future work are presented.

1.2. Related Work

Crack detection in road pavements has been a hot topic of research over the years due to its importance, high level of complexity, and challenging characteristics. It has seen a remarkable evolution and a huge variety of proposals based on different methodologies. What most differentiates the available methodologies using a camera sensor is the strategy used to extract features from images that allow for identifying pavement surface distresses [4].

In the scientific literature, two main categories of methods can be identified: (i) those based on traditional image processing; (ii) those using machine learning techniques. Traditional image processing techniques can be further divided into three subcategories: (i) threshold segmentation; (ii) edge detection; and (iii) region growing [5].

Crack detection using threshold segmentation assumes that pixels belonging to a crack are darker compared to its surroundings [6]. These methods start with a pixel-based analysis to determine whether they belong to a crack or to the background, where selecting a suitable threshold value is the key to good method performance. For instance, Oliveira and Correia [7] identified dark pixels belonging to potential cracks using the dynamic thresholding technique that comprised two steps. The first threshold was able to label image pixels into "image background" and "potentially belonging to cracks". Then, after dividing the resulting segmented image into non-overlapping blocks of a given dimension, they were finally classified into "crack" and "non-crack" after applying a second threshold determined based on the set of entropy values obtained from the analysis of the pixel intensities of each non-overlapping image block (as many entropy values as the number of non-overlapping blocks in the image).

Edge detection methods are based on identifying the edges of regions of the image presenting pavement surface distress, enabling the outlying of their contours [6]. Various edge detection operators include Sobel, Roberts, Prewitt, and Canny operators. However, using a single operator can hardly reach the desired results. Hence, many researchers used edge detection operators with other techniques to improve crack detection performance [5]. Ayenu-Prah and Attoh-Okine [8] proposed a road crack detection method, combining a bi-dimensional empirical mode decomposition (BEMD) with Sobel edge detection with BEMD, extending the original empirical mode decomposition [9] to remove noise from the input signal efficiently.

Region-growing strategies aim to gather similar pixels together to form regions. Then, the characteristics of pavement surface defect regions can be estimated. Li et al. [10] proposed an automatic crack detection method, the F* Seed growth Algorithm (FoSA), to improve the detection of discontinuous and blurred cracks. The FoSA is an extension of the F* method [11], exploiting a seed-growing strategy (as illustrated in Figure 1) to eliminate the requirement that start and endpoints should be known or pre-defined in advance. The global search area is reduced to local to improve search efficiency. However, the results of FoSA may be impaired by the lighting conditions of the images to be analyzed.

Figure 1. Illustration of the seed-growing strategy described in [10] for different seed-growing radius.

More recently, through the enormous development of machine learning techniques, deep learning has dominated proposals for the detection of cracks in pavement surface images taken during road surveys, as well as their classification, by allowing a computer to automatically learn new characteristics and classification rules based on a sufficiently large set of representative training examples. Adopting a supervised learning approach requires that the training set contains images labelled with information about the presence and location of cracks, a task that typically requires a substantial previous effort by human experts to produce this labelling information and that can bring out the subjectivity of human analysis.

Methods based on deep learning have achieved great success in many related areas, such as image classification, object detection, and image segmentation. Therefore, deep learning methods, mainly using convolutional neural networks (CNN), have also been proposed for automatic crack detection and/or crack-type classification.

Crack detection and crack-type classification can be seen as image classification problems, and two different approaches can be followed: (i) assigning a label to the whole input image; or (ii) dividing the image into blocks and classifying each block as belonging to a particular class. The labels to be considered can be binary ("crack" or "non-crack") or else indicate the type of crack that was detected (longitudinal, transverse, and crocodile skin, among others).

For example, Lei et al. [12] proposed a deep learning-based algorithm that divides pavement surface images of size 3264 × 2248 pixels into smaller blocks of size 99 × 99 pixels (keeping the 3 RGB channels). They used a CNN to perform a binary classification of each image block according to the probability of whether or not it contains cracks. The CNN used by the authors was ConvNet, and its architecture is illustrated in Figure 2.

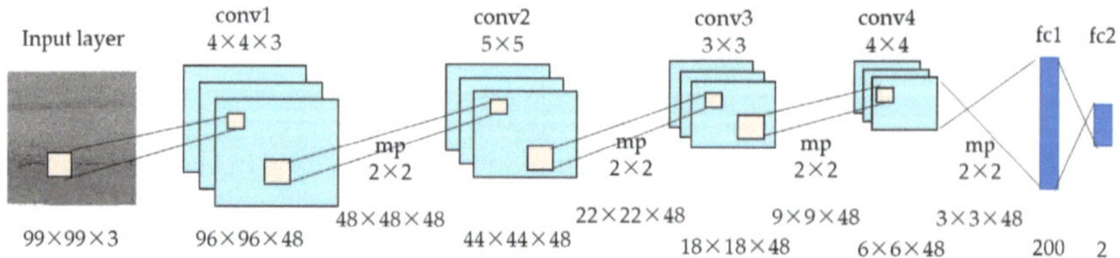

Figure 2. Illustration of the ConvNet architecture presented in [12].

Aslan et al. [13] implemented a CNN that classifies several types of surface distress present in images of road pavements. The proposed CNN architecture (Figure 3) consists of two convolutional units with 32 filters, each applied before a 16-neuron fully-connected layer and a softmax classification layer with four output neurons. The developed model was trained through a balanced dataset that includes four types of pavement surface defects: (i) longitudinal crack; (ii) transversal crack; (iii) alligator crack; and (iv) pothole.

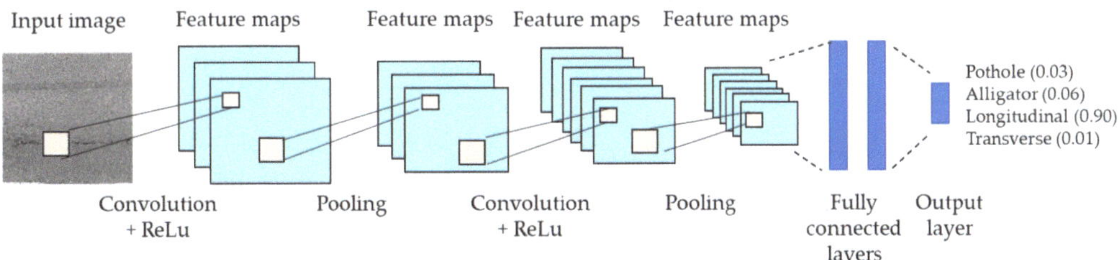

Figure 3. Illustration of the CNN architecture proposed in [13].

Anand et al. [14] proposed an autonomous real-time system with an associated camera to detect cracks and potholes in images of road pavement surfaces. The method, denoted Crack-pot, uses traditional image processing techniques, such as edge detection, to locate and generate the bounding boxes of the detected objects, combined with a CNN employed as a classification model.

Jenkins et al. [15] proposed a fully convolutional neural network [16] to classify all pixels presented in pavement surface images. The network architecture proposed by the authors is an encoder–decoder architecture based on the U-Net model [17]. In this type of architecture, the layers belonging to the encoder reduce the size of the input image through down-sampling operations and learn a high density of lower-level feature maps. The layers belonging to the decoder then map the encoded features back to their original resolution, using pooling layers to perform up-sampling operations efficiently. The network ends with a classification layer to assign an individual label to each pixel. During the evaluation of this method, the authors concluded that the training data were scarce and contained a low diversity of cracks, which meant that the performance of the proposed classification system could still improve, given the network's capabilities.

Zhang et al. [18] proposed a fully convolutional network for per-pixel semantic segmentation/detection. The authors stated that their system, CrackGAN, aims to solve issues found in published research works dealing with the crack detection problem in images of pavement surfaces: the one known as "All black", which occurs in FCN-based pixel-level crack detection when using partially accurate ground truths, where the network treats all the pixels as belonging to the image background, along with the data imbalance issue concerning the network's training step. Their system processed images from three publicly available image datasets, but most of them were captured by very sophisticated imaging systems, namely laser imaging systems. Although the proposed approach achieved state-of-the-art performance, the calculation of the pavement surface distress level was not performed, nor the classification of distress types.

Liu et al. [19] proposed a feature fusion encoder–decoder network (FFEDN) with two novel modules to improve the crack detection accuracy by enhancing the representation capability of tiny cracks and reducing the influence of background interference. The FFEDN system processed images from three publicly available image datasets mainly captured by cell phones, and published results showed that the proposed system outperformed all the eight methods used for comparison. Nevertheless, all the images presented in the published work are free of artifacts in the pavement surface, such as white lane markings (continuous or interrupted), oil spots, and shadows cast by objects near the road edges, which does not allow us to assess on the system performance in processing images with such artifacts.

A recently published research paper by Ma et al. [20] addressed problems regarding data acquisition (image and video) and the defect counting of road pavement surfaces, aiming at increasing the efficiency and accuracy of traditional detection systems. The authors developed a generative adversarial network (PCGAN) to generate realistic crack images, aiming to solve the problem of small amounts of training data, an important issue when using deep neural networks. Regarding crack detection, an improved version of a regression-based object detection and median flown model was developed (YOLO-MF) that allows for obtaining the number of cracks in an image/video captured by an imaging sensor carried by an unmanned aerial vehicle. Although the published performance metrics are auspicious, the images shown in the paper exhibit very prominent cracks that appear to be easily detectable. On the other hand, the images presented in the paper did not show some often-visible artifacts on road pavement surfaces, such as oil spots or even white lane markings. Additionally, the authors concluded that their method presented difficulties in dealing with small and complex pavement distresses and detecting extensive alligator cracks.

CrackW-Net was the methodology developed by Han et al. [21], namely a skip-level round-trip sampling block structure with the implementation of convolutional neural networks aimed at segmenting cracks in images of the pavement surface at the pixel level. The authors tested their proposed approach with only two image datasets, namely the Crack500 [22] and a self-built dataset. Processed images were free from well-known artifacts (oil spots and white lane markings, among others), with most of them exhibiting simple and fairly noticeable cracks. CrackW-Net training costs were compared to FCN [23], U-Net [24], and ResU-Net [25], which were significantly higher concerning the number of neural network parameters and the training speed. The authors concluded that extensive validations are still needed before the use of CrackW-Net becomes feasible.

As highlighted, numerous studies have been devoted to deep learning in crack detection and classification due to convolution neural networks' excellent feature representation capabilities. The approach presented in this paper follows this trend, proposing, however, a segmentation deep neural network to identify crack regions and a classification deep neural network to classify detected cracks into types to handle more complex images of road pavement surfaces that frequently present noticeable artifacts' visible and more complex crack shapes and textures.

2. Materials and Methods

This section describes the proposed methodology for the developed crack detection and classification system. It consists of five main steps, as represented by the generic architecture presented in Figure 4.

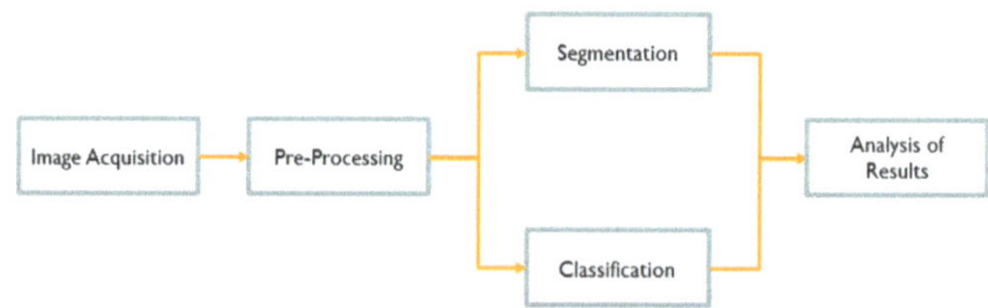

Figure 4. Generic architecture of the proposed system.

The five main steps of the proposed system consist of the following:

1. Image Acquisition—This step deals with the acquisition of the road pavement images to be analyzed. A set of representative images was considered to create a dataset for usage along the development of the crack detection and classification system.
2. Preprocessing—This step performs a set of image manipulations and transformations, notably for normalization purposes.
3. Segmentation—This step creates an image map identifying which image pixels are affected by cracking.
4. Classification—This step assigns a crack type classification for each detected crack.
5. Analysis of Results—This step is responsible for combining the results produced by the segmentation and classification steps.

The proposed system captures pavement surface images using a conventional 2D camera. Each road is analyzed considering segments of a pre-specified length (e.g., 100 m). The output consists of road pavement crack segmentation maps and a list containing the type of cracking that the system assigned to each road segment. The analysis of results module then outputs a set of tables summarizing the calculated information, including a distressing percentage per road segment, which provides valuable help for the quick identification of road segments that require careful analysis by experts when planning road maintenance operations.

2.1. Image Acquisition

The proposed system includes a simple image acquisition system consisting of a conventional 2D camera that can be easily mounted on a regular vehicle. The system was provided by Tecnofisil [3], and an illustration of its setup is included in Figure 5; for the purpose of this work, the camera was set up on the back of a small van and mounted to guarantee the desired slant angle towards the ground.

Figure 5. Image acquisition system consisting of a conventional 2D camera mounted on the back of a vehicle.

The image sensor used is a Teledyne Lumenera Lt16059H high-performance camera [26], which has a 16 Megapixel Charge-Coupled Device (CCD) sensor and an electronic shutter, that can capture high-quality and high-speed images. The maximum resolution of this camera is 4864×3232 pixels. A sample acquired image is shown in Figure 6, where it is possible to see that the road pavement surface region of interest to be analyzed, corresponding to the lane where the van is travelling, appears with a trapezoidal shape marked in red in the figure.

Figure 6. Sample image acquired by the proposed system. The area of interest for analysis is delimited in red.

Image acquisition needs to ensure that the entire width of the road lane is captured. Additionally, the acquisition system should not skip any portion of the road. For this reason, it was set up so that consecutive images have some overlap in the captured regions of interest. The setup of the image acquisition system requires setting the following parameters:

- Camera slant angle;
- Camera focal length;
- Camera frame rate;
- Camera resolution;
- Vehicle speed.

Image acquisition tests were conducted on Portuguese highways, considering an initial slant angle of 30° towards the ground and camera focal length values of 16 and 28 mm, corresponding to the minimum and maximum values supported by the imaging sensor, respectively. Figure 7 shows sample images captured with these parameters. Please note that the present work is mostly focused on the analysis of highways and other roads for circulating at relatively high speeds and not so much for very twisty roads.

(a) (b)

Figure 7. Images captured with a slant angle of 30° and a camera focal length of: (**a**) 16 mm; (**b**) 28 mm.

As expected, when considering a short focal length, as illustrated in Figure 7a, a more extensive road area is captured due to a wider angle of view, although the details of the pavement surface are smaller due to the lower magnification. On the other side, with a longer focal length, as illustrated in Figure 7b, a smaller road area is captured, but the pavement surface details are more visible. This second setting would require the driver to keep the vehicle in the middle of the road lane, which can be challenging depending on the road trajectory.

After the initial tests, it was decided to adopt a camera focal length of 22 mm, which ensures an easy capture of the entire road lane width, with good focus and detail in the region of interest of the pavement surface. A sample image acquired in these conditions is presented in Figure 6.

The camera slant angle was kept at 30° towards the pavement surface because, for higher angles, the images could suffer from shadow effects cast by the vehicle itself, and the van's roof could also appear. On the other hand, for angles less than 30°, the pavement surface closest to the van was no longer captured, but the region of interest would be captured at a greater distance and, consequently, at a lower resolution.

The selection of the remaining three parameters, namely camera frame rate, camera resolution, and vehicle speed, should be performed in such a way as to ensure that an overlap area exists between the regions of interest of consecutive images to make sure that there is no loss of pavement surface area in the analysis. The camera offers a trade-off between frame rate and resolution; with the maximum resolution of 4864 × 3232 pixels, the camera can operate at a maximum of 3 frames per second. For higher frame rates, the resolution should be lower. For instance, reducing to a quarter of the maximum resolution (2432 × 1616 pixels) increases the frame rate to 8 frames per second. With this in mind, several tests were conducted with images captured on a 3 km stretch of a major Portuguese road to decide the values of the above-mentioned parameters. A summary of the tests conducted is presented in Table 1, with the overlap area of the pavement surface between consecutive images included in the last column. The length of the region of interest was set to 6.5 m.

Table 1. Tests conducted to help set the values of the camera frame rate, camera resolution, and vehicle speed. The last column shows the overlap between the regions of interest in consecutive images.

Trial	Speed (km/h)	Frame Rate	Resolution	Overlapping Region (m)
1	50	3	4864 × 3232	2/6.5
2, 3, 4	70	3	4864 × 3232	not guaranteed
5, 6	70	8	2432 × 1616	4.5/6.5
7, 8	90	8	2432 × 1616	3.5/6.5

Table 1 shows the results for four combinations of speed, frame rate, and resolution parameters. During these tests, it was noticed that the camera frame rate underwent minor variations during the acquisition, meaning that even if the frame rate is set at 3 or 8 frames per second, it will vary, typically in the range of ±0.1–0.3 frames per second. These oscillations influence the overlapping region achieved between consecutive images. The technical team setting up the project decided that an overlap of 3.5 m, with a region of interest of length of 6.5 m, would be adequate.

A camera resolution of 2432 × 1616 pixels was considered sufficient for crack detection when using the previously defined values of a 30° slant angle and a focal distance of 22 mm. The value of the camera frame rate (and consequently of the time lapse between successive shots) was adjusted according to the vehicle speed, as summarized in Table 2, to achieve a 3.5 m overlap between the regions of interest of consecutively captured images.

Table 2. Tradeoff between vehicle speed and camera frame rate to achieve a 3.5 m overlap between consecutively captured images.

Speed (km/h)	Frame Rate (fps)	Time Lapse (ms)
50	4.6	217
70	6.5	154
90	8	120

The proposed system image acquisition can be used to acquire images for training, validation, and testing sets; however, other images extracted from public datasets were used to train the neural networks that compose the proposed system to increase image diversity in training. As automatic crack detection in pavement surface images is an emergent topic that has been investigated over the years, several image datasets of pavement surfaces are publicly available, allowing for the validation and testing of new models and systems. They differ by various parameters, such as the sensor with which the images are captured and the scale used, among others. The public datasets that were used in this work are as follows:

- CrackForest: Contains 118 crack images of a road pavement surface in Beijing that are 480 × 320 pixels in size. The sensor used was the camera of an iphone5 [27].

- AigleRN: Contains 38 pre-processed grayscale images of a road pavement surface in France. Half of these images are of size 991 × 462 pixels, and the other half are of size 311 × 462 pixels [28].
- CrackTree260: Contains 260 images of size 800 × 600 pixels, captured by an area-array camera under visible light illumination conditions [29].
- CRKWH100: Contains 100 images of a road pavement surface of size 512 × 512 pixels, captured by a line array camera under visible light illumination conditions [30].
- CrackLS315: Contains 315 images of size 512 × 512 pixels, captured by a line array camera under laser illumination [31].

These five public datasets have differences in lighting conditions, texture, pavement type, and other characteristics, ensuring image diversity in the neural network training process. Along with image diversity, it is also appropriate to include images of the training set, such as those that the neural networks analyze in their applications to improve neural network performance. Accordingly, it was considered advantageous to include images of Portuguese road pavements in the training step of the classification model, namely some images also obtained during the survey of the "IP3".

It should be noted that, in order to be able to use images of different sizes, coming from several public datasets, for the purposes of the training, validation, and testing of the neural networks used by the proposed system, those images were first resized to match the input size expected by each of the neural networks used, notably considering an input size of 480 × 320 for the segmentation network and an input size of 512 × 512 for the classification network.

2.2. Preprocessing

Image pre-processing includes two main tasks: (i) identifying the region of interest of the captured pavement surface image and converting its trapezoidal shape to a rectangular shape; (ii) creating a more extensive image of the pavement surface area for distress analysis by compositing several consecutive reshaped regions of interest. This second task is performed because, from the point of view of the road analysis experts, observing surface distress in images of road lane sections longer than 6.5 m in length provides more relevant results.

In the first task, all images were converted to grayscale and then subjected to a perspective transformation, as illustrated in Figure 8. This task was accomplished using the functions getPerspectiveTransform() and warpPerspective() available in the OpenCV library [32].

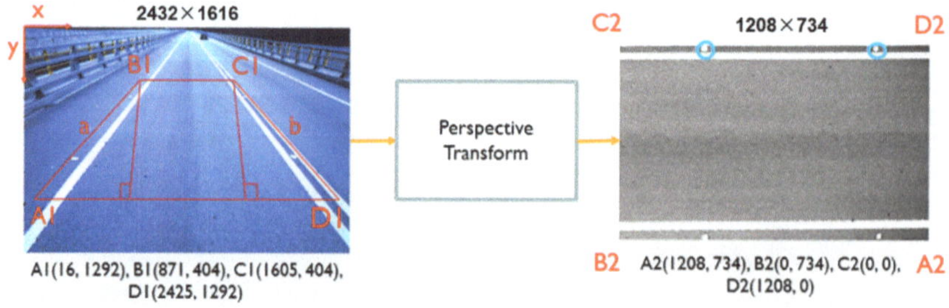

Figure 8. Selection of the region of interest (**left**) and the output of perspective transformation (**right**).

In order to define the region of interest, a 20% lower cut and a 25% uppercut were made relative to the image's vertical axis, fixing the vertical (y) value of points A1 and D1 and points B1 and C1, respectively. The 20% lower cut ensures that the white lane markings are visible at the bottom of the region of interest. On the other hand, the 25% uppercut is necessary since only a portion of the entire pavement surface that is visible in the images can be used for analysis because its upper part presents a less detailed pavement surface (less prominent distresses may not even be visible) compared to the lower part.

Once the height of the trapezium relative to the image's vertical axis is fixed, its width is set to cover the width of the road lane. However, since it can be challenging to keep the vehicle carrying the image acquisition system in the road lane's center, the width of the trapezium was slightly increased. Thus, margins were created for the white lane markings to remain in the region of interest, even when vehicles deviate from the road lane's center. With this in mind, lines were drawn parallel to the white lane markings along the image. By intersecting them with the horizontal lines that define the trapezium's height already specified by the lower and uppercuts, it was possible to determine the image region of interest for the acquired images (Figure 8).

As illustrated in Figure 8, the region of interest is given by a rectangle with height equal to the length of the segment defined by points B1-C1, namely 734 pixels, while the rectangle's width was set equal to the shorter length of the segments defined by points A1-B1 and D1-C1, namely 1208 pixels $\left(\sqrt{(x_{D1} - x_{C1})^2 + (y_{D1} - y_{C1})^2} \right)$. After defining the dimensions of the region of interest in pixels, it was essential to match them to the true dimensions in meters relative to the pavement surface. To perform this, two blue circles were marked in the image as shown on the right side of Figure 8, whose real distance was known to be 3.5 m. Taking this measurement as a reference, the actual dimensions considered for the region of interest were 6.5 m long by 3.5 m wide.

Since two neural networks were used in the proposed system, independently and articulated in parallel, the second pre-processing step comprised two independent paths to conveniently prepare the images for the segmentation and classification steps.

The images resulting from the first pre-processing step (a sample is shown on the right side of Figure 8) only required a resize operation to be processed by the classification network, which expects a smaller image as input to keep reasonable values of required memory and processing time.

For the segmentation step, images of road lane sections approximately 18.5 m long by 3.5 m wide were created from those resulting from the first pre-processing step. An overlapping area between two consecutive images was essential in forming such sections. For this purpose, two options are presented: (i) construction of a panoramic image; (ii) image concatenation. In the first alternative, a panorama was created using the functions createStitcher() and stitch() from the OpenCV library. Figure 9 presents a panoramic image sample.

Figure 9. Sample of a panoramic image.

To build a panorama, the images must share an overlapping area featuring a reasonable number of salient points, also known as key points, which are used as reference points to successfully join two or more images into one. However, pavement surface images often exhibit a uniform structure and texture, a most noticeable feature when the road pavement is free of distress, making it very difficult to identify the key points. Therefore, the panoramic construction of a road lane section may fail, and if it does, it is crucial to have an alternative, such as image concatenation, which is a feasible solution because the overlap area between consecutive images is known. An image concatenation sample for the same road lane section shown in Figure 9 is presented in Figure 10.

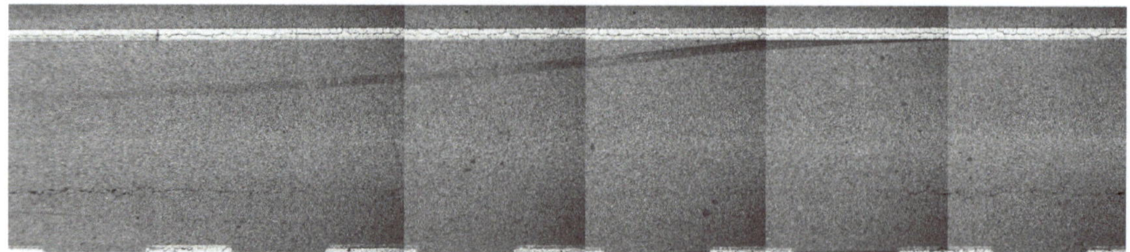

Figure 10. Road section generated through the image concatenation method.

Analyzing the two alternatives, it was decided that panoramic construction would be the first option, and in case of failure, the image concatenation would ensure the generation of the respective road lane section as a fallback solution.

2.3. Segmentation

This step segments the road lane sections generated in the preprocessing step, whose proposed architecture is shown in Figure 11. The road lane sections are submitted to a segmentation neural network, whose architecture is identical to the U-Net [17], illustrated in Figure 12. The difference between U-Net and the proposed neural network for the segmentation task is the dimension of the input image, corresponding to 480×320 pixels. A dataset was built with images from five public datasets, as described in Section 2, to train the proposed segmentation neural network. Furthermore, the parameters used for training were Categorical Crossentropy Loss and Adam optimizer with a learning rate of 0.0001.

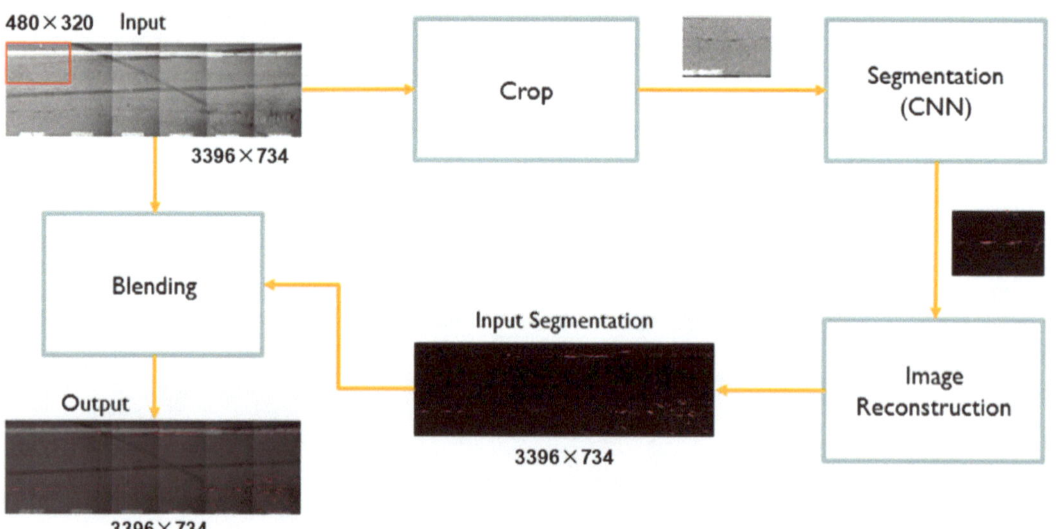

Figure 11. Road lane section processing during the segmentation step.

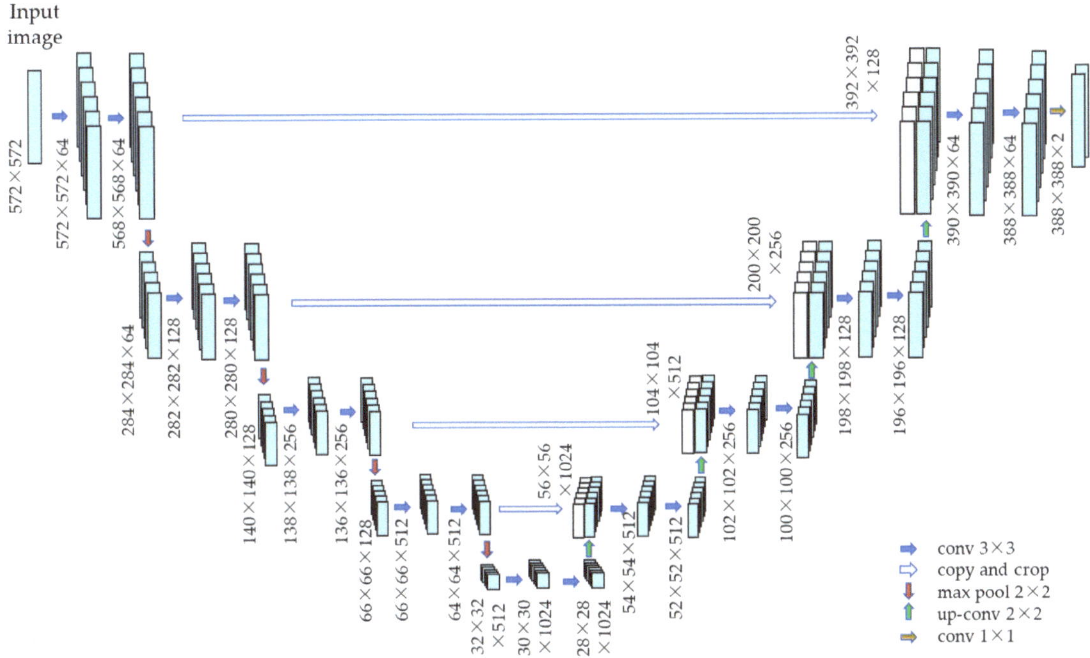

Figure 12. Illustration of the U-Net architecture proposed in [24].

In Figures 13–15, it is possible to observe a road lane section, the corresponding segmentation, and the merging of the two previous images, respectively.

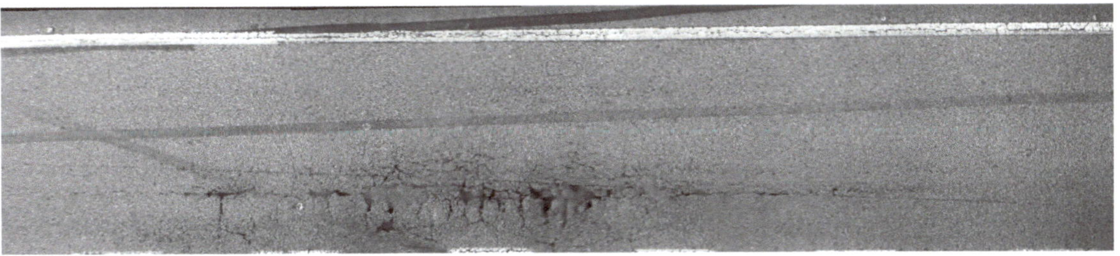

Figure 13. Sample of a road lane section.

Figure 14. Road lane section segmented by the segmentation neural network.

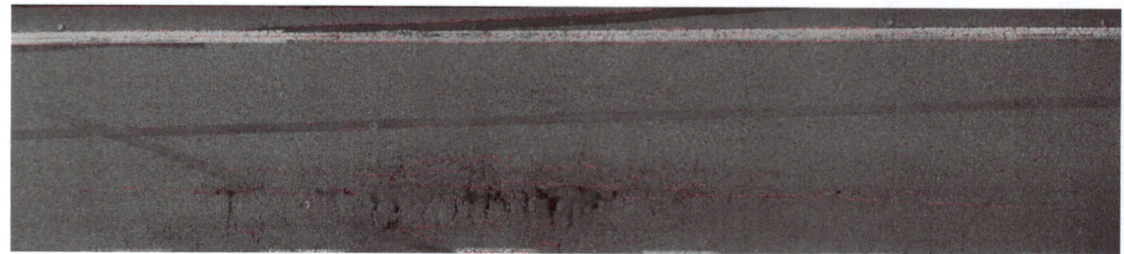

Figure 15. The result of merging the road section with its segmentation.

Figures 14 and 15 show that the segmentation method performs satisfactorily, distinguishing the cracks from the rest of the road pavement surface. However, the model is influenced by white lane markings and tire tracks on the pavement, segmenting some pixels that do not belong to cracks.

2.4. Classification

The classification step aims at label assignment regarding the type of crack present in each image coming from the pre-processing step. The core of this step is a multi-class classification neural network (Figure 16), whose proposed architecture is detailed in Table 3. The output of the classification model is a vector of probabilities. Each vector element contains the probability of a particular label being assigned to the image under analysis, and the one with the highest probability is chosen. The possible labels considered for classification are: (i) alligator crack; (ii) longitudinal crack; (iii) transverse crack; and (iv) non-crack. Figure 17 shows samples of each possible label.

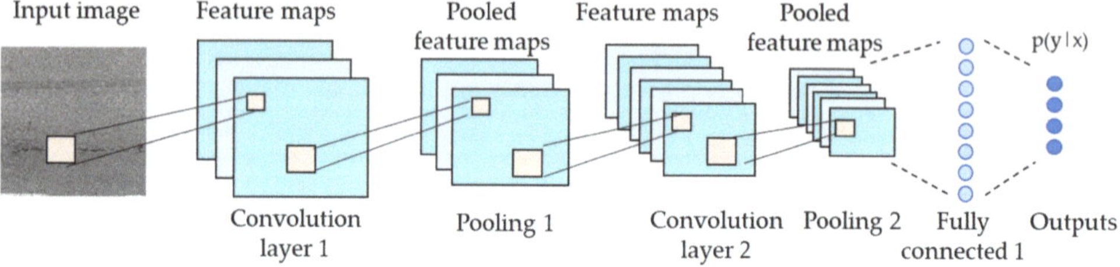

Figure 16. Generic architecture of a multi-class classification neural network.

Table 3. Proposed architecture for the classification network.

Layer	Kernel	Activation Function	Output
Convolutional	5×5	ReLU	(512, 512, 64)
MaxPooling	2×2	-	(256, 256, 64)
Convolutional	5×5	ReLU	(256, 256, 128)
MaxPooling	2×2	-	(128, 128, 128)
Convolutional	5×5	ReLU	(128, 128, 256)
MaxPooling	2×2	-	(64, 64, 256)
GlobalMaxPooling	-	-	256
Dense	-	ReLU	64
Dense	-	Softmax	4

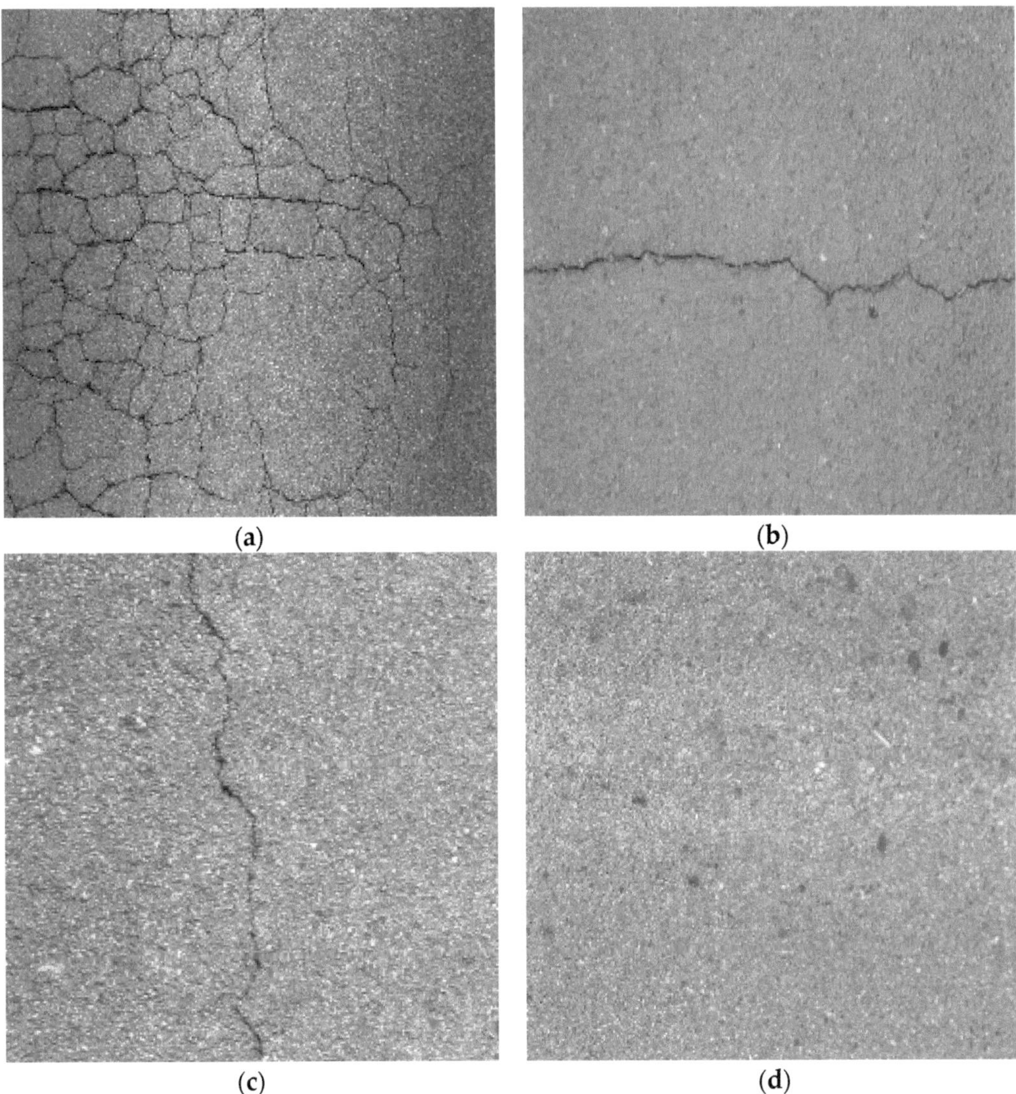

Figure 17. (**a**) Alligator crack; (**b**) longitudinal crack; (**c**) transverse crack; and (**d**) non-crack samples.

A balanced dataset was built to train the proposed classification neural network by collecting images from the five public datasets presented in Section 2 and by selecting some images captured during a survey of a major road in Portugal, the "IP3", between the locations of "Souselas" and "Viseu" (a 2 km route section). Thus, 600 images were considered for training, where each of the four classes are represented by 150 images.

The parameters for training the classification model were number of epochs, 30; batch size, 20; optimizer Adam with a learning rate of 0.001; and Categorical Crossentropy Loss, which is described by Equation (1), where $\hat{y}_i$ is the ith scalar value in the model output, y_i is the corresponding target value, and the output size is the number of scalar values in the model output:

$$Loss = -\sum_{i=1}^{outputsize} y_i \cdot \log \hat{y}_i.$$

(1)

2.5. Crack Percentage and Analysis of Results

The crack percentage is obtained by dividing the affected area by the total area of a road segment. The cracking level characterization adopted in this paper is structured in three severity levels, as described in Table 4, following the proposal of road analysis experts of the company.

Table 4. Cracking level characterization as proposed by the company's road analysis experts.

Severity	Description	Affected Area
Level 1	Isolated but noticeable crack (<2 mm)	0.5 m × affected length
Level 2	Open and/or branched longitudinal or transverse cracks	2 m × affected length
Level 3	Alligator cracks	Road lane width × affected length

The first severity level (Level 1) was not considered because it is not guaranteed to detect cracks less than 2 mm in width on pavement surface images captured by the proposed system since a low-cost image sensor was used.

The classification model is also able to distinguish between longitudinal, transverse, and alligator cracks, assigning them a severity level according to the criteria listed in Table 5.

Table 5. Criteria for assigning crack severity labels according to the crack type.

Label	Severity	Affected Area
Longitudinal	Level 2	2 m × 3 m
Transverse	Level 2	2 m × 3.5 m
Alligator	Level 3	3.5 m × 3 m
Non-crack	-	0

Since the proposed system does not incorporate any methodology to measure the crack length, the size of the affected pavement surface area was considered according to the following: (i) if a crack is classified as longitudinal, its maximum length is equal to the length of the non-overlapping region of the image, i.e., 3 m; (ii) if a crack is classified as transverse, its maximum length is equal to the road lane width, i.e., 3.5 m; (iii) if a crack is classified as of alligator type, the affected area corresponds to the non-overlapping area of the image, i.e., 3.5 m wide by 3 m long. It should be noted that the cracking estimate obtained in this way tends to overestimate the actual crack value observed by visual inspection.

Finally, it should be noted that, if there is a large overlap between consecutive images, the same crack will be detected twice. To avoid counting the same crack twice, in a sequence of n classifications of a given type of cracking, only $n - 1$ are considered for the affected area calculation. As an example of calculating the crack percentage, consider the analysis of a 100 m road segment, represented by a sequence of four images containing a longitudinal cracking. The crack percentage contribution of this sequence is obtained by applying three times the longitudinal crack criteria presented in Table 5, where 2 m corresponds to the affected road width considered by the company when a longitudinal crack is detected, and 3 m is the length of the non-overlapping region in each image, as described above. Therefore, the affected area is $3 \times (2 \times 3) = 18$ m^2 and dividing by the total area of the road segment $(3.5 \times 100 = 350$ m$^2)$ and dividing by the total area of the road segment $(3.5 \times 100 = 350$ m$^2)$, resulting in a crack percentage of $(18/350) \times 100 = 5.1\%$.

The overlap between consecutive images allows the implementation of an anomaly detection methodology and the association of an uncertainty value to the crack percentage result computed by the proposed system. For instance, a possible anomaly corresponds to a crack classification in the middle of two non-crack classifications and vice-versa (a non-crack classification in the middle of two crack classifications). For detected anomalies, the user can choose to visualize the corresponding images and correct whatever is necessary.

When an isolated classification is detected, its contribution to uncertainty depends on the crack type that the label translates. In this way, it is necessary to calculate the maximum influence of each classification hypothesis on the crack percentage. In this case of study, it is intended to calculate the crack percentage in road sections whose width is 100 m. Accordingly, the maximum uncertainty associated with a given image, depending on the four classification hypotheses, is shown in Table 6.

Table 6. Uncertainty associated with each classification hypothesis in road sections with 100 m width.

Label	Affected Area	Uncertainty (%)
Longitudinal	2 m × 3 m	1.7
Transverse	2 m × 3.5 m	2
Alligator	3.5 m × 3 m	3
Non-crack	0	0

Although it was not developed in this work to promote the robustness and accuracy of the system, it would be interesting to compare the results of the segmentation neural network with the results of the classification neural network since there are cases (described in Section 4) where the classification model may fail to classify an image as containing cracks, while the segmentation model was able to detect them. Combining the results of both neural networks, the system's robustness and precision would increase. Another way to improve the accuracy in determining cracking percentages would be to incorporate a methodology to measure the length of the cracks. This measurement could be made with the result of crack segmentation; that is, from the segmentation result, the length in pixels of the crack would be obtained and, subsequently, the conversion of the crack length from pixels to meters could be made, enabling a more detailed analysis and a better calculation of the crack percentage.

3. Results

The Google Colab platform was utilized to carry out the experimental work described in this article, by means of a PC with an Intel(R) Xeon(R) CPU @ 2.30 GHz, NVIDIA-SMI 470.74 GPU, and a maximum provided RAM of 12 GB. The programming language chosen was Python 3.7.12, using the Tensorflow [33] and Keras [34] libraries.

A comparison between the analysis carried out by road analysis experts and the one resulting from the developed system was carried out to evaluate the system's performance in a real case scenario, namely during the inspection of the "IP3" pavement surface between the locations of "Souselas" and "Viseu". The crack percentages' results of the analyzed road lane sections are presented in Table 7.

Table 7. Crack percentages: FT 2—crack severity level 2; FT 3—crack severity level 3; T—Tecnofisil results; S—system results.

Road Segment [Start (m), End (m)]	FT 2 %		FT 3 %		Total %		S—Uncertainty (%)	S—Total with Uncertainty (%)
	T	S	T	S	T	S		
[75,900; 76,000]	0	0	0	0	0	0	0	0
[76,000; 76,100]	0	0	0	0	0	0	0	0

Table 7. *Cont.*

Road Segment [Start (m), End (m)]	FT 2 %		FT 3 %		Total %		S—Uncertainty (%)	S—Total with Uncertainty (%)
	T	S	T	S	T	S		
[76,200; 76,300]	41	26	0	3	41	29	4.7	[24.3; 33.7]
[76,300; 76,400]	54	5.1	0	0	54	5.1	1.7	[3.4; 6.8]
[76,400; 76,500]	0	1.7	0	0	0	1.7	1.7	[0; 3.4]
[76,500; 76,600]	0	7.1	0	0	0	7.1	5.4	[1.7; 12.5]
[76,600; 76,700]	18	7.1	0	0	18	7.1	3.4	[3.7; 10.5]
[76,700; 76,800]	57	15.7	0	3	57	18.7	7.1	[11.6; 25.8]
[76,900; 77,000]	18	27.7	0	0	18	27.7	3.4	[24.3; 31.1]

4. Discussion

By analyzing the results in Table 7, it was observed that the crack percentages presented by road analysis experts and the proposed system were similar in most road lane sections. In sections without cracking identified by road analysis experts, the proposed system gave a good response, presenting null or low percentages, as was the case of the following road lane sections: [75,900; 76,000], [76,000; 76,100], [76,400; 76,500], and [76,500; 76,600].

Then, by observing the five road lane sections with cracking identified by road analysis experts, for three of them, the proposed system came close to the percentage obtained by road analysis experts, namely in the following sections: [76,200; 76,300], [76,600; 76,700], and [76,900; 77,000]. The remaining two road lane sections with cracks, more specifically the sections [76,300; 76,400] and [76,700; 76,800], were critical cases in which the percentages obtained by the developed system presented a significant discrepancy, as in both cases the developed system erroneously presented a low percentage. Carrying out a more detailed analysis of these two road lane sections, with the help of the labels that the classification model assigned, it was concluded that the model did not identify the longitudinal cracking found within the white lane markings and sometimes in the road pavement. Hence, the classification model labeled images containing longitudinal cracking as non-crack images. Sample images corresponding to these two road lane sections are shown in Figures 18 and 19.

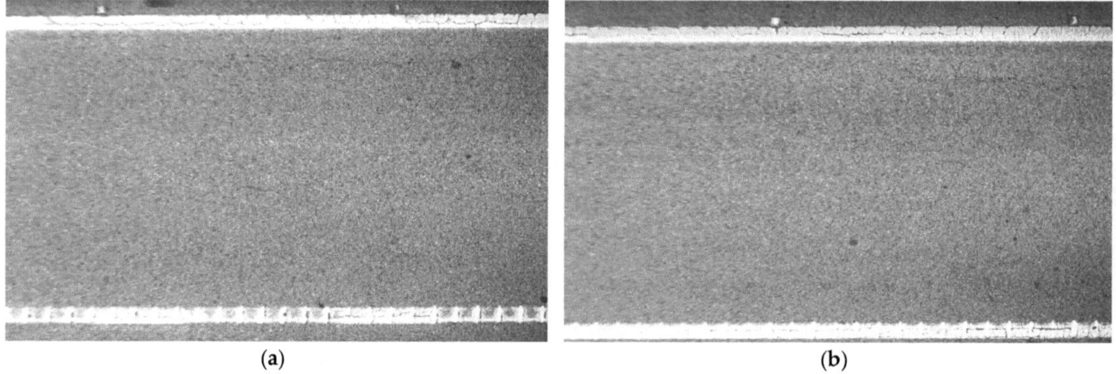

(a) (b)

Figure 18. Images relating to the section [76,300; 76,400]: (**a**) longitudinal crack in the pavement and within the upper white lane marking (the system classified the image as non-crack); (**b**) longitudinal crack in the pavement (the system classified the image as non-crack).

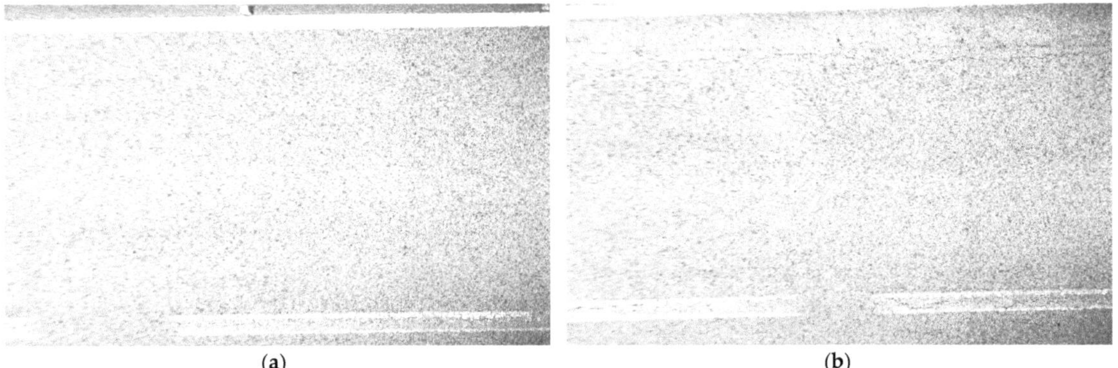

(a)

(b)

Figure 19. Images relating to the section [76,700; 76,800]: (**a**) longitudinal crack within the lower white lane marking (the system classified the image as non-crack); (**b**) longitudinal crack in the pavement and within the lower white lane marking (the system classified the image as non-crack).

By observing the images presented in Figures 18 and 19, it can be seen that the classification model has difficulties in identifying longitudinal cracks within the white lane markings. Sometimes, when the crack is within the white lane markings, it is at an early stage, and it can also be misinterpreted as flaws in the paint due to its wear and tear. Furthermore, the cracks in these images have a reduced width (less than 2 mm) and are barely visible, being below the detection threshold at which the classification model can detect cracking. One of the reasons why these cracks go unnoticed by the classification model is due to the resize procedure performed in the pre-processing step, as this transformation can attenuate the cracking present in the images. It should be noted that, although the classification model did not detect the longitudinal cracking in Figure 18, it was successfully detected by the segmentation model. Therefore, it would be interesting to associate the results of both networks to increase the system's robustness.

The situations highlighted in the previous analysis were cases marked by failures of the proposed system. However, it would also be interesting to compare these cases with some in which the system performed correct classifications to observe the differences. Therefore, in Figure 20, two sample images are presented, which were well classified by the classification model.

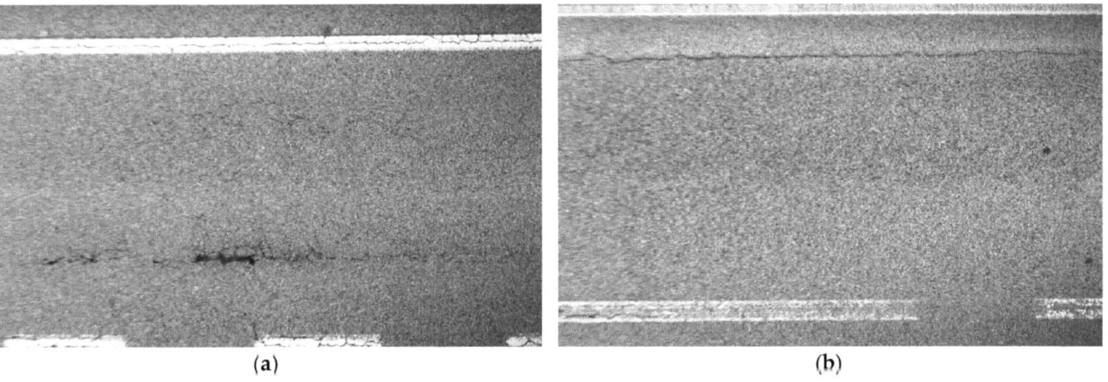

(a)

(b)

Figure 20. Correct classifications performed by the classification model: (**a**) image from section [76,200; 76,300] with a longitudinal crack in the pavement surface and within the white lane markings (the system classified the image as longitudinal crack); (**b**) image from section [76,900; 77,000] with a longitudinal crack in the pavement surface (the system classified the image as longitudinal crack).

Comparing the failures of the proposed system (samples shown in Figures 18 and 19) with cases in which the system correctly classified the longitudinal cracking present in the images (samples shown in Figure 20) verified that the cracking present in the latter sample is sharper and more noticeable. This comparison proves the existence of a detection threshold from which the classification model fails to detect less noticeable cracks, a system outcome that is related to the crack's width, i.e., thin cracks are less noticeable and thus less likely to be detected by the developed system than thick cracks.

Concluding the analysis, since the system is not totally reliable, the system's results should serve as a warning for the subsequent manual analysis of the road sections where possible problems were marked.

5. Conclusions

This paper proposed a low-cost automatic crack detection and classification system. By capturing images of road pavements, the proposed system provides the cracking percentage that characterizes the level of pavement surface degradation. Overall, the proposed system presented promising results, showing the potential to help the experts analyze and inspect road pavements. Moreover, correcting and improving some particularities of the system in its steps can become a very reliable and precise tool. With the use of the proposed system, the analysis and inspection of road pavements can give rise to faster procedures.

The proposed system has advantages and disadvantages. One of its advantages is that if the results are dubious, the system allows the user to visualize their origin. In this case, the system outputs a file with the labels assigned to the images corresponding to each road lane section, either in the case of classification or segmentation methods. This way, identifying errors and correcting the results becomes easier for the system's users. Another advantage is the processing time, which corresponds roughly to ten minutes per kilometer of the analyzed pavement surface.

In conclusion, this work proposed a low-cost and straightforward system that applies what was developed in the scientific community to obtain results from the image analysis of road pavement surfaces. It aimed to bridge the gap that exists between research work and what is demanded by the business world in the automatic detection of cracks in road pavement surfaces.

The proposed system has some margin for improvement as discussed in the following.

The formation of road lane sections could be improved through a more robust and controllable panorama creation function. Another alternative to improve the image concatenation could be to carry out preprocessing to standardize the pixel intensities presented by the images. Consequently, the transition between consecutive images would be smoother, minimizing the possibility of detecting false cracks in the union between concatenated images.

The robustness of the results could be improved by combination of the results provided by the segmentation neural network and those of the classification neural network, as the networks exploit different features, and therefore, the fusion of their results will lead to improved crack detection performance. Regarding image segmentation, adaptive learning approaches should be considered in further developments of the proposed system [35–38].

Another improvement for determining cracking percentages would be to incorporate the measurement of crack lengths. This measurement can easily be obtained from the crack segmentation result.

Finally, it would also be interesting to associate a global position system (GPS) with the acquisition system. In this way, it would be possible to match the GPS position of the acquisition vehicle to each captured image, allowing it to properly synchronize the vehicle speed with the consecutive camera's shutter releases. This strategy would guarantee that all the road lane sections would be correctly surveyed. Moreover, further developments of the system may consider a different image acquisition system, e.g., a camera mounted on a drone supported by photogrammetric techniques, which allows for overcoming the problem of georeferencing the images.

Author Contributions: Conceptualization, D.I.; methodology, D.I. and P.C.; software, D.I.; validation, P.C. and P.O.; formal analysis, D.I, P.C., P.O. and H.O.; investigation, D.I. and P.C.; resources, P.O.; data curation, D.I. and P.O.; writing—original draft preparation, D.I.; writing—review and editing, P.C. and H.O.; visualization, P.C and H.O.; supervision, P.C. and P.O.; funding acquisition, P.C. and P.O. All authors have read and agreed to the published version of the manuscript.

Funding: This work was partly funded by FCT/MEC under the project UIDB/50008/2020.

Data Availability Statement: Data are available from the authors upon reasonable request and with the permission of Pedro Oliveira (pedro.oliveira@tecnofisil.pt) from Tecnofisil.

Acknowledgments: The authors would like to thank Instituto de Telecomunicações and Tecnofisil, for the opportunity to work together with and for the help provided throughout this project.

Conflicts of Interest: The authors declare no conflict of interest.

References

1. Mohan, A.; Poobal, S. Crack detection using image processing: A critical review and analysis. *Alex. Eng. J.* **2018**, *57*, 787–798. [CrossRef]
2. Ouyang, A.; Luo, C.; Zhou, C. Surface distresses detection of pavement based on digital image processing. In *Computer and Computing Technologies in Agriculture IV*; Springer: Berlin/Heidelberg, Germany, 2011; pp. 368–375.
3. Tecnofisil. Tecnofisil–Consultores de Engenharia. Available online: https://tecnofisil.pt/ (accessed on 21 July 2022).
4. Oliveira, H.; Correia, P.L. Automatic Road Crack Detection and Characterization. *IEEE Trans. Intell. Transp. Syst.* **2013**, *14*, 155–168. [CrossRef]
5. Cao, W.; Liu, Q.; He, Z. Review of Pavement Defect Detection Methods. *IEEE Access* **2020**, *8*, 14531–14544. [CrossRef]
6. Waseem Khan, M. A Survey: Image Segmentation Techniques. *Int. J. Future Comput. Commun.* **2014**, *3*, 89–93. [CrossRef]
7. Oliveira, H.; Correia, P.L. Automatic road crack segmentation using entropy and image dynamic thresholding. In Proceedings of the 17th European Signal Processing Conference (EUSIPCO 2009), Glasgow, UK, 24–28 August 2009; pp. 622–626.
8. Ayenu-Prah, A.; Attoh-Okine, N. Evaluating Pavement Cracks with Bidimensional Empirical Mode Decomposition. *EURASIP J. Adv. Signal Process.* **2008**, *2008*, 861701–861708. [CrossRef]
9. Wu, Z.; Huang, N.E. A study of the characteristics of white noise using the empirical mode decomposition method. *Phys. Eng. Sci.* **2004**, *460*, 1597–1611. [CrossRef]
10. Li, Q.; Zou, Q.; Zhang, D.; Mao, Q. FoSA: F* Seed-growing Approach for crack-line detection from pavement images. *Image Vis. Comput.* **2011**, *29*, 861–872. [CrossRef]
11. Tarjan, R.E. *Data Structures and Network Algorithms*; Society for Industrial and Applied Mathematics: Philadelphia, PA, USA, 1983.
12. Zhang, L.; Yang, F.; Daniel Zhang, Y.; Zhu, Y.J. Road crack detection using deep convolutional neural network. In Proceedings of the 2016 IEEE International Conference on Image Processing (ICIP), Phoenix, AZ, USA, 25–28 September 2016; pp. 3708–3712.
13. Aslan, O.D.; Gultepe, E.; Ramaji, I.J.; Kermanshachi, S. Using Artifical Intelligence for Automating Pavement Condition Assessment. In Proceedings of the International Conference on Smart Infrastructure and Construction 2019 (ICSIC), Cambridge, UK, 8–10 July 2019; pp. 337–341.
14. Anand, S.; Gupta, S.; Darbari, V.; Kohli, S. Crack-pot: Autonomous Road Crack and Pothole Detection. In Proceedings of the 2018 Digital Image Computing: Techniques and Applications (DICTA), Canberra, Australia, 10–13 December 2018; pp. 1–6.
15. David Jenkins, M.; Carr, T.A.; Iglesias, M.I.; Buggy, T.; Morison, G. A Deep Convolutional Neural Network for Semantic Pixel-Wise Segmentation of Road and Pavement Surface Cracks. In Proceedings of the 2018 26th European Signal Processing Conference (EUSIPCO), Rome, Italy, 3–7 September 2018; pp. 2120–2124.
16. Kang, K.; Wang, X. Fully Convolutional Neural Networks for Crowd Segmentation. *arXiv* **2014**, arXiv:1411.4464.
17. Ronneberger, O.; Fischer, P.; Brox, T. U-Net: Convolutional Networks for Biomedical Image Segmentation. In *Medical Image Computing and Computer-Assisted Intervention–MICCAI 2015*; Lecture Notes in Computer Science; Springer: Berlin/Heidelberg, Germany, 2015; pp. 234–241.
18. Zhang, K.; Zhang, Y.; Cheng, H.-D. CrackGAN: Pavement Crack Detection Using Partially Accurate Ground Truths Based on Generative Adversarial Learning. *IEEE Trans. Intell. Transp. Syst.* **2021**, *22*, 1306–1319. [CrossRef]
19. Liu, C.; Zhu, C.; Xia, X.; Zhao, J.; Long, H. FFEDN: Feature Fusion Encoder Decoder Network for Crack Detection. *IEEE Trans. Intell. Transp. Syst.* **2022**, *23*, 15546–15557. [CrossRef]
20. Ma, D.; Fang, H.; Wang, N.; Zhang, C.; Dong, J.; Hu, H. Automatic Detection and Counting System for Pavement Cracks Based on PCGAN and YOLO-MF. *IEEE Trans. Intell. Transp. Syst.* **2022**, *23*, 22166–22178. [CrossRef]
21. Han, C.; Ma, T.; Huyan, J.; Huang, X.; Zhang, Y. CrackW-Net: A Novel Pavement Crack Image Segmentation Convolutional Neural Network. *IEEE Trans. Intell. Transp. Syst.* **2022**, *23*, 22135–22144. [CrossRef]

22. Yang, F.; Zhang, L.; Yu, S.; Prokhorov, D.; Mei, X.; Ling, H. Feature Pyramid and Hierarchical Boosting Network for Pavement Crack Detection. *IEEE Trans. Intell. Transp. Syst.* **2020**, *21*, 1525–1535. [CrossRef]
23. Long, J.; Shelhamer, E.; Darrell, T. Fully convolutional networks for semantic segmentation. In Proceedings of the 2015 IEEE Conference on Computer Vision and Pattern Recognition (CVPR), Boston, MA, USA, 7–12 June 2015; pp. 3431–3440.
24. U-Net Architecture. Available online: https://lmb.informatik.uni-freiburg.de/people/ronneber/u-net/ (accessed on 23 July 2022).
25. Diakogiannis, F.I.; Waldner, F.; Caccetta, P.; Wu, C. ResUNet-a: A deep learning framework for semantic segmentation of remotely sensed data. *ISPRS J. Photogramm. Remote Sens.* **2020**, *162*, 94–114. [CrossRef]
26. Camera Teledyne Lumenera lt16059h. Available online: https://www.znjtech.cn/resources/datasheets/teledyne-lumenera/lt16059h-datasheet.pdf (accessed on 23 July 2022).
27. CrackForest Dataset. Available online: https://github.com/cuilimeng/CrackForest-dataset (accessed on 23 July 2022).
28. AigleRN. Available online: http://telerobot.cs.tamu.edu/bridge/Datasets.html (accessed on 23 July 2022).
29. CrackTree260. Dataset and Ground Truth. Available online: https://1drv.ms/f/s!AittnGm6vRKLyiQUk3ViLu8L9Wzb (accessed on 23 July 2022).
30. CRKWH100. Dataset, Ground Truth. Available online: https://1drv.ms/f/s!AittnGm6vRKLglyfiCw_C6BDeFsP (accessed on 23 July 2022).
31. CrackLS315. Dataset, Ground Truth. Available online: https://1drv.ms/u/s!AittnGm6vRKLg0HrFfJNhP2Ne1L5?e=WYbPvF (accessed on 23 July 2022).
32. Bradski, G. The OpenCV Library. *Dr. Dobb's J. Softw. Tools* **2000**, *120*, 122–125.
33. TensorFlow: Large-Scale Machine Learning on Heterogeneous Systems. Available online: https://www.tensorflow.org/ (accessed on 23 July 2022).
34. Keras: Deep Learning for Humans. Available online: https://github.com/fchollet/keras (accessed on 23 July 2022).
35. Guo, P.; Zhang, W.; Li, X.; Zhang, W. Adaptive Online Mutual Learning Bi-Decoders for Video Object Segmentation. *IEEE Trans. Image Process.* **2022**, *31*, 7063–7077. [CrossRef] [PubMed]
36. Kerdvibulvech, C. A methodology for hand and finger motion analysis using adaptive probabilistic models. *J. Embed. Syst.* **2014**, *2014*, 18. [CrossRef]
37. Yang, F.; Yuan, X.; Ran, J.; Shu, W.; Zhao, Y.; Qin, A.; Gao, C. Accurate Instance Segmentation for Remote Sensing Images via Adaptive and Dynamic Feature Learning. *Remote Sens.* **2021**, *13*, 4774. [CrossRef]
38. Bourouis, S. Adaptive variational model and learning-based SVM for medical image segmentation. In Proceedings of the ICPRAM 2015–4th International Conference on Pattern Recognition Applications and Methods, Lisbon, Portugal, 10–12 January 2015; pp. 149–156.

remote sensing

MDPI

Article

Road Condition Detection and Emergency Rescue Recognition Using On-Board UAV in the Wildness

Chang Liu [1,2] and Tamás Szirányi [1,2,*]

1   Department of Networked Systems and Services, Budapest University of Technology and Economics, Műegyetem rkp. 3, H-1111 Budapest, Hungary
2   Machine Perception Research Laboratory of Institute for Computer Science and Control (SZTAKI), Kende u. 13–17, H-1111 Budapest, Hungary
*   Correspondence: sziranyi.tamas@sztaki.hu; Tel.: +36-20-999-3254

Abstract: Unmanned aerial vehicle (UAV) vision technology is becoming increasingly important, especially in wilderness rescue. For humans in the wilderness with poor network conditions and bad weather, this paper proposes a technique for road extraction and road condition detection from video captured by UAV multispectral cameras in real-time or pre-downloaded multispectral images from satellites, which in turn provides humans with optimal route planning. Additionally, depending on the flight altitude of the UAV, humans can interact with the UAV through dynamic gesture recognition to identify emergency situations and potential dangers for emergency rescue or re-routing. The purpose of this work is to detect the road condition and identify emergency situations in order to provide necessary and timely assistance to humans in the wild. By obtaining a normalized difference vegetation index (NDVI), the UAV can effectively distinguish between bare soil roads and gravel roads, refining the results of our previous route planning data. In the low-altitude human–machine interaction part, based on media-pipe hand landmarks, we combined machine learning methods to build a dataset of four basic hand gestures for sign for help dynamic gesture recognition. We tested the dataset on different classifiers, and the best results show that the model can achieve 99.99% accuracy on the testing set. In this proof-of-concept paper, the above experimental results confirm that our proposed scheme can achieve our expected tasks of UAV rescue and route planning.

Keywords: UAV rescue; wildness; road condition detection; emergency gesture recognition; sign for help; route planning

Citation: Liu, C.; Szirányi, T. Road Condition Detection and Emergency Rescue Recognition Using On-Board UAV in the Wildness. *Remote Sens.* **2022**, *14*, 4355. https://doi.org/10.3390/rs14174355

Academic Editors: Valerio Baiocchi, Alessandro Mei and Xianfeng Zhang

Received: 29 July 2022
Accepted: 30 August 2022
Published: 2 September 2022

Publisher's Note: MDPI stays neutral with regard to jurisdictional claims in published maps and institutional affiliations.

1. Introduction

With the development of artificial intelligence technology, machine vision technology is being applied to many aspects of human life. For example, in precision agriculture, it can be used for crop farming [1]; in transportation, the use of machine vision technology enables the detection and counting of vehicles in highway scenes [2]; in human daily life, it can also detect pedestrians in surveillance videos [3] and human falls [4], etc. Due to the flexibility and portability of drones and other features, unmanned aerial vehicle (UAV) vision has also been developed to a certain extent. Drone vision technology [5] is also increasingly being noticed and applied by the research community as well as industry. Especially in the field of wilderness rescue, drones will be even more advantageous, as they can reach places that are inaccessible to humans, which makes them extremely convenient for wilderness rescue [6,7]. In wilderness environments, where the GPS, mobile, radio, or road path is not fully covered and where a few people are present or congregate nearby, there is no proper infrastructure. Indeed, most of the wilderness is forested, mountainous, or unexplored, so when one or more people are in these environments for whatever reason, there is a certain potential danger. For example, for people who love hiking [8] and for those who are lost in the wilderness, a timely drone rescue and possible path monitoring are necessary when they are in danger.

Drones flying at low latitudes can detect in real-time what is happening on the ground, and to some extent enable search and rescue in the event of a natural disaster [9,10]. Similarly, remote sensing data from satellites [11–13] at high latitudes can also provide multiple channels of information to humans. Satellite data can be downloaded in advance and is constantly updated if the satellite is working. It is usually of a lower resolution than the image information captured by low altitude drones, but satellite data can provide information in multiple bands [14], which is not possible with most drones. Satellite remote sensing images play an irreplaceable role in the field of earth observation, and the different combinations of multiple channels can provide mankind with a wealth of ground-based information. Satellite remote sensing has been used in many fields, such as food security in agriculture [15], water body detection [16–19], segmentation of remote sensing images [20], land-use land-cover classification [21,22], change detection [23–25], environmental monitoring [26], etc. Based on geographical location information, we can fuse satellite information with real-time information captured by drones at low latitudes, both of which complement each other to provide greater assistance to humans. Of course, with the development of drone technology, a part of the drone is designed to carry multispectral camera to capture information of different bands, but such sensors are far from the bands captured by satellites.

The research background of this work is for a field environment with a poor network and bad weather conditions. When humans are in the wild without the network, owing to a lack of active antenna towers and/or cable infrastructure, they can encounter some unexpected problems and must seek help from the outside world; at this time, the drone rescue is particularly important. Drones will be more efficient compared to rescue workers. In addition to the above-mentioned network problems, wilderness environments also feature poor road conditions [27]; due to the sparse population, many roads in the wilderness are not well built, are usually composed of bare soil, or there may even be no roads that can allow people to walk. At this time, if there is a rainstorm and other bad weather, then the soil roads will become very muddy, which is extremely unfavorable to human use. Considering the poor road conditions in most of the wilderness, coupled with the effects of bad weather, it is particularly important that drones can detect road conditions and plan optimal routes for humans in distress. Even if humans have maps downloaded in advance on their phones in the above environment, there is no guarantee that the downloaded maps will match 100% with the actual roads encountered, especially in the wild, as many roads are not included in the maps [28].

Based on the above research background, the main goals of this work are as follows:

- The UAV plans an optimal route for human navigation in real-time in wildness environments with poor mobile networks and bad weather;
- With the local road quality ensuring the probable walking or driving speed, multispectral images are combined with weather information to estimate the walking speed of pedestrians, and the throughput of the road should also be evaluated;
- Different roads are assigned weights by geometric features, such as road length and width on the road extraction map, as well as estimated pedestrian speed and road throughput;
- Human–UAV interaction in low latitudes by adjusting the flight altitude of the UAV to facilitate the recognition of emergency hand gesture signals and potential hazards. e.g., injured or tired persons, or vehicles with technical defects.

The main challenges to be solved to achieve the above goals are as follows:

- The drones use extracted road network maps with different weights to navigate the user in real-time and plan an optimal route that meets the needs by weighted searching algorithms;

- The classification of normalized difference vegetation index (NDVI) values by multi-spectral images allows the differentiation of different road types. The walking speed of pedestrians on the road and the road throughput can be obtained by certain analyses;
- Road extraction allows us to obtain the length, width, and connectivity characteristics of a road. Pedestrian detection and tracking techniques allow for a theoretical evaluation of the road's throughput. Combined with weather information and road surface materials, the walking speed of pedestrians can be estimated. In short, these are the main basis for road priority assignment;
- The model for human–UAV interaction needs to be accurate, stable, and robust enough when there is a problem with navigation or when a human is in danger and needs to be rescued. Different models need to be compared in order to select the best and most reliable one.

In this proof-of-concept paper, we propose a technique to detect road conditions and plan the best route for humans in a field environment using UAV vision and multispectral imagery based on road extraction [29]. We use the road extraction techniques and modify the shortest path searching algorithms that have been proposed and widely used. The type of road material is detected based on multispectral information from satellites or UAVs, and the walking speed of pedestrians on different roads and the throughput of the road are evaluated in combination with weather information to plan an optimal route for humans. Based on the different flight heights of drones, drones at high altitudes can plan the optimal route for humans, and drones at low altitudes can perform human–UAV interaction. Users can interact with the drones through dynamic "Attention" and "Cancel" gestures [30] when humans encounter potential dangers or unexpected situations on the ground in the wild, such as when the best route planned according to the drones is wrong, the actual situation is that there is no more road to walk on ahead, or the user is threatened by terrorists, etc. At this point the drone can recognize emergency rescue signals and re-route the user or send rescuers directly. However, this work is limited by the accuracy of road extraction techniques, which is the main reason why we establish low-altitude human–drone interaction to assist humans in real-time. The road throughput and the assignment of road weights have not been carried out automatically.

The main innovations and contributions of this paper are as follows:

1. The surface material of the roads in the wildness environment is classified using an immediate UAV-boarded multispectral camera or, additionally, the latest available satellite-based multispectral camera images. The main distinction was made between bare soil roads and concrete or gravel roads in good condition;
2. When flying at high altitude, the UAV analyzed the road conditions based on the detected road surface materials and weather information, evaluated the throughput of different roads, and assigned different weights to the roads extracted from the map for optimal route planning;
3. When the UAV is flying at low altitudes, human–UAV interaction is possible. Based on the media-pipe hand landmarks, the "ok, good, v-sign, sign for help" dataset is created to identify emergency distress signals, so that corresponding measures can be taken to re-route or send rescuers;
4. The fusion of low-altitude drone imagery and high-altitude satellite imagery allows for a wider range of search and rescue information. Drones use their flexibility for high-altitude road condition detection and extraction, low-altitude human–UAV interaction, and emergency hazard identification to maximize their ability to help humans in distress.

In the subsequent sections, Section 2 focuses on the related work, including a description of the technical background and the relevant datasets used, as well as the process of data processing and the parts that have been published and completed. Section 3 is the main implementation approach, introducing the whole system flowchart, detection of road surface materials, analysis of road conditions, and evaluation of road throughput, followed by the collection of emergency rescue dataset and training of the models. Section 4 shows the experimental results of different parts of the whole system, including the results of the detection and analysis of road conditions in the wilderness environment when the UAV is flying at high altitudes, and the results of the recognition of emergency rescue signals by the UAV at low altitudes, as well as the evaluation of the model. Sections 5 and 6 contain a discussion section, the conclusions of this work, and an outlook for future work.

2. Related Work

2.1. Technical Background

The technical background in which this work is implemented is a GPU-equipped UAV with a NVIDIA Jetson AGX Xavier developer kit [31], along with a Parrot Sequoia+ Multispectral Camera [32]. A standalone on-board system is very important, since in the wilderness we do not have network to rely on. From Sabir Hossain's experiments [33] on different GPU systems, it was evident that the Jetson AGX Xavier was powerful enough for this proposed system. As shown in Figure 1, the UAV can process the video information captured by the camera in real-time. The Parrot Sequoia+ multispectral camera is chosen because it is multispectral and can capture four channels of information. The details of the specifications are shown in Table 1.

Figure 1. Drone with on-board GPU and multispectral camera.

Table 1. Jetson AGX Xavier and Parrot Sequoia+ multispectral camera specification.

Jetson AGX Xavier [31]		Parrot Sequoia + Multispectral Camera [32]	
GPU	512-core Volta GPU with Tensor Cores	**Sensor**	Multispectral sensor + RGB camera
CPU	8-core ARM v8.2 64-bit CPU, 8MB L2 + 4MB L3	**Multispectral sensor**	4-band
Memory	32GB 256-Bit LPDDR4x | 137GB/s	**RGB resolution**	16 MP, 4608 $\times$ 3456 px
Storage	32GB eMMC 5.1	**Single-band resolution**	1.2 MP, 1280 $\times$ 960 px
DL Accelerator	(2$\times$) NVDLA Engines	**Multispectral bands**	Green (550 $\pm$ 40 nm) Red (660 $\pm$ 40 nm) Red-edge (735 $\pm$ 10 nm) Near-infrared (790 $\pm$ 40 nm)
Vision Accelerator	7-way VLIW Vision Processor	**Single-band shutter**	Global
Encoder/Decoder	(2$\times$) 4Kp60 | HEVC/(2$\times$) 4Kp60 | 12-Bit Support	**RGB shutter**	Rolling
Size	105 mm $\times$ 105 mm $\times$ 65 mm	**Size**	59mm $\times$ 41mm $\times$ 28mm
Deployment	Module (Jetson AGX Xavier)	**Weight**	72 g (2.5 oz)

In addition to information captured by UAV cameras with multispectral sensors, satellite information is also involved. Currently, the European Space Agency (ESA) [34] or United States Geological Survey (USGS) [35] has many publicly available datasets that users can download and use for free, and the channel information provided by satellites is more and more extensive. As such, UAV equipped with GPU can operate in real-time without relying on the local network. Although its flight time is affected by battery life, etc., humans are working on solving this problem, and the battery life can be predicted [36], meaning that stable energy can be provided during the disaster [37]. Thus, in special cases for emergency rescue, the flight time can be controlled according to demand. In our proposed drone system, the endurance of a GPU-equipped UAV with an NVIDIA Jetson AGX Xavier developer kit, along with a Parrot Sequoia+ multispectral camera, is around half an hour. The endurance of the drones will be improved in our system if the battery capacity is increased and if the drones are flown with rotary wings. Another solution for the constraint of the drone endurance is that drones can also work together in the wilderness, with different drones performing different tasks and collaborating with each other through communication to carry out rescue work. Indeed, UAV can revisit the wilderness after returning from the charging station, or a replacement team is activated, where the returning unit checks the proposed roads, thus, taking advantage of the return time. The training of the model for the human–computer interaction part is carried out on the ground station computer, which is equipped with an NVIDIA GeForce GTX Titan GPU and an Intel(R) Core (TM) I7-5930k CPU.

2.2. Related Datasets and Data Processing

The dataset presented in Figure 2 was captured by a drone equipped with a Parrot Sequoia+ multispectral camera, located in Biatorbágy, Hungary. The latitude of Biatorbágy is 47.470682, and the longitude is 18.820559. Biatorbágy (Wiehall-Kleinturwall, in German) is a town in Pest County, Budapest Metropolitan Area, Hungary [38]. From Figure 2, we can see that the image captured by the drone contains four bands, which are green, near-infrared, red-edge, and red. This UAV dataset was collected in 2017, and current satellite imagery from Google Maps [39] is also shown in Figure 2. Some changes in houses and vegetation can be clearly seen, but several of the main roads are unchanged.

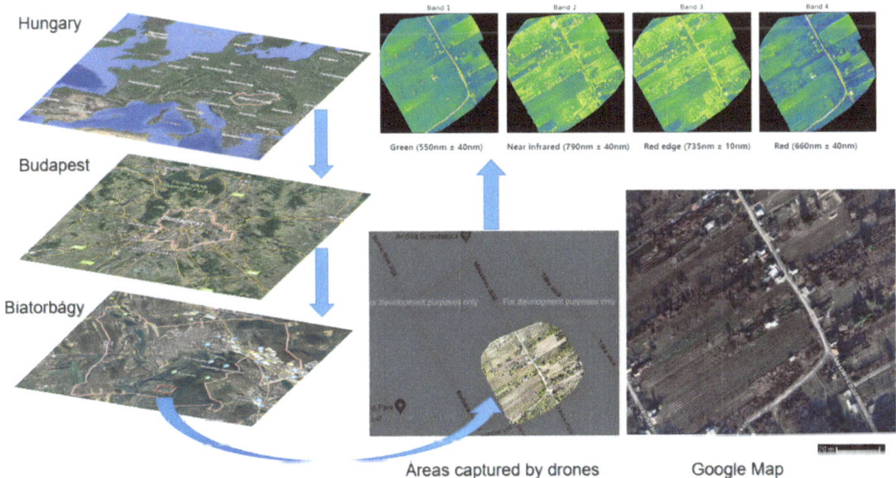

Figure 2. UAV with Parrot Sequoia+ multispectral camera dataset (The location is Biatorbágy, Budapest, Hungary).

GeoEye's OrbView-3 satellite (2003 to 2007) was among the world's first commercial satellites to provide high-resolution imagery from space. OrbView-3 collected 1 m panchromatic (black and white) or 4 m multispectral (color) imagery at a swath width of 8 km for both sensors. One meter imagery enables more accurate viewing and mapping of houses, automobiles, and aircraft, and makes it possible to create precise digital products. Four meter multispectral imagery provides color and near-infrared (NIR) information to further characterize cities, rural areas, and undeveloped land from space. Imagery from the OrbView-3 satellite complements existing geographic information system (GIS) data for commercial, environmental, and national security customers [40]. The data downloaded in this paper are from USGS (https://earthexplorer.usgs.gov/, accessed on 1 May 2022) [35], located in Australia, and the coordinates are −35.038926, 138.966428. Specific satellite information and information on the satellite datasets used in this paper are shown in Table 2.

Table 2. Orbview-3 Specifications and GeoEye's OrbView-3 satellite Dataset Attribute.

Orbview-3 Specifications [37]			Dataset (BirdWood, Adelaide, Australia) Attribute [35]	
Imaging Mode	**Panchromatic**	**Multispectral**	**Entity ID**	3V070304M0001619071A52 0001900252M_001655941
Spatial Resolution	1 m	4 m	**Acquisition Date**	2007/03/04
Imaging Channels	1 channel	4 channels	**Map Projection**	GEOGRAPHIC
Spectral Range	450–900 nm	450–520 nm (blue) 520–600 nm (green) 625–695 nm (red) 760–900 nm (NIR)	**Date Entered**	2011/11/10

In this work, we mainly used data from the Birdwood neighborhood. Birdwood is a town near Adelaide, South Australia. It is in the local government areas of the Adelaide Hills Council and the Mid Murray Council [41]. Figure 3 shows the specific information of the OrbView-3 satellite dataset used in this work, containing information in four different bands, namely blue, green, near-infrared, and red, as well as satellite information from the OrbView-3 satellite near Birdwood and satellite image information from Google Maps. Combining the information from the satellite dataset in Table 2, the area covered by the satellite is in Australia, and our study site is in the town of Birdwood in the city of Adelaide. Specific scale information and the location of the geographical coordinates of the study site, the town of Birdwood, can also be found in Figure 3. We selected the entire satellite image of Birdwood because the location is near a forested area, far from the urban center, and is mostly unoccupied in the wilderness, with both good and bad road conditions, which is consistent with the background of this work. The arrows in Figure 3 show the data selection process, and, finally, we captured the same areas on Google Maps for comparative display. Figure 3 also shows the four bands of the entire satellite map.

In the data processing phase, the input to the proposed system is the video sequence captured by the camera of the airborne UAV in real-time. When the available UAV does not have a camera with multispectral sensors, then the multispectral satellite data downloaded in advance will play a crucial role; when the UAV's camera has multispectral sensors, then the multispectral satellite information acquired in advance is available to be fused and used, and the two complement each other. Thus, whether through the drone data source or satellite data source, multispectral information can be accessed and used by humans. Since the UAV is equipped with an NVIDIA Jetson AGX Xavier, it can perform the processing of the video sequences captured by the camera in real-time into image sequences, which can be RGB images or multispectral images. Here, RGB three-channel images will be used as input for the road extraction part, and multispectral images will be used as input when detecting road conditions. Figures 2 and 3 show the experimental data areas from UAVs and satellites used in this task, respectively. The data types and features of the specific UAV and satellite referenced experimental areas are shown in Table 3.

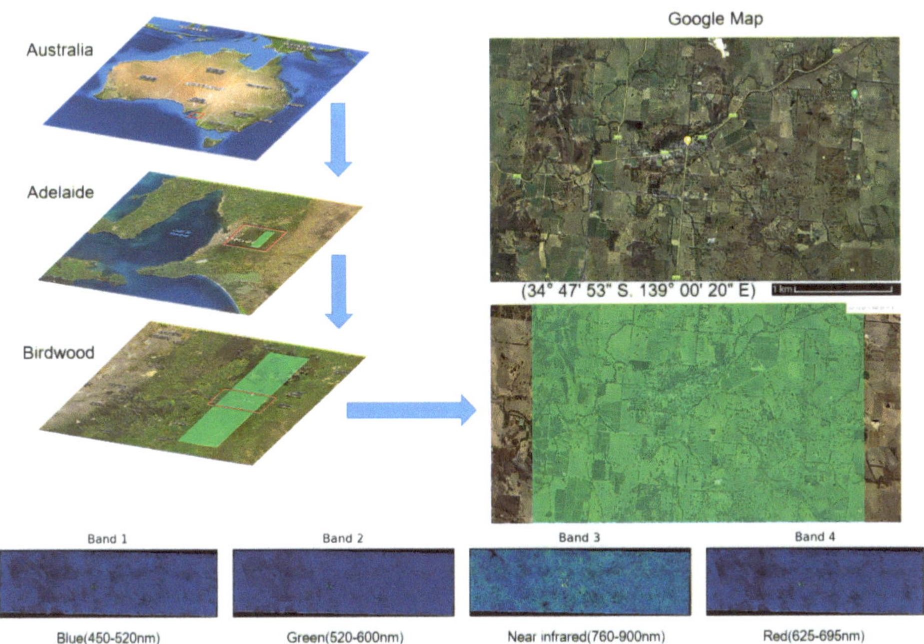

Figure 3. GeoEye's OrbView–3 satellite Dataset (The location is Birdwood, Adelaide, Australia).

Table 3. The data types and features of the UAV and satellite referenced experimental areas.

Data Types and Features	UAV Dataset	Satellite Dataset
RGB Images Resolution	15,735 × 14,355	7202 × 2151
Multispectral Images	Red (660 ± 40 nm) Near-infrared (790 ± 40 nm)	Red (625–695 nm) Near-infrared (760–900 nm)
Cloud Cover	0	0
Coverage area	Small	large

2.3. Road Extraction and Optimal Route Planning

This subsection describes the work that has been carried out in road extraction and optimal route planning. For the task of road extraction, there are already many publicly available datasets collected by satellite that can be used, such as the Massachusetts Road Dataset [42] and DeepGlobe Dataset [43], and different road extraction networks have been proposed for these datasets, such as U-Net [44], Seg-Net [45], LinkNet [46], etc. In this work, considering our research background and the problem to be solved, we chose D-LinkNet [47], which won the first place in the Deep Globe Road Extraction Challenge. In addition to its better precision, recall, and F1 than other networks on the same dataset, D-LinkNet can handle road characteristics, such as narrowness, connectivity, complexity, and long span to some extent by expanding the sensory field and assembling multi-scale features in the central part while preserving detailed information, which is needed in our research work. For optimal route planning, road connectivity must be considered, because the routes are continuous rather than intermittent. D-LinkNet is a LinkNet with a pretrained encoder and dilated convolution for high resolution satellite imagery road extraction. D-LinkNet can perform road extraction for the input three-channel RGB images, regardless of whether they are UAV images or satellite images. It can perform road segmentation well, labeling the roads as foreground and the other parts of the image as background. The DeepGlobe dataset also fits the context of our work, which is the field environment.

D-LinkNet can solve the road connectivity problem well, and the trained model can be used directly on the on-boarded UAV.

In the optimal route planning part, we have proposed and published the corresponding solution [48]. According to the map of road extraction, different weights are assigned to different roads. The weights are assigned by the road condition information, mainly including the road surface material, the length and width of the road, the walking speed of pedestrians on that road, and the estimated road throughput combined with the weather information, etc. Finally, the road map with weights is used for route planning by the A^* algorithm [49], and the best route can be planned for the user based on the road network map with different road priorities. Of course, users are required to report their starting location and the location of the destination they want to go to. Some optimal route planning results with complex road networks can be found in [48].

3. Methodology

3.1. Proposed System

The main workflow of the proposed system is to extract the road network by processing and analyzing the video streams captured in real-time using a UAV. The road surface condition is then detected in combination with pre-downloaded weather information and satellite information to assess the pedestrian speed and the road throughput. Next, the evaluation of pedestrian speed and road throughput is combined with the length and width of the road from the road extraction section to prioritize the different roads in the extracted road network. The final optimal route planning is based on the start location and destination provided by the user. When the user encounters a problem or potential danger, the user can attract the drone's attention through dynamic emergency gestures, and the drone can communicate with the user at low altitudes by adjusting its flight altitude, thus, providing maximum assistance to humans in distress.

The flow chart of the entire system is shown in Figure 4, where satellite data and weather data can be downloaded and analyzed and processed in advance. With the continuous development of weather forecasting, the information obtained from weather forecasts is becoming more and more accurate and can be relied upon [50]. As described in the introduction to the first section of this paper, the background of this work is that of a human being in a poorly networked and weathered field environment. The input to the system is a video sequence captured by a UAV flying in real-time, and the GPU-equipped UAV can process the video sequence to obtain a resolution of $1024 \times 1024 \times 3$ RGB images. If the UAV camera has a multispectral sensor, then the system will get four different channels of image information, namely green, near-infrared, red-edge, and red. The above steps prepared the way for the next road extraction and road condition analysis. The UAV at this stage was flown at an altitude of 50 m or more, and the RGB map obtained through data processing had a range of 512 m $\times$ 512 m.

The RGB image with a resolution of $1024 \times 1024 \times 3$ is used as the input of the road extraction network. With the road extraction by D-LinkNet [47], we can obtain information on the length and width of the road. Based on the satellite image information corresponding to the area where the UAV flies and the weather information downloaded in advance, we can determine whether there is bad weather, such as heavy rain, dusty weather, or a snowstorm, etc. This weather can lead to problems, such as a muddy bare soil road surface, low air visibility, or smooth road surfaces becoming icy, snowy, sandy, or water-covered, which can make them difficult for pedestrians to walk on. At this point, the material of the road surface is particularly important. Having a better condition road material is beneficial for human walking in rainy weather if the low visibility of the road caused by nearby bare soil is not considered in windy weather. On the contrary, if the road surface is composed of bare soil, then it will be difficult for humans to walk after heavy rain. Roads made of concrete or gravel are not so muddy, so the weather has a great influence depending on the different road materials used in an area. For the detection of road material information, we use multispectral information from satellites or drones to assess the normalized difference vegetation index [51,52] and, depending on the classification of the normalized difference vegetation index, the road material is detected, since the normalized difference vegetation index values are different between bare soil and roads with concrete or gravel cover.

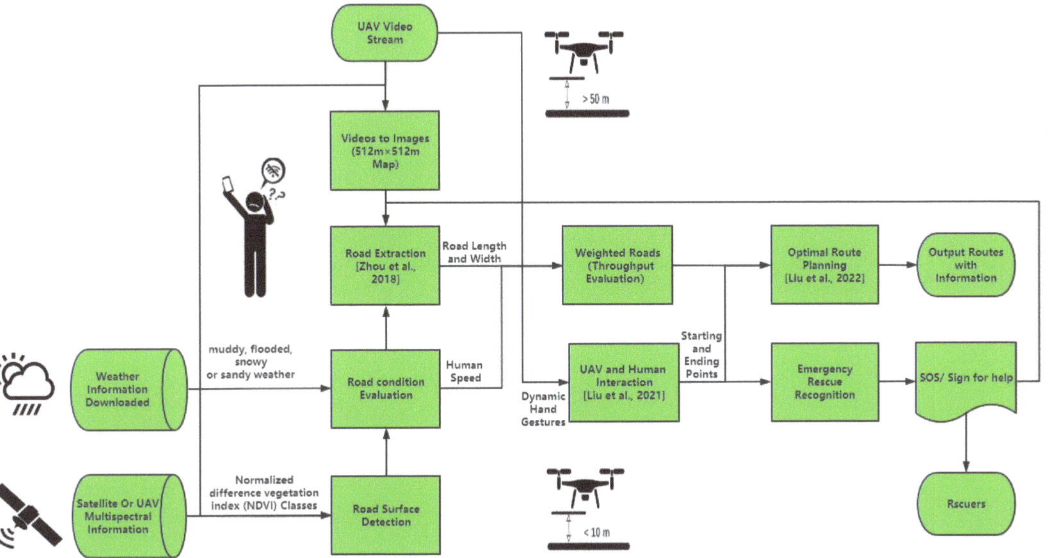

Figure 4. Flowchart of the proposed whole system [30,47,48].

For a road, we define two basic measures over the geometrical parameters of width and slope, namely walkability (the speed of pedestrians) and transferability (the number of pedestrians able to enter or exit the road during the analysis period). Combining weather information and road surface material information, we can estimate the speed of pedestrians. With pedestrian detection and tracking we can evaluate the throughput of the road [53,54] and finally combine the road length and width information to assign priority to the road. The road map with the weights can then be used to plan the best route using the A^* algorithm [48]. The system will eventually output an optimal route for the current situation to the user, thus, enabling the UAV to help humans navigate in the field.

Since, in the real world, there may be emergency situations, such as a human being encountering a threat or a route going wrong, and other special situations, communication between the drone and the human being is crucial. The drone can adjust its flight altitude, and when it lowers its flight altitude to less than 10 m, it can complete the interaction process with the human being on the ground through dynamic attention and cancellation gestures [30]. The drone will accurately identify potential dangers and emergencies, reroute humans in distress, or call SOS directly to request the intervention of rescuers.

3.2. Road Surface Detection and Road Throughput Evaluation

The normalized difference vegetation index (NDVI) is a simple graphical indicator that is often used to analyze RS measurements and assess whether or not the target being observed contains green healthy vegetation. This index will be referred to by the abbreviation NDVI throughout the rest of the paper. The NDVI [51,52,55] is derived from the red: near-infrared reflectance ratio, where NIR and RED are the amounts of near-infrared and red light, respectively, reflected by the vegetation and captured by the sensor of the satellite. The formula is based on the fact that chlorophyll absorbs RED, whereas the mesophyll leaf structure scatters NIR. Thus, NDVI values range from −1 to +1, where negative values correspond to an absence of vegetation [52]. The NDVI quantifies vegetation by measuring the difference between near-infrared (NIR) (which the vegetation strongly reflects) and red light (which the vegetation absorbs/has a low reflectance) [56]. In our Section 2.2, both the UAV dataset and the satellite dataset are present in the near-infrared band and the red band. The formula for NDVI is given as follows in Equation (1), where NIR is near-infrared light and Red is visible red light:

$$NDVI = (NIR - Red)/(NIR + Red) \tag{1}$$

The value of the NDVI will always fall between -1 and $+1$. Values between -1 and 0 indicate dead plants, or inorganic objects, such as stones, roads, and houses. On the other hand, NDVI values for live plants range between 0 to 1, with 1 being the healthiest and 0 being the least healthy [57]. Bare soils range from about 0.2–0.3 [58]. Based on the above presentation of the different NDVI values, we can determine that the value of NDVI for a road not covered by vegetation should be negative. However, if the road consists of bare soil only, then the NDVI values will be distributed in the range of 0.2–0.3 [58]. This effectively distinguishes the roads covered by concrete or gravel from those covered by bare soil.

Before evaluation of road throughput, we can estimate the human walking speed on different kinds of roads based on weather information and road surface material information. For rainy weather or just after a heavy rain, the speed of a human walking in the bare soil road is about 1.07 m/s [59]. The asphalt road or stone road will be more friendly to humans and, thus, human walking speed in rainy weather on the asphalt or stone road is about 1.2 m/s [60]. For dusty weather, which brings a reduction in air visibility, walking on asphalt-covered roads will be less dusty and faster than on bare soil-covered roads. For the snowy weather, a smooth road surface will speed up the human walking speed, so the road material is no longer the main factor affecting the human walking speed. For no adverse weather conditions, the human walking speed on different roads is the same. The speed can also be effectively estimated from other parts of the remotely sensed areas, where the situation is more definite.

After evaluating the speed of humans walking on different types of roads, the drones can see that humans are moving within and between different roads. In a limited time, for the 10 min duration of the drone analysis, the drone can evaluate the throughput of different roads and, depending on the results of the road extraction, different throughputs are obtained for different roads. Equation (2) is the definition of the road throughput. In our previous work, the UAV can count the number of pedestrians in real-time [61]. Throughput is defined as the number of distinct people able to enter or exit the road during the analysis period. To aid its interpretation, the throughput should be divided into five components according to whether or not the people entered, exited, never entered, or never exited the road during the analysis period [54]. The five categories are described as follows and shown in Figure 5. In Figure 5 we can see that a total of five categories of people are included, and each person in each category is given a specific ID. The ID of the same person is kept constant when entering and leaving the road, and the technique of assigning IDs to different people and tracking them can be found in our previous work in [61].

$$Through\text{-}put = N/T \tag{2}$$

where N is the number of distinct people able to enter or exit the road during the analysis period, and T is the time during the analysis period (e.g., 10 min). The five categories are as follows:

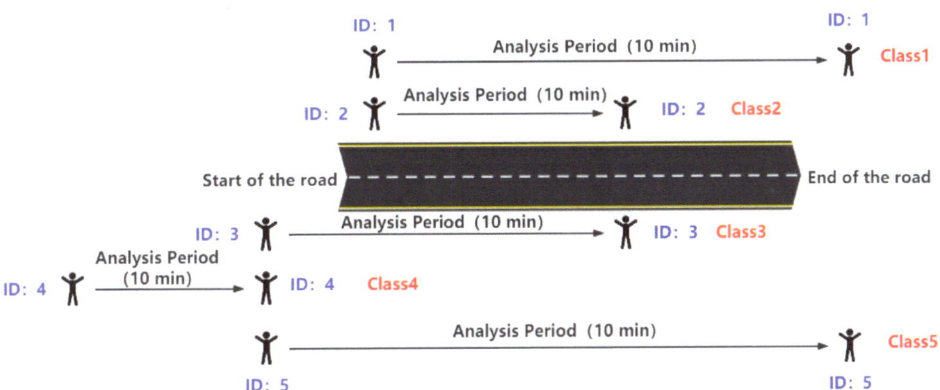

Figure 5. The five categories used to assess the road throughput.

1. Human Class 1—people that were present at the start of the analysis period and were able to successfully exit the road before the end of the analysis period;
2. Human Class 2—people that were present at the start of the analysis period but were unable to successfully exit the road before the end of the analysis period;
3. Human Class 3—people that were able to enter the road during the analysis period but were unable to successfully exit the road before the end of the analysis period;
4. Human Class 4—people that tried to enter the road during the analysis period but were completely unsuccessful;
5. Human Class 5—people that entered during the analysis period and were able to successfully exit the road before the end of the analysis period.

The percent of incomplete trips will be the sum of human classes 1, 2, 3, and 4, divided by the sum of all human classes (1 + 2 + 3 + 4 + 5). Generally, higher throughputs and lower percentages of incomplete trips are desired since they reflect the productivity of the road [54]. The throughput of the road and the length and width geometrical characteristics of the road directly affect the road weights that are assigned, that is, the priority assignment problem.

The task of optimal route planning based on roads with different weights has been solved in [48]. The road weight assignment in [48] is manual and based only on the road connectivity and length–width geometric properties, and does not make an analysis and evaluation of pedestrian walking speed, road surface material, road throughput, etc. This work complements the prioritization evaluation and weight assignment part of different roads in the previous road map networks. It is important to emphasize that in the route planning and navigation section, we are assuming that humans can report the starting location and destination for the UAV.

3.3. Emergency Rescue Recognition

The main flaw of drones flying at altitudes above 50 m, as is responsible for navigation, is that they are not able to see the situation on the ground in real-time and cannot meet the needs of the user in the current situation, so it is necessary to adjust the flight height to carry out human–UAV interaction. There are many uncertainties in the wilderness environment, so the drone flying at low altitudes below 10 m can communicate with the user through dynamic gestures [30] and, once there is an emergency, it is possible for the drone to make a timely response and take measures.

The Signal for Help [62] is a single-handed gesture that can be used to alert others that people feel threatened and need help. Originally, the signal was created as a tool to combat the rise in domestic violence cases around the world linked to self-isolation measures that were related to the COVID-19 pandemic. The signal is performed by holding one hand up with the thumb tucked into the palm, then folding the four other fingers down, symbolically trapping the thumb by the rest of the fingers. It was designed intentionally as a single continuous hand movement, rather than a sign held in one position, so it could be made easily visible. As this gesture is widely spread and popularized, it is increasingly known and used as a signal for potential hazard identification in this work, and we created a dataset of the gesture by mixing the gesture into some common human gestures, and the detail of the dataset is shown in Table 4.

Table 4. Emergency rescue recognition dataset.

Name	Number of Data	Hand Gestures
ok	4232	
v-sign	3754	
Good	4457	
SignForHelp (dynamic)	3504	1. Palm to camera and tuck thumb 2. Trap thumb

The datasets were collected using a 1080P 160° fisheye surveillance camera module for Raspberry Pi on the 3DR SOLO UAV system. Three people from our lab participated in UAV emergency rescue gesture dataset collection, the genders were two males and one female, aged between 25 and 30 years old. We collected data for each gesture in different orientations to cover as many situations as possible, making our dataset more generalizable. Table 4 shows the details of the UAV emergency rescue recognition dataset. The acquisition of this dataset was based on media-pipe hand landmarks, where we extracted the positional information of 21 key points on the human hand and saved them in a CSV file. The data extracted for each hand gesture was taken from a different person separately, and the final amount of data for each gesture is given in Table 4.

Figure 6 shows the main process of emergency rescue recognition, which is carried out in advance at the ground base station, and the final obtained model will be deployed directly on the UAV for use. The dataset collection is based on media-pipe hand landmarks [63,64], and the first process is the extraction of 21 hand key points. We collected four gestures from different people in our lab, which are ok, v-sign, good and sign for help. This extracted hand key-point data were stored in a CSV file, after which 70% of the entire dataset was used as the training set and 30% was used as the test set to be tested on different classifiers. We tested on the following five different classifiers: logistic regression [65], ridge classifier [66], random forest classifier [67], gradient boosting classifier [68], and deep neural network [69]. The classifier with the highest model accuracy at the end will win and be deployed for use on the UAV.

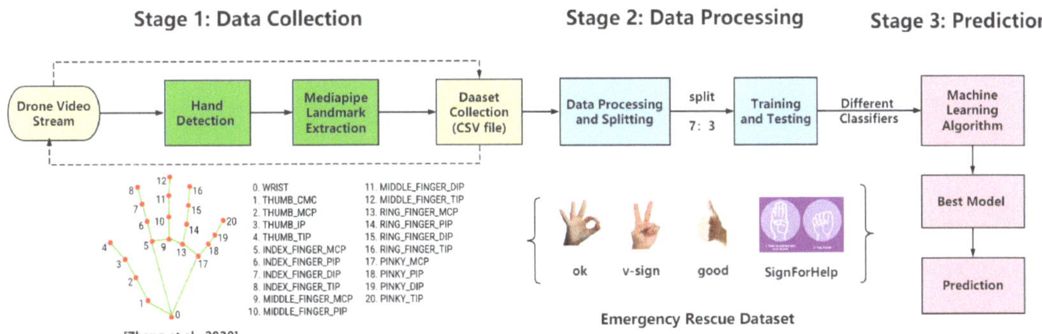

Figure 6. Architecture of the emergency rescue recognition method [64].

The architecture of the emergency rescue recognition method is divided into three main stages. The first stage is data collection, where the 21 key points of the hand are continuously presented from the video stream of the UAV input and stored in a CSV file for later model training. The second stage is the processing of the collected data and the separation of the data set into a training set and a test set. The third stage is the training of the different classifiers and the prediction of the results.

4. Experiments and Results

This section presents the experimental results of this work and contains two main subsections. The first, Section 4.1, presents the results of the detection of road surface materials on different datasets and shows the results of optimal route planning on different types of roads as well. The second part, Section 4.2, is the evaluation of a human–machine interaction model for UAVs in low altitude flight and a demonstration of simulated emergency rescue recognition in wildness.

4.1. Road Surface Condition Detection and Optimal Route Planning Results

This subsection shows the results of the road surface material detection and demonstrates the navigation task of optimal path planning. For the UAV dataset presented in Section 2.2, we obtained the results shown in Figures 7 and 8 with the proposed system. Figure 7 contains the RGB images taken by the UAV and the resulting NDVI images obtained by multi-channel image calculation. Normally, the range of NDVI is –1 to +1, but in this UAV dataset, we show the NDVI results as −0.19 to 0.94, in order to clearly distinguish some road surface materials. The NDVI image is well illustrated in Figure 7, and by comparing the analysis with the RGB image, we can see that the red area from −0.19 to 0.25 is largely uncovered by vegetation and mostly inorganic material. The orange area from 0.25 to 0.45 is mainly bare soil, while the yellow or green part above 0.45 is covered by vegetation. We zoom in on the part containing the road to show it at the bottom of Figure 7, and we can see that the main road in the middle is red, which means that the road surface is composed of non-soil gravel material, while the narrow road on the right is close to orange in color, and we can see that the surface material is mainly composed of bare soil. In this way, the different road composition materials can be effectively distinguished by the NDVI.

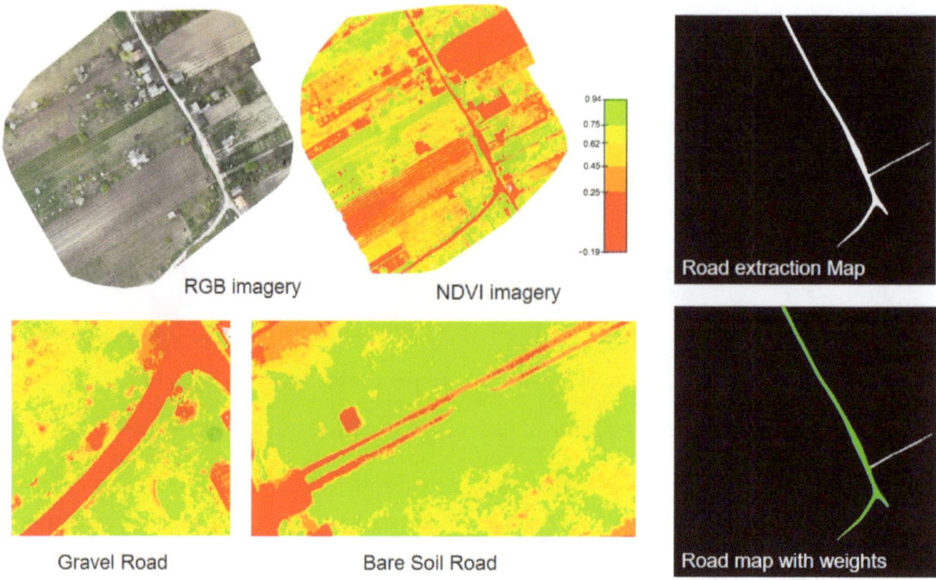

Figure 7. Road surface condition detection and road extraction with weights results for the UAV dataset.

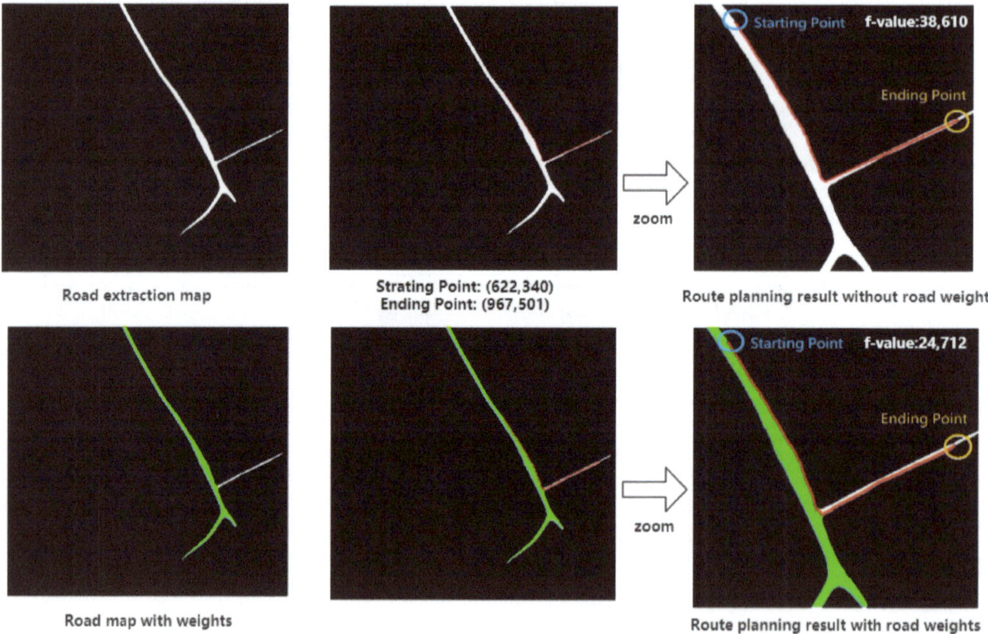

Figure 8. Optimal route planning results for the road maps with and without weight assignment in the UAV dataset.

The material composition of the road surface can have an impact on the walking speed of pedestrians under different weather conditions. According to the analysis and description inside the methodology in Section 3, we can evaluate the pedestrian speed by walking on different road surfaces under different weather conditions. In this UAV dataset, if the user encounters rainy weather, then the road consisting of narrow bare soil is very muddy and the walking speed of pedestrians is slower, but the walking speed of pedestrians on the main road consisting of gravel is faster. After road extraction, we can obtain the geometric features of the length and width of different roads. Applying the previously deployed yolo3-tiny model [61], we can evaluate the transferability of the road and, thus, obtain the throughput of that road. It is important to note here that the throughput of the road in the real case is not tested and, therefore, is not shown accordingly in the results section. The road surface material, the length and width of the road, the walking speed of the pedestrians, and the throughput of the road are combined and evaluated to give priority to the extracted road network. Of these, road throughput according to Equation (2) is used as the main criterion and basis. The results of road extraction and road priority assignment for the UAV dataset can also be found in Figure 7. The road network included in this dataset is relatively simple and not very complex due to the relatively small area included in the UAV dataset. More complex cases and optimal route planning results can be found in our published paper [48].

Figure 8 shows the best intelligent navigation route provided by the UAV based on the starting and ending position coordinates reported by the user. It can be clearly seen that in the two path planning comparison plots on the right side of Figure 8, the value of f in the plot with road priority assignment is reduced, where f indicates the cost of the shortest path at the pixel scale. Each pixel is considered a node, except that these nodes are given different weights. Here, $f(n)$ is the sum of $g(n)$ and $h(n)$, where n is the next node on the path, $g(n)$ is the cost of the path from the start node to the n node, and $h(n)$ is a heuristic function that estimates the cost of the cheapest path from n to the goal. Details of how the modified A^* algorithm finds the minimum value of f can be found in our previously published work [48]. The minimum value of f corresponding to the best route, found using the search algorithm, is 38,610. On the contrary, if the road is given priority, the minimum value of f corresponding to the best route found by the search algorithm is reduced to 24,712. This greatly improves efficiency for the user. As can be seen in the path planning in Figure 8, the optimal route search algorithm makes maximum use of the roads marked in green, because green roads represent roads in good condition, while white roads represent roads in poor condition. To show that the algorithm for optimal path planning works well, we used the results of a more complex road network from our previous work [48] in Figure 9 to show that the UAV is able to plan the best route when there are multiple paths to choose from. The f-*value* of the optimal path obtained by the best route search for the road map without priority assignment in Figure 9 is 4733, and the f-*value* of the best path planned for the road map with the priority road, on the contrary, is 1044. It should be noted that the image used in Figure 9 is from the DeepGlobe Road Extraction dataset [40], and the images are RGB three-channel.

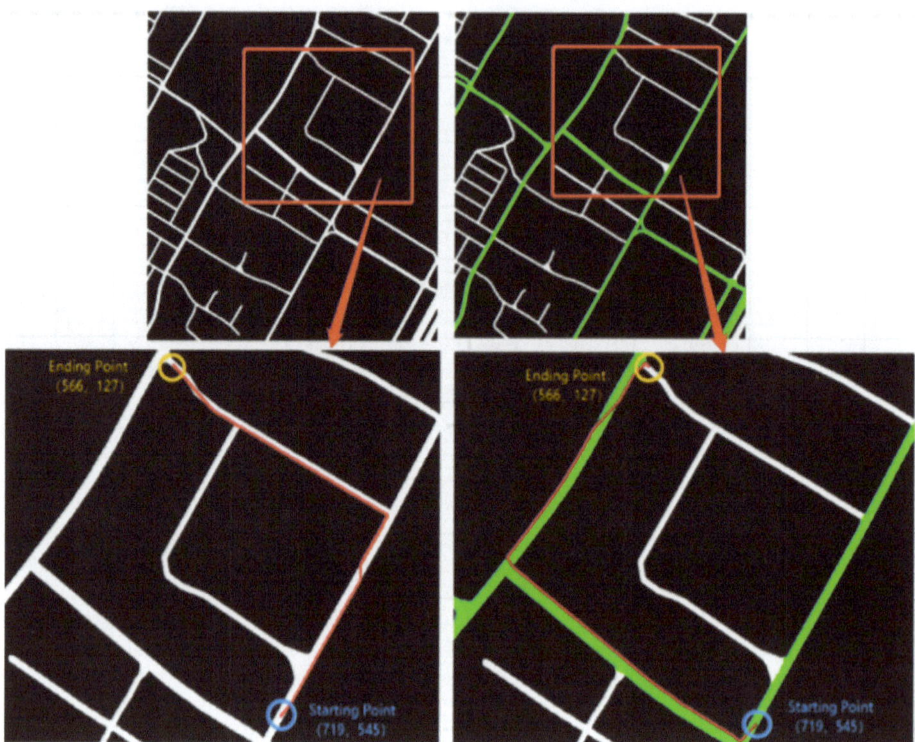

Figure 9. Optimal route planning results for the complex road maps with and without weight assignment in the previous work [48].

For the satellite dataset presented in Section 2.2, we also performed the corresponding tests. Figure 10 shows the NDVI index results for the whole satellite image and the results after classifying the NDVI. Here, similar to the processing of the UAV dataset described above, we shifted the NDVI value results from −1.0 to +1.0 to −0.2 to +1.0, displaying them with the purpose of distinguishing more clearly the material of the road surface. Based on the different values of NDVI we classified them into three categories. Those with NDVI values less than 0.2 are classified as being in good condition, which is generally inorganic in this case, and for road areas, where the surface material is composed of non-soil. On the other hand, NDVI values in the range of 0.2 to 0.4 are classified as being in a bare soil condition, and NDVI values greater than 0.4 are classified as vegetation. These three categories are shown in gray, tomato red, and green, respectively. These results can be found in Figure 10. We can see that, for roads, those with pavements consisting of bare soil are shown in tomato color, while those with pavements consisting of concrete and in good condition are shown in gray. In summary, roads with different road materials are well distinguished, which also provides the basis for the subsequent road priority assignment.

Figure 11 zooms in on the NDVI classification results of Figure 10, and we can see that the entire satellite image undergoes constant regional zooming, which allows a good distinction between the road conditions in different conditions. The entire satellite image was selected to show the results for the suburban area near Birdwood, but the rest of the road information is also well differentiated and can be compared with the actual situation in Google Maps. In fact, the classification of the whole satellite image is consistent with the information observed by Google Maps satellites as far as the places covered by vegetation are concerned. If you zoom in on the area, the information for the road is also the same as that observed by Google Maps, so the different road surface materials are effectively distinguished from each other. The last zoomed area in Figure 11 shows that the road network is clear; some roads are shown in tomato color which means that the roads are composed of bare soil, while some roads are shown in gray color which means that the surface of these roads is composed of non-bare soil, such as gravel or asphalt, and they are not very muddy after bad weather, especially after heavy rain.

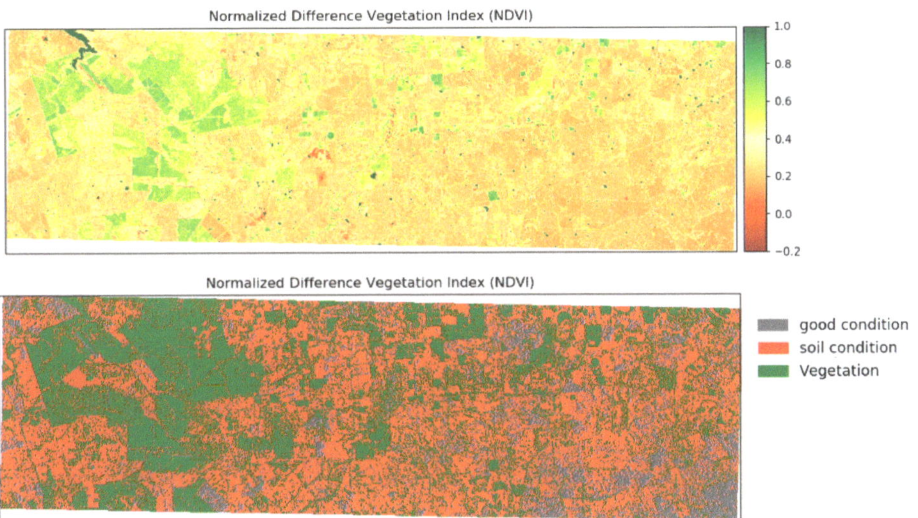

Figure 10. NDVI index and NDVI classes results for GeoEye's OrbView–3 satellite dataset.

Figure 11. Road surface condition detection and classification results for GeoEye's OrbView–3 satellite dataset by zooming in on white boxes.

Near the Birdwood area, we performed the detection and classification of the road surface material in a small area, as well as the extraction and prioritization of the roads in the area, and the related results are shown in Figure 12. According to the classification of road surface materials of the different roads shown in Figure 11, we can evaluate the walking speed of pedestrians under the influence of bad weather. After heavy rain, the road surface composed of bare soil is muddy, which is unfavorable for pedestrians to walk on. On the contrary, if the road composition is non-muddy, such as concrete or gravel, then the road is less muddy and the walking speed of pedestrians is about 1.2 m/s, which is faster than the walking speed on the muddy road. The walking speed of pedestrians also affects the throughput of that road, so it is necessary to evaluate the speed. The results of road extraction allow us to obtain the two geometric characteristics of road length and width. Similarly, by deploying the yolo3-tiny model we can analyze the throughput of each road in 10 min. Here, the corresponding results are not shown, due to the lack of real cases to test. Finally, the road weights are assigned by considering the composition of the road material, the walking speed of pedestrians, the length and width of the road, and the throughput of the road. Here, green represents roads with priority and white represents roads without priority, consistent with our previous work [48]. Theoretically, as shown in Figure 5, road throughput is obtained by counting and tracking the number and location of pedestrians on each road in real-time at low altitude by a GPU-equipped UAV [61]. The road surface material, the estimation of pedestrian speed, the road length and road width, and the road throughput assessment are the main basis for assigning road weights.

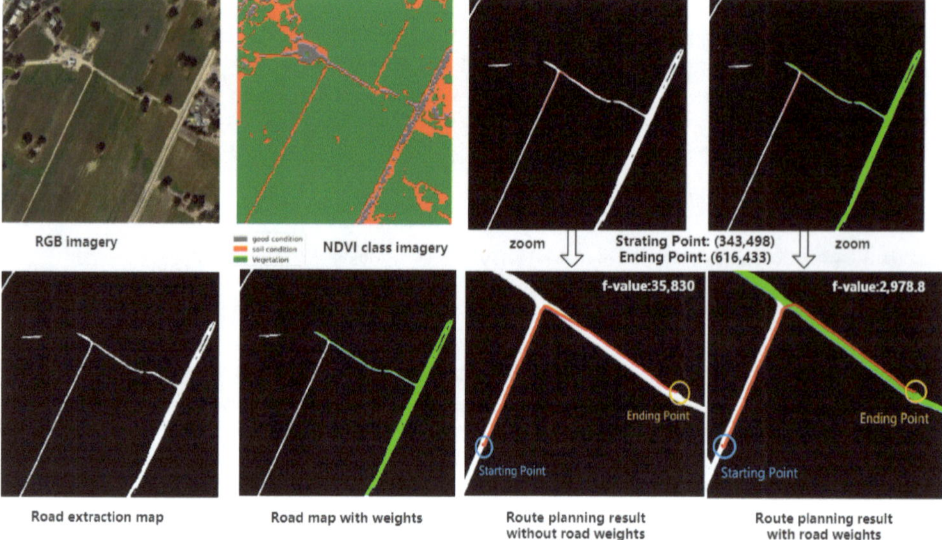

Figure 12. Results of road extraction and optimal path planning on the GeoEye's OrbView–3 satellite dataset.

The results of the best route planning are also shown in Figure 12, where we can see that the minimum value f from the final route search is different based on the same starting position and ending position coordinates reported by the user. The blue circle in the figure represents the starting location, the yellow circle represents the destination, and the red line is the best route searched by the algorithm. We can see that the value of f in the best route search result is 35,830 when no priority is assigned to the roads and, on the contrary, that the value of f is reduced to 2978.8 when roads have priority, which means that the route is faster and more efficient than the previous one. The maps shown here for road extraction are simple. The relatively complex road extraction results and the corresponding optimal route planning results can be found in [48], which also shows that the A^* algorithm with weights is effective. Since the extraction result of D-LinkNet does not match the real situation 100%, we can see that, in the road extracted image, a section of the road with bare soil on the surface is missing compared to the actual RGB image, which may cause the user to encounter some emergency situations, such as the route planned by the UAV appearing to have no road in front of it, or the route planned by the UAV is best in the model, but the user actually found a better road to walk on. These situations require the drone to communicate with the user in real-time, and emergency rescue situations are possible to re-plan the routes for the users.

4.2. Human–UAV Interaction for Emergency Rescue Recognition

Based on the dataset created in Table 4, we split the entire dataset into 70% and 30% portions, with the 30% serving as the testing set. The testing set for validating the accuracy of the model contains 4784 data points, i.e., 30% of the entire data set was randomly selected to validate the accuracy of the model. Compared to other hand gesture recognition methods, such as using 3D convolutional neural networks [70], we finally chose the hand key points as the basic feature for emergency rescue recognition. The reason is that the features of the hand key points are concise, intuitive, and easy to distinguish between different hand gestures. In contrast, 3DCNN is both time-consuming and struggles to train large neural networks. As for the classifiers, we tested on five different classifiers, namely logistic regression [65], ridge classifier [66], random forest classifier [67], gradient boosting classifier [68], and deep neural network [69]. The accuracies obtained for these classical and commonly used classifiers on the testing set of our dataset are shown in Table 5. We have retained six decimal digits for the accuracy results. We can see that the DNN results are the highest, which is why we chose to train the model with DNN at the base station and finally deploy it on the drone. The model accuracy of the DNN can reach 99.9% on the testing set, and this accuracy is crucial for the research part of this work, as the drones need to be ready to accurately identify when the user is in an emergency or when there is potential danger around them.

Table 5. Accuracy of emergency rescue recognition dataset on different classifiers.

Classifiers	Accuracy on Testing Dataset
Logistic Regression [65]	99.5193%
Ridge Classifier [66]	98.5789%
Random Forest Classifier [67]	99.6865%
Gradient Boosting Classifier [68]	99.5402%
Deep Neural Network [69]	99.9164%

The DNN model has been programmed using Keras Sequential API in Python and compiled using Keras with a TensorFlow backend. There are four layers with batch normalization behind each one and 128, 64, 16, and 4 units in each dense layer sequentially. The total number of parameters included is 18,260, of which 17,844 are trainable parameters and 416 are non-trainable parameters. The last layer of the model is with Softmax activation and 4 outputs. The categorical cross-entropy loss function is utilized because of its suitability to measure the performance of the fully connected layer's output with Softmax activation. Adam optimizer with an initial learning rate of 0.0001 is utilized to control the learning rate. Figure 13 shows the changes in accuracy and loss of the DNN model throughout the training process. We can see that the model stabilizes in accuracy and loss after 20 epochs of training and, after 100 epochs of training, the model achieves an accuracy of 99.99% on the training set and 99.92% on the testing set. Figure 14 shows the confusion matrix of the model on the testing set, which is an evaluation of the DNN model, and we can see that the predictions are concentrated on the diagonal, meaning that most of them are accurately predicted. The processing time of the DNN model was computed using the start a timer clock function in Python code. The real running time of the emergency rescue recognition framework is around 20 ms. In human interaction, the FPS value is maintained at around 5, which is in accordance with the real motion.

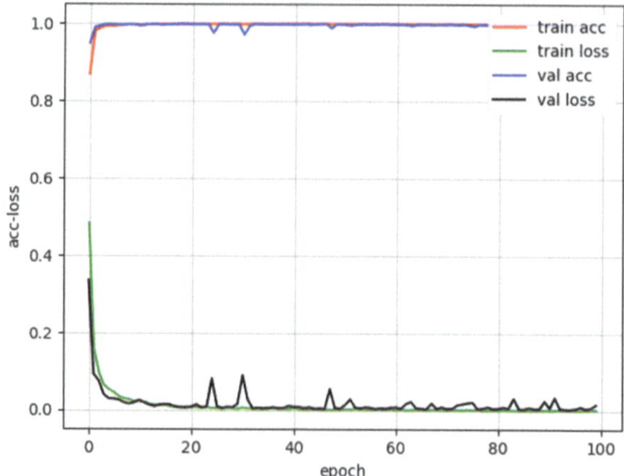

Figure 13. Evaluation of emergency rescue recognition DNN model.

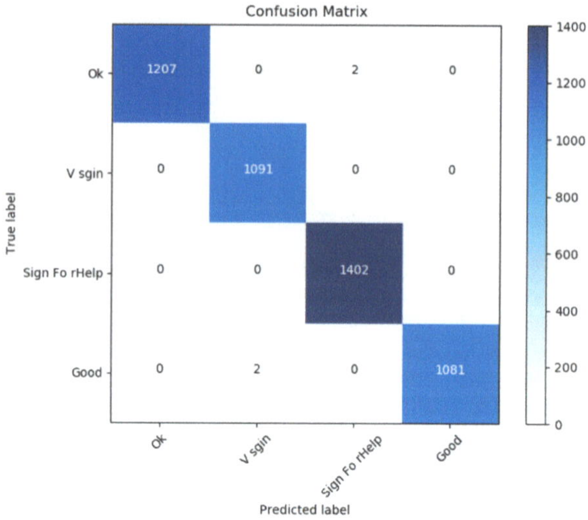

Figure 14. Confusion matrix with predicted labels on *X*-axis and true labels on the *Y*-axis tested in the testing set.

Figure 15 shows a demonstration of emergency rescue recognition. While the UAV was flying at an altitude of less than 10 m, we tried to simulate a field environment and performed a demonstration of emergency rescue recognition. It can be seen that different hand gestures are well predicted and, due to the high accuracy of the model, the model is very sensitive to the recognition of switching between different gestures, even dynamic sign for help gestures. This part of the function also assists the drone in its navigation tasks. When a human is following the route planned by the drone and suddenly encounters a situation where there is no road ahead or a better alternative route is available, then the human can interact with the drone in real-time and the drone will re-route for the human. The background environment in Figure 15 is a simulated field environment in the laboratory, where you can see the presence of some plants of about the same height as a human in the background, and object detection [61] was implemented in our previous work to give an early warning when a human is present. The first row of three images in Figure 15 shows the following normal static gestures: good, v-sign, and ok, which means that everything is normal from the user's point of view, and we can see that the categories and probabilities of the gestures are well presented in the top right corner. The second row shows the dynamic emergency gesture (sign for help), where a warning is immediately given so that the drone can take the appropriate rescue measures, re-route for humans, or directly call a rescuer. It is important to note that the main reason we did not choose voice interaction here was due to the background of the paper, where the field environment combined with the bad weather and external noise was not conducive to verbal interaction. Therefore, gesture recognition is one of the best possible solutions. Later, UAV may interact with sharp sounds or projected phrase/signals on the ground.

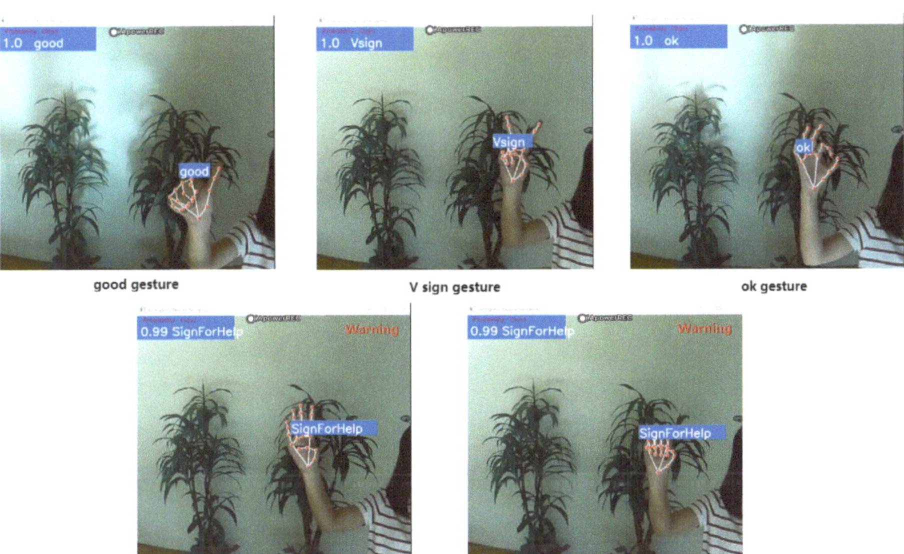

Figure 15. Demonstration of the results of simulated wilderness environments for emergency rescue recognition.

5. Discussion

The main concept of this work is to propose a road condition detection and emergency rescue recognition system for people in wilderness environments with poor network and bad weather conditions. The system utilizes the flexibility of UAVs and the collaboration between low altitude UAVs and high altitude satellites to provide the most timely assistance and monitor the best routes for humans in distress. Firstly, the system's upfront input is satellite data and weather data, and the real-time input is a video sequence captured by the drone camera. For drones equipped with multispectral camera sensors, in addition to the real-time video sequences, we can also access four channels of multispectral images. If the UAV does not have a multispectral sensor, then the detection of road material is completely dependent on the multispectral image information from the satellite. Next comes the data processing of the captured video sequences by the UAV. The UAV equipped with a GPU can segment the video sequences into image sequences in real-time, where RGB images are used as input for road extraction, and hyperspectral images are used as input for road material detection. The main backbone technologies of this work are as follows: the NDVI values are classified by multispectral images for the purpose of detecting the material composition of the road, i.e., bare soil road and non-bare soil gravel road can be well differentiated. The information on the different material compositions of the road surface is combined with the weather information entered in advance to estimate the speed of pedestrians walking on the different materials of the road surface. Based on the yolo3-tiny model previously deployed on the UAV, the throughput of each road can be evaluated. D-LinkNet is the backbone technical support for road extraction and, based on the extracted roads, we can obtain geometric characteristics, such as the length and width of the different roads. Combining all that has been described, the pedestrians walking speed, the throughput of the road, the length and width of the road, and the connectivity are all used as the main basis for giving priority to the road. Finally, roads in the road network are given different priorities and the A^* algorithm with weights is applied to the starting and ending positions reported by the user for optimal route planning and ultimately for the purpose of navigation. As the optimal route planning is limited by the accuracy of the road extraction results, it is necessary for the UAV to adjust the flight altitude and have real-time human–UAV interaction with the user. Thus, a dataset for emergency rescue recognition was built and a reliable DNN model with high accuracy was trained. The model is primarily used for emergency rescue or to adjust navigation routes for the user.

The road condition detection in this work is different from the traditional road condition detection methods [71–74]. In this work, road condition detection is achieved by fusing a pre-trained road extraction model from a relevant dataset with NDVI information from different bands obtained from remote sensing information. Road extraction using D-LinkNet is then combined with advance weather information downloads to obtain the width of the road, the length of the road, the speed of pedestrians, the surface material of the road, and the throughput of the road, ultimately enabling the detection of road conditions. Most of the existing methods rely on various sensors, such as Wi-Fi, GPS, accelerometers, microphones, GSM antennae, in-vehicle standard sensors, laser sensors, stereo cameras, RGB cameras etc., which are usually deployed on ground vehicles, ground robots, or smartphones; however, the platform used as the data source for this paper is a drone with multispectral camera. The UAV platform is well-matched for aerial reconnaissance, and it has an unhindered large field of view. It allows for navigation through difficult terrains and facilitates safe and quick inspections. Some of the above-mentioned sensors will not work well in the wilderness without a network, and it is important to consider that these sensors do not affect the endurance of the drone once it is deployed. The endurance of a GPU-equipped UAV with an NVIDIA Jetson AGX Xavier developer kit, along with a Parrot Sequoia+ multispectral camera is around half an hour in our proposed system. With the development of artificial intelligence technology, the data sets collected by different sensors on different platforms are becoming more and more available, and it is in the area of autonomous driving that most applications in road detection are being made. Classification of road surface materials and the prediction of the damage to the road surface are being achieved with a high accuracy [75–79], but the above platforms are limited by a smaller field of view compared to drones, and the use cases are different.

The innovation of using drones for emergency rescue recognition at low altitude is the use of the latest dynamic gesture recognition technology for potential hazards (such as sign for help), avoiding interference from the external environment in the field. There are many options for using UAVs for emergency rescue, and vision technology is one of the best solutions to avoid interference from the drone rotors or external environmental noise to the voice recognition technology for search and rescue [80]. The emergency rescue recognition system proposed in this paper is a pre-trained model deployed on a UAV that runs in real-time which can then return the results to the ground base station as a projected signal or a sharp sound. There are many special and specific applications for drones for emergency rescue, such as searching for people [81], for man overboard rescue scenarios [82], etc. Most of these emergency rescue approaches incorporate machine learning methods to propose some new models with high accuracy from already existing datasets or use the collected datasets to train specified models to be deployed for use on UAVs. In this work, the core of the emergency rescue approach is to gain the attention of the drone through local dynamic gestures in order to help the user, such as in cases of re-routing. This dataset was collected in our lab and divided 7 to 3 into a training set and a testing set, respectively. The models were trained on each of the four most-used classifiers, and the winning DNN model was finally deployed for use on the drone.

The limitations of this paper are discussed based on the experimental results of the whole system as follows. In the optimal path planning section, the user is required to actively report their starting position and destination to the UAV, a process that requires the UAV to fly at low altitude, and an interaction which must be completed through dynamic interactive gestures [30]. Due to the lack of real wilderness tests, the evaluation of the walking speed of pedestrians on different roads and the road throughput is not shown in the results section, but according to our previous work proposal [61], it is theoretically possible to count and track the number and location of people in real-time by deploying a yolo3-tiny model on the UAV. The road network in the UAV dataset used in this work is relatively simple. Of course, many publicly available UAV datasets with complex road networks already exist, but these are collected with cameras that do not use multispectral sensors on board, so they do not meet the needs of road surface material detection and classification. For satellite datasets, there are many and mostly free human-accessible databases, such as the sentinel-1 dataset and sentinel-2 dataset, which are publicly available and have many bands. It is worth noting that when selecting these datasets, we need to avoid datasets with cloud cover. The system proposed in this work is limited by the accuracy of road extraction techniques, which is the reason for creating the emergency rescue recognition dataset and training the highest accuracy model. Real-time interaction at low altitude is necessary when a human is following the route planned by the UAV and there is a special situation, such as there being no road ahead or a better alternative route available where the UAV can re-route for the user. Finally, because the endurance of the UAV proposed in this paper is limited (around half an hour), the battery life of the UAV should be well predicted and controlled and, if necessary, swarms of UAVs can also be added, with different UAVs performing different tasks to ensure the successful completion of road condition detection and emergency rescue recognition through cooperation.

The cooperation between drones and satellites is particularly important when extreme situations are encountered in the field, for example, when roads are flooded. Even if the drone is equipped with a multispectral camera, the satellite data can be downloaded in advance and used together. By downloading the satellite information in advance, we can obtain geographical information about the area, such as the terrain, road networks, rivers, etc. At the same time, research in related fields, such as water level detection by drones during floods [83] and flood detection [84], can be considered for deployment and application to drones. Similarly, reference can be made to water segmentation in very high-resolution aerial and satellite imagery [85], which indicates a problem at the same location if the results do not match the road and river information downloaded in advance from the satellite data. Solutions for river detection and flood detection can be found in our previous work [20] using the fusion MRF method. Humans in such extreme conditions will need to wait for rescue workers to be transported by boat.

Considering people from different cultures, we use hand gestures which have general meanings, so that in each context the human can communicate receiving the feedback (e.g., audio message or projected text on the surface) from the UAV board. In short, the results that can be achieved and applied in this work are the detection of road conditions and the accurate recognition of emergency signals by UAVs for human–machine interaction.

6. Conclusions

The main contributions and novelties of this paper are as following. Firstly, the information provided by the UAV itself or by the combination of both UAV and satellite enables the detection of road surface materials and use multispectral information to differentiate between different road types (bare soil roads and non-bare soil roads). Secondly, the walking speed of pedestrians on different roads was estimated in combination with the constituent materials of the road surface and weather information, and the throughput of the road was also assessed, which complements and theoretically supports our previous road weights assignment work [48]. Thirdly, drones flying at high altitudes can plan the best route for humans in distress; drones flying at low altitudes can communicate freely with humans on the ground in real-time and provide accurate recognition of emergency rescue information. Drones maximize the multi-faceted ways in which they can help users in distress. Finally, the information provided by low-altitude drones is fused with hyperspectral image information provided by high-altitude satellites to provide humans with a wider range of search and rescue information. The main challenges have been solved, as follows:

- Drones can plan the optimal routes in real-time for humans in the wilderness with poor network conditions and bad weather;
- Different road types are well differentiated, the walking speed of pedestrians is estimated, and the throughput of the road can theoretically be evaluated;
- In the road extraction network map, the priority of the different roads is given on a sufficient basis, which refines our previous work;
- Here, UAVs flying at low latitudes can perform human–machine interaction tasks very accurately.

In short, the novel features of this proof-of-concept paper are as follows:

- Estimating the soil type from multispectral information;
- Estimating the road quality from the weather and soil information;
- Finding the paths/roads on the terrain;
- Weighting the paths by walking/driving quality measure;
- Finding the optimal routes based on the weighted route-map.

This paper has shown a tool-set for a future emergency related situation. In future work, if emergency rescue situation simulation is allowed, we will test this system in a field environment. We will also try to collect UAV multispectral datasets with complex road networks. Finally, for the A^* algorithm of path planning, we will try to use and compare other related path planning search algorithms. Different algorithms for route search with weights will be further compared and optimized.

Author Contributions: C.L. designed and implemented the whole system and wrote and edited the paper. She also created the emergency rescue dataset with her colleagues. T.S. had the initial idea of this project and worked on the paper, who supervised and guided the whole project. All authors have read and agreed to the published version of the manuscript.

Funding: The research was supported by the Ministry of Innovation and Technology NRDI Office within the framework of the Autonomous Systems National Laboratory Program and by the Hungarian National Science Foundation (NKFIH OTKA) No. K139485. This research was also funded by Stipendium Hungaricum scholarship and China Scholarship Council.

Data Availability Statement: The Satellite dataset can be downloaded from this link https://earthexplorer.usgs.gov/ (accessed on 1 May 2022) and the UAV dataset presented in this paper are available on request from the authors.

Acknowledgments: The work is carried out at Institute for Computer Science and Control (SZTAKI), Hungary and the authors would like to thank their colleague László Spórás for the technical support and for the helpful scientific community of Machine Perception Research Laboratory.

Conflicts of Interest: The authors declare no conflict of interest.

References

1. Mavridou, E.; Vrochidou, E.; Papakostas, G.A.; Pachidis, T.; Kaburlasos, V.G. Machine vision systems in precision agriculture for crop farming. *J. Imaging* **2019**, *5*, 89. [CrossRef] [PubMed]
2. Song, H.; Liang, H.; Li, H.; Dai, Z.; Yun, X. Vision-based vehicle detection and counting system using deep learning in highway scenes. *Eur. Transp. Res. Rev.* **2019**, *11*, 51. [CrossRef]
3. Varga, D.; Szirányi, T. Robust real-time pedestrian detection in surveillance videos. *J. Ambient Intell. Humaniz. Comput.* **2017**, *8*, 79–85. [CrossRef]
4. Panahi, L.; Ghods, V. Human fall detection using machine vision techniques on RGB–D images. *Biomed. Signal Process. Control* **2018**, *44*, 146–153. [CrossRef]

5. Kanellakis, C.; Nikolakopoulos, G. Survey on computer vision for UAVs: Current developments and trends. *J. Intell. Robot. Syst.* **2017**, *87*, 141–168. [CrossRef]
6. Kashino, Z.; Nejat, G.; Benhabib, B. Aerial wilderness search and rescue with ground support. *J. Intell. Robot. Syst.* **2020**, *99*, 147–163. [CrossRef]
7. Alsamhi, S.H.; Almalki, F.A.; AL-Dois, H.; Shvetsov, A.V.; Ansari, M.S.; Hawbani, A.; Gupta, S.K.; Lee, B. Multi-drone edge intelligence and SAR smart wearable devices for emergency communication. *Wirel. Commun. Mob. Comput.* **2021**, *2021*, 6710074. [CrossRef]
8. Heggie, T.W.; Heggie, T.M. Dead men hiking: Case studies from the American wilderness. *Med. Sport.* **2012**, *16*, 118–121. [CrossRef]
9. Mishra, B.; Garg, D.; Narang, P.; Mishra, V. Drone-surveillance for search and rescue in natural disaster. *Comput. Commun.* **2020**, *156*, 1–10. [CrossRef]
10. Alsamhi, S.H.; Shvetsov, A.V.; Kumar, S.; Shvetsova, S.V.; Alhartomi, M.A.; Hawbani, A.; Rajput, N.S.; Srivastava, S.; Saif, A.; Nyangaresi, V.O. UAV Computing-Assisted Search and Rescue Mission Framework for Disaster and Harsh Environment Mitigation. *Drones* **2022**, *6*, 154. [CrossRef]
11. Harris, R. *Satellite Remote Sensing—An Introduction*; Routledge Kegan & Paul: London, UK, 1987.
12. Patino, J.E.; Duque, J.C. A review of regional science applications of satellite remote sensing in urban settings. *Comput. Environ. Urban Syst.* **2013**, *37*, 1–17. [CrossRef]
13. Lo, C. *Applied Remote Sensing*; Taylor & Francis: Abingdon, UK, 1986.
14. Zhu, L.; Suomalainen, J.; Liu, J.; Hyyppä, J.; Kaartinen, H.; Haggren, H. *A Review: Remote Sensing Sensors—Multi-Purposeful Application of Geospatial Data*; IntechOpen: London, UK, 2018; pp. 19–42.
15. Karthikeyan, L.; Chawla, I.; Mishra, A.K. A review of remote sensing applications in agriculture for food security: Crop growth and yield, irrigation, and crop losses. *J. Hydrol.* **2020**, *586*, 124905. [CrossRef]
16. Yang, X.; Qin, Q.; Grussenmeyer, P.; Koehl, M. Urban surface water body detection with suppressed built-up noise based on water indices from Sentinel-2 MSI imagery. *Remote Sens. Environ.* **2018**, *219*, 259–270. [CrossRef]
17. Özelkan, E. Water body detection analysis using NDWI indices derived from landsat-8 OLI. *Polish J. Environ. Stud.* **2020**, *29*, 1759–1769. [CrossRef]
18. Yuan, K.; Zhuang, X.; Schaefer, G.; Feng, J.; Guan, L.; Fang, H. Deep-learning-based multispectral satellite image segmentation for water body detection. *IEEE J. Sel. Top. Appl. Earth Obs. Remote Sens.* **2021**, *14*, 7422–7434. [CrossRef]
19. Dang, B.; Li, Y. MSResNet: Multiscale residual network via self-supervised learning for water-body detection in remote sensing imagery. *Remote Sens.* **2021**, *13*, 3122. [CrossRef]
20. Sziranyi, T.; Shadaydeh, M. Segmentation of remote sensing images using similarity-measure-based fusion-MRF model. *IEEE Geosci. Remote Sens. Lett.* **2014**, *11*, 1544–1548. [CrossRef]
21. Talukdar, S.; Singha, P.; Mahato, S.; Pal, S.; Liou, Y.A.; Rahman, A. Land-use land-cover classification by machine learning classifiers for satellite observations—A review. *Remote Sens.* **2020**, *12*, 1135. [CrossRef]
22. Castelluccio, M.; Poggi, G.; Sansone, C.; Verdoliva, L. Land use classification in remote sensing images by convolutional neural networks. *arXiv* **2015**, arXiv:1508.00092.
23. Peng, D.; Zhang, Y.; Guan, H. End-to-end change detection for high resolution satellite images using improved UNet++. *Remote Sens.* **2019**, *11*, 1382. [CrossRef]
24. Asokan, A.; Anitha, J.J.E.S.I. Change detection techniques for remote sensing applications: A survey. *Earth Sci. Inform.* **2019**, *12*, 143–160. [CrossRef]
25. Szirányi, T.; Zerubia, J. Multilayer Markov Random Field Models for Change Detection in Optical Remote Sensing Images. *ISPRS J. Photogramm. Remote Sens.* **2015**, *107*, 22–37.
26. Li, J.; Pei, Y.; Zhao, S.; Xiao, R.; Sang, X.; Zhang, C. A review of remote sensing for environmental monitoring in China. *Remote Sens.* **2020**, *12*, 1130. [CrossRef]
27. Laurance, W.F.; Clements, G.R.; Sloan, S.; O'connell, C.S.; Mueller, N.D.; Goosem, M.; Venter, O.; Edwards, D.; Phalan, B.; Balmford, A.; et al. A global strategy for road building. *Nature* **2014**, *513*, 229–232. [CrossRef] [PubMed]
28. Ciepłuch, B.; Jacob, R.; Mooney, P.; Winstanley, A.C. Comparison of the accuracy of OpenStreetMap for Ireland with Google Maps and Bing Maps. In Proceedings of the Ninth International Symposium on Spatial Accuracy Assessment in Natural Resuorces and Enviromental Sciences, Leicester, UK, 20–23 July 2010; p. 337.
29. Wang, W.; Yang, N.; Zhang, Y.; Wang, F.; Cao, T.; Eklund, P. A review of road extraction from remote sensing images. *J. Traffic Transp. Eng.* **2016**, *3*, 271–282. [CrossRef]
30. Liu, C.; Szirányi, T.A. Gesture Recognition for UAV-based Rescue Operation based on Deep Learning. In Proceedings of the International Conference on Image Processing and Vision Engineering (IMPROVE 2021), Anchorage, AL, USA, 19–22 September 2021; pp. 180–187.
31. Jetson AGX Xavier Developer Kit. NVIDIA Developer. 2018. Available online: https://developer.nvidia.com/embedded/jetson-agx-xavier-developer-kit (accessed on 4 July 2022).
32. Parrot Sequoia+. SenseFly. 2018. Available online: https://www.sensefly.com/camera/parrot-sequoia/ (accessed on 4 July 2022).
33. Hossain, S.; Lee, D.-J. Deep learning-based real-time multiple-object detection and tracking from aerial imagery via a flying robot with GPU-based embedded devices. *Sensors* **2019**, *19*, 3371. [CrossRef]

34. Esa.int. ESA—Home. 2019. Available online: https://www.esa.int/ (accessed on 4 July 2022).
35. USGS. Science for a Changing World. Available online: https://www.usgs.gov/ (accessed on 4 July 2022).
36. Mansouri, S.S.; Karvelis, P.; Georgoulas, G.; Nikolakopoulos, G. Remaining useful battery life prediction for UAVs based on machine learning. *IFAC Pap.* **2017**, *50*, 4727–4732. [CrossRef]
37. Saif, A.; Dimyati, K.; Noordin, K.A.; Shah, N.S.M.; Alsamhi, S.H.; Abdullah, Q. August. Energy-efficient tethered UAV deployment in B5G for smart environments and disaster recovery. In Proceedings of the 2021 1st International Conference on Emerging Smart Technologies and Applications (eSmarTA), Sana'a, Yemen, 10–12 August 2021; IEEE: New York, NY, USA, 2021; pp. 1–5.
38. Wikipedia Contributors. Biatorbágy. Wikipedia, Wikimedia Foundation. 2021. Available online: https://en.wikipedia.org/wiki/Biatorb%C3%A1gy (accessed on 5 July 2022).
39. Google. Google Maps. 2022. Available online: www.google.com/maps (accessed on 5 July 2022).
40. USGS EROS Archive—Commercial Satellites—OrbView 3. U.S. Geological Survey. Available online: www.usgs.gov/centers/eros/science/usgs-eros-archive-commercial-satellites-orbview-3 (accessed on 5 July 2022).
41. Birdwood. Wikipedia. 2021. Available online: https://en.wikipedia.org/wiki/Birdwood (accessed on 5 July 2022).
42. Mnih, V. *Machine Learning for Aerial Image Labeling*; University of Toronto: Toronto, ON, Canada, 2013.
43. Demir, I.; Koperski, K.; Lindenbaum, D.; Pang, G.; Huang, J.; Basu, S.; Hughes, F.; Tuia, D.; Raskar, R. Deepglobe 2018: A challenge to parse the earth through satellite images. In Proceedings of the IEEE Conference on Computer Vision and Pattern Recognition Workshops, Lake City, UT, USA, 18–22 June 2018; pp. 172–181.
44. Ronneberger, O.; Fischer, P.; Brox, T. U-net: Convolutional networks for biomedical image segmentation. In Proceedings of the International Conference on Medical Image Computing and Computer-Assisted Intervention, Singapore, 18–22 September 2015; Springer: Cham, Switzerland, 2015; pp. 234–241.
45. Badrinarayanan, V.; Kendall, A.; Cipolla, R. Segnet: A deep convolutional encoder-decoder architecture for image segmentation. *IEEE Trans. Pattern Anal. Mach. Intell.* **2017**, *39*, 2481–2495. [CrossRef]
46. Chaurasia, A.; Culurciello, E. Linknet: Exploiting encoder representations for efficient semantic segmentation. In Proceedings of the 2017 IEEE Visual Communications and Image Processing (VCIP), St. Petersburg, FL, USA, 10–13 December 2017; IEEE: New York, NY, USA, 2017; pp. 1–4.
47. Zhou, L.; Zhang, C.; Wu, M. D-LinkNet: LinkNet with pretrained encoder and dilated convolution for high resolution satellite imagery road extraction. In Proceedings of the IEEE Conference on Computer Vision and Pattern Recognition Workshops, Lake City, UT, USA, 18–22 June 2018; pp. 182–186.
48. Liu, C.; Szirányi, T. UAV Path Planning based on Road Extraction. In Proceedings of the International Conference on Image Processing and Vision Engineering (IMPROVE 2021), Brussels, Belgium, 16–17 June 2022; pp. 202–210.
49. Goto, T.; Kosaka, T.; Noborio, H. On the heuristics of A* or A algorithm in ITS and robot path-planning. In Proceedings of the 2003 IEEE/RSJ International Conference on Intelligent Robots and Systems (IROS 2003) (Cat. No. 03CH37453), Las Vegas, NV, USA, 27–31 October 2003; IEEE: New York, NY, USA, 2003; Volume 2, pp. 1159–1166.
50. Abhishek, K.; Singh, M.; Ghosh, S.; Anand, A. Weather forecasting model using artificial neural network. *Procedia Technol.* **2012**, *4*, 311–318. [CrossRef]
51. Running, S.W. Estimating terrestrial primary productivity by combining remote sensing and ecosystem simulation. In *Remote Sensing of Biosphere Functioning*; Springer: New York, NY, USA, 1990; pp. 65–86.
52. Myneni, R.B.; Hall, F.G.; Sellers, J.; Marshak, A.L. The interpretation of spectral vegetation indexes. *IEEE Trans. Geosci. Remote Sens.* **1995**, *33*, 481–486. [CrossRef]
53. Papageorgiou, M.; Diakaki, C.; Dinopoulou, V.; Kotsialos, A.; Wang, Y. Review of road traffic control strategies. *Proc. IEEE* **2003**, *91*, 2043–2067. [CrossRef]
54. Definition, Interpretation, and Calculation of Traffic Analysis Tools Measures of Effectiveness—6.0 Recommended MOEs. Available online: https://ops.fhwa.dot.gov/publications/fhwahop08054/sect6.htm (accessed on 1 June 2022).
55. Pettorelli, N.; Vik, J.O.; Mysterud, A.; Gaillard, J.M.; Tucker, C.J.; Stenseth, N.C. Using the satellite-derived NDVI to assess ecological responses to environmental change. *Trends Ecol. Evol.* **2005**, *20*, 503–510. [CrossRef] [PubMed]
56. Gupta, V.D.; Areendran, G.; Raj, K.; Ghosh, S.; Dutta, S.; Sahana, M. Assessing habitat suitability of leopards (*Panthera pardus*) in unprotected scrublands of Bera, Rajasthan, India. In *Forest Resources Resilience and Conflicts*; Elsevier: Amsterdam, The Netherlands, 2021; pp. 329–342.
57. Kraetzig, N.M. 5 Things to Know about NDVI (Normalized Difference Vegetation Index). UP42 Official Website. Available online: https://up42.com/blog/tech/5-things-to-know-about-ndvi#:~{}:text=The%20value%20of%20the%20NDVI (accessed on 6 July 2022).
58. UW-Madison Satellite Meteorology. Available online: https://profhorn.meteor.wisc.edu/wxwise/satmet/lesson3/ndvi.html (accessed on 6 July 2022).
59. Gast, K.; Kram, R.; Riemer, R. Preferred walking speed on rough terrain: Is it all about energetics? *J. Exp. Biol.* **2019**, *222*, jeb185447. [CrossRef] [PubMed]
60. Mohamed, O.; Appling, H. Clinical assessment of gait. *Orthot. Prosthet. Rehabil.* **2020**, *4*, 102–144.
61. Liu, C.; Szirányi, T. Real-time human detection and gesture recognition for on-board UAV rescue. *Sensors* **2021**, *21*, 2180. [CrossRef]
62. Signal for Help. Wikipedia. 2020. Available online: https://en.wikipedia.org/wiki/Signal_for_Hel (accessed on 1 July 2021).
63. Mediapipe. Hands. Available online: https://google.github.io/mediapipe/solutions/hands.html (accessed on 1 June 2021).

64. Zhang, F.; Bazarevsky, V.; Vakunov, A.; Tkachenka, A.; Sung, G.; Chang, C.L.; Grundmann, M. Mediapipe hands: On-device real-time hand tracking. *arXiv* **2020**, arXiv:2006.10214.

65. Wright, R.E. *Logistic Regression*; APA: Washington, DC, USA, 1995.

66. Singh, A.; Prakash, B.S.; Chandrasekaran, K. A comparison of linear discriminant analysis and ridge classifier on Twitter data. In Proceedings of the 2016 International Conference on Computing, Communication and Automation (ICCCA), Greater Noida, India, 29–30 April 2016; IEEE: New York, NY, USA, 2016; pp. 133–138.

67. Pal, M. Random forest classifier for remote sensing classification. *Int. J. Remote Sens.* **2005**, *26*, 217–222. [CrossRef]

68. Natekin, A.; Knoll, A. Gradient boosting machines, a tutorial. *Front. Neurorobot.* **2013**, *7*, 21. [CrossRef]

69. Canziani, A.; Paszke, A.; Culurciello, E. An analysis of deep neural network models for practical applications. *arXiv* **2016**, arXiv:1605.07678.

70. Carreira, J.; Zisserman, A. Quo vadis, action recognition? A new model and the kinetics dataset. In Proceedings of the IEEE Conference on Computer Vision and Pattern Recognition, San Juan, PR, USA, 17–19 June 1997.

71. Chugh, G.; Bansal, D.; Sofat, S. Road condition detection using smartphone sensors: A survey. *Int. J. Electron. Electr. Eng.* **2014**, *7*, 595–602.

72. Castillo Aguilar, J.J.; Cabrera Carrillo, J.A.; Guerra Fernández, A.J.; Carabias Acosta, E. Robust road condition detection system using in-vehicle standard sensors. *Sensors* **2015**, *15*, 32056–32078. [CrossRef] [PubMed]

73. Jokela, M.; Kutila, M.; Le, L. Road condition monitoring system based on a stereo camera. In Proceedings of the 2009 IEEE 5th International Conference on Intelligent Computer Communication and Processing, Cluj-Napoca, Romania, 27–29 August 2009; IEEE: New York, NY, USA, 2009; pp. 423–428.

74. Ranyal, E.; Sadhu, A.; Jain, K. Road condition monitoring using smart sensing and artificial intelligence: A review. *Sensors* **2022**, *22*, 3044. [CrossRef] [PubMed]

75. Xie, Q.; Hu, X.; Ren, L.; Qi, L.; Sun, Z. A Binocular Vision Application in IoT: Realtime Trustworthy Road Condition Detection System in Passable Area. In *IEEE Transactions on Industrial Informatics*; IEEE: New York, NY, USA, 2022.

76. Gupta, A.; Anpalagan, A.; Guan, L.; Khwaja, A.S. Deep learning for object detection and scene perception in self-driving cars: Survey, challenges, and open issues. *Array* **2021**, *10*, 100057. [CrossRef]

77. Chun, C.; Ryu, S.K. Road surface damage detection using fully convolutional neural networks and semi-supervised learning. *Sensors* **2019**, *19*, 5501. [CrossRef] [PubMed]

78. Wang, D.; Liu, Z.; Gu, X.; Wu, W.; Chen, Y.; Wang, L. Automatic Detection of Pothole Distress in Asphalt Pavement Using Improved Convolutional Neural Networks. *Remote Sens.* **2022**, *14*, 3892. [CrossRef]

79. Rateke, T.; Justen, K.A.; Von Wangenheim, A. Road surface classification with images captured from low-cost camera-road traversing knowledge (rtk) dataset. *Rev. De Inf. Teórica E Apl.* **2019**, *26*, 50–64. [CrossRef]

80. Yamazaki, Y.; Tamaki, M.; Premachandra, C.; Perera, C.J.; Sumathipala, S.; Sudantha, B.H. Victim detection using UAV with on-board voice recognition system. In Proceedings of the 2019 Third IEEE International Conference on Robotic Computing (IRC), Naples, Italy, 25–27 February 2019; IEEE: New York, NY, USA, 2019; pp. 555–559.

81. Castellano, G.; Castiello, C.; Mencar, C.; Vessio, G. Preliminary evaluation of TinyYOLO on a new dataset for search-and-rescue with drones. In Proceedings of the 2020 7th International Conference on Soft Computing & Machine Intelligence (ISCMI), Stockholm, Sweden, 14–15 November 2020; IEEE: New York, NY, USA, 2020; pp. 163–166.

82. Cafarelli, D.; Ciampi, L.; Vadicamo, L.; Gennaro, C.; Berton, A.; Paterni, M.; Benvenuti, C.; Passera, M.; Falchi, F. MOBDrone: A Drone Video Dataset for Man OverBoard Rescue. In Proceedings of the International Conference on Image Analysis and Processing, Bangkok, Thailand, 16–17 November 2022; Springer: Cham, Switzerland, 2022; pp. 633–644.

83. Rizk, H.; Nishimur, Y.; Yamaguchi, H.; Higashino, T. Drone-based water level detection in flood disasters. *Int. J. Environ. Res. Public Health* **2021**, *19*, 237. [CrossRef]

84. Tanim, A.H.; McRae, C.; Tavakol-Davani, H.; Goharian, E. Flood Detection in Urban Areas Using Satellite Imagery and Machine Learning. *Water* **2022**, *14*, 1140. [CrossRef]

85. Zhang, Z.; Lu, M.; Ji, S.; Yu, H.; Nie, C. Rich CNN Features for water-body segmentation from very high-resolution aerial and satellite imagery. *Remote. Sens.* **2021**, *13*, 1912. [CrossRef]

Article

Automatic Detection of Pothole Distress in Asphalt Pavement Using Improved Convolutional Neural Networks

Danyu Wang [1,2], Zhen Liu [1,2], Xingyu Gu [1,2,*], Wenxiu Wu [3], Yihan Chen [1,2] and Lutai Wang [1,2]

[1] Department of Roadway Engineering, School of Transportation, Southeast University, Nanjing 211189, China
[2] National Demonstration Center for Experimental Road and Traffic Engineering Education, Southeast University, Nanjing 211189, China
[3] Jinhua Highway Administration Bureau, Jinhua 321000, China
* Correspondence: guxingyu1976@seu.edu.cn; Tel.: +86-025-52091362

Abstract: To realize the intelligent and accurate measurement of pavement surface potholes, an improved You Only Look Once version three (YOLOv3) object detection model combining data augmentation and structure optimization is proposed in this study. First, color adjustment was used to enhance the image contrast, and data augmentation was performed through geometric transformation. Pothole categories were subdivided into P1 and P2 on the basis of whether or not there was water. Then, the Residual Network (ResNet101) and complete IoU (CIoU) loss were used to optimize the structure of the YOLOv3 model, and the K-Means++ algorithm was used to cluster and modify the multiscale anchor sizes. Lastly, the robustness of the proposed model was assessed by generating adversarial examples. Experimental results demonstrated that the proposed model was significantly improved compared with the original YOLOv3 model; the detection mean average precision (mAP) was 89.3%, and the F1-score was 86.5%. On the attacked testing dataset, the overall mAP value reached 81.2% (−8.1%), which shows that this proposed model performed well on samples after random occlusion and adding noise interference, proving good robustness.

Keywords: pavement distress; pothole detection; YOLOv3; data augmentation; robustness

Citation: Wang, D.; Liu, Z.; Gu, X.; Wu, W.; Chen, Y.; Wang, L. Automatic Detection of Pothole Distress in Asphalt Pavement Using Improved Convolutional Neural Networks. *Remote Sens.* **2022**, *14*, 3892. https://doi.org/10.3390/rs14163892

Academic Editors: Valerio Baiocchi, Alessandro Mei and Xianfeng Zhang

Received: 30 June 2022
Accepted: 8 August 2022
Published: 11 August 2022

Publisher's Note: MDPI stays neutral with regard to jurisdictional claims in published maps and institutional affiliations.

1. Introduction

Pavement surface distresses including potholes pose a significant threat to driving safety. Therefore, rapid detection and intelligent maintenance of potholes are prerequisites for pavement management [1]. However, conventional pothole detection mainly relies on manual methods [2], which suffer from strong subjectivity, high cost, a long cycle, and which are not conducive to rapid detection. The emergence of digital image processing technology (DIP) makes the digital detection of pavement potholes possible, which is a method of processing images through denoising, enhancing, restoring, segmenting, and extracting features using a computer, including edge detection [3], threshold segmentation [4], and morphological processing [5]. These methods are widely used in pavement distress detection.

Although DIP has made good progress in pothole detection [6], it cannot achieve the automatic detection of potholes accurately. Fortunately, with the progress of deep learning (DL) [7], especially the emergence of convolutional neural networks (CNNs) in 2012 [8], DL models represented by CNNs have gradually shown their superiority in object detection because they can quickly and accurately detect objects by automatically extracting features [9]. Subsequently, CNNs have been extensively used in remote sensing [10], the healthcare industry [11], and other fields [12,13]. Furthermore, they are also applied in scenarios such as transportation infrastructure [14,15], pedestrian detection [16], unmanned driving [17], and pavement distress detection [18]. At present, there are two kinds of object detection algorithms in identifying pavement distress: one-stage algorithms and two-stage algorithms.

On the one hand, two-stage algorithms first generate the proposal region, and then perform CNN-based classification and identification of the region. Representatives include the region-based CNN (R-CNN) series [19], R-FCN [20], and SPP-Net [21]. Nie et al. [22] applied the faster R-CNN model to detect pavement distress; these experimental results represented a significant breakthrough, but the dataset used was small. Then, Pei et al. [23] further refined the faster R-CNN model by modifying the anchor size and combining it with the VGG-16 network based on the expanded dataset, whose final detection accuracy was 89.97%. Song et al. [24] further considered the distress conditions for different pavements and proposed a method based on the faster R-CNN model to identify various distresses accurately. In conclusion, these studies failed to achieve rapid detection because the two-stage algorithm has numerous network parameters, which makes the period of training and testing time-consuming.

On the other hand, the emergence of one-stage object detection models addressed the issues of low efficiency in the two-stage models. One-stage algorithms directly set sizes of the multiscale anchor, integrating classification and regression into one step instead of region proposal, which improves the model's speed. Cao et al. [25] proposed a single-shot multi-box detector-based (SSD) object detection method on airport pavement with high detection accuracy (89.3%). However, the sizes of anchors in the SSD model were manually determined by data distribution, and a method for automatic anchor generation was urgently requested. The emergence of You Only Look Once version three (YOLOv3) solves this problem [26]. As one of the classic one-stage object detection models, YOLOv3 adopts the K-Means algorithm to generate anchors of three scales, which can realize multiscale detection. Zhu et al. [27] used the YOLOv3 model to detect pavement distress, and they verified the feasibility of YOLOv3 by comparing it with several mainstream object detection models. YOLOv3 has a high comprehensive performance, but the detection accuracy is not good enough to meet the maintenance needs.

Many studies have improved the detection performance of the YOLOv3 model, and the existing improvement methods mainly include data augmentation, network structure adjustment, improvement of the loss function, and hyperparameter optimization [28]. Liu et al. [29] expanded the number of samples in the dataset through data augmentation to avoid overfitting, which improved the generalization ability of the YOLOv3 model. Bochkovskiy et al. [30] combined the SPP-Net and PANet networks to replace the original FPN network in YOLOv3, enabling the network to fully extract small object features. Tang et al. [31] employed the distance between the anchor and the cluster center point to define the loss function, which was more reasonable for evaluating the loss of the model. However, the above studies were aimed at cracks, and this approach has rarely been applied to potholes. In addition, the enhancement methods are not comprehensive enough, lacking the consideration of the actual scene. Meanwhile, the robustness of the model has not been evaluated.

Therefore, a method of combining data augmentation and YOLOv3 structure optimization is proposed to study pavement pothole detection. The main contributions of this study are as follows:

(1) The pothole dataset was contrast-enhanced by color adjustment, and geometric transformation was adopted to expand the number of samples. A data augmentation strategy suitable for potholes was proposed to train DL models.

(2) The ResNet101 network was used to improve the feature extraction network of YOLOv3. Complete intersection over union (CIoU) was applied to measure the loss of the proposed model. The anchor sizes were modified by the K-Means++ algorithm. An object detection model applicable for potholes was established.

(3) Adversarial samples of potholes were generated by random occlusion and adding noise before testing, which verified the robustness of this model.

The remainder of this study is organized as follows: the methodology is described in Section 2, including the image preprocessing methods, YOLOv3 network structure, robustness analysis, and evaluation index. The experimental results and analysis are presented in Section 3. Conclusions are drawn in Section 4.

2. Methodology

2.1. Pavement Pothole Dataset

The pavement distress images were collected from typical provincial highways in Zhejiang Province. The dataset was captured using a mobile mapping system using an onboard high-definition camera (HD camera), as illustrated in Figure 1. Figure 1 illustrates the acquisition process of gray images and local details of camera (the front and the back). The detection field of the camera takes the maximum detection width of the single lane as 3750 mm. If two cameras are selected, the lateral visual field of one camera must be at least 2000 mm. Two cameras were selected here, and the camera's information is listed in Table 1. Camerlink interface meets the demand of high-speed image acquisition, and it has the advantages of strong anti-interference ability and low power consumption. CMOS cameras are more suitable for high-speed acquisition with low cost and low power consumption. One obtained image is shown on the right of Figure 1.

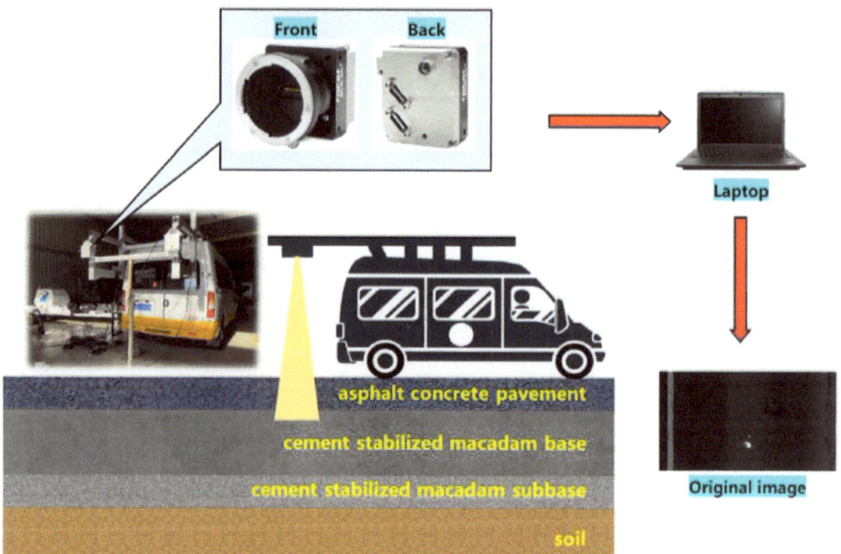

Figure 1. Mobile mapping system used to collect pavement images.

Table 1. Technical parameters of the high-definition camera.

Camera Features	Details	Camera Features	Details
Type	Basler raL2048-80km	Power supply requirements (typical value)	3 W
Interface	Camera Link	Type of light-sensitive chips	CMOS
Resolution ratio	3854 px × 2065 px	Size of light-sensitive chips	14.3 mm

The original gray image was 3854 by 2065 pixels. A total of 500 pavement images were collected. After the filtration, 300 images were used for model training, and 100 images were chosen for testing. Two color adjustment methods were used to improve the image brightness. Then, four types of geometric transformation were performed on the preprocessed images. Considering the small scale of the dataset, the risk of information leakage [32], and the low accuracy of model performance, the dataset was divided into a training dataset, validation dataset, and testing dataset according to the ratio of 6:2:2 [33]. Details are shown in Table 2.

Table 2. Detailed division information of image dataset of pavement potholes.

Dataset	Training Dataset	Validation Dataset	Testing Dataset
Image with potholes	480	160	160
Image without potholes	720	240	240
Total	1200	400	400

Potholes on rainy days accumulate water, while potholes on sunny days are mostly dry. Therefore, the potholes with accumulated water and dry surfaces were marked as P1 and P2, respectively. All labeling work was performed using LabelImg software (v1.8.6) [34].

2.2. Pothole Data Pre-Processing

To adjust the brightness of the dataset, two types of color adjustment method were used. The contrast and sharpness of the image were modified. The performance of DL models depends on the size and the quality of the dataset. Therefore, data augmentation was used to address the above issues [35]. Four geometric augmentation methods were adopted [36]. Details are described below.

2.2.1. Color Adjustment

The pavement pothole images were three-channel color images. Each color pixel featured the three components of red, green, and blue, referring to the color image at a specific spatial position and, thus, forming a vector to describe the image. For the processing of color images, two color augmentation operations (contrast and sharpness) were used [37].

To solve the problem of low contrast caused by the small gray-level range of the pothole image, contrast augmentation was used to enlarge the gray-level range of the pothole image and make the image clearer [38]. Using the hue–saturation–intensity (HSI) color model, the probability smoothing method was conducted for the components of intensity and saturation, which converted the intensity and saturation components into a uniform distribution. The calculation formulas of the intensity and the saturation components are shown in Equation (1) and Equation (2), respectively.

$$y_{1k} = F(x_{1k}) = P\{x_I \leq x_{Ik}\} = \sum_{m=0}^{k} f(x_{Im}) = \sum_{m=0}^{k} P\{x_I \leq x_{Ik}\}, \tag{1}$$

$$y_{st} = F(x_{1k}|x_{Ik}) = \sum_{m=0}^{t} f(x_{Sm}|x_{Ik}) = \sum_{m=0}^{t} \frac{P\{x_I = x_{Ik}, x_s = x_{Sm}\}}{P\{x_I = x_{Ik}\}}, \tag{2}$$

where $k = 0, 1, \ldots, L - 1$ and $t = 0, 1, \ldots, M - 1$; L and M represent discrete levels of intensity and saturation, respectively. $X = (x_H, x_S, x_I)^T$ is a vector of color pixels representing each image. $F(\cdot)$ is a probability function, and $F(Z) = F(x_I, x_S) = P\{x_I \leq x_I, x_S \leq x_S\}$.

Sharpness eliminates the blurring of the pothole image by increasing the contrast of the pixels in the neighborhood, making the pothole object clearer. Laplace sharpness adds gradient values (Laplace operator) to the pothole image [39]. The enhancement method based on the Laplace operator is shown in Equation (3).

$$g(x,y) = f(x,y) - \begin{bmatrix} \nabla^2 R(x,y) \\ \nabla^2 G(x,y) \\ \nabla^2 B(x,y) \end{bmatrix}, \tag{3}$$

where $g(x, y)$ is the pothole image after sharpness, $f(x, y)$ is original pothole image, and $\nabla^2 R(x, y)$, $\nabla^2 G(x, y)$, and $\nabla^2 B(x, y)$ are the Laplace operators of each component (red, green, and blue) of the color images, respectively.

2.2.2. Geometric Transformation

As a traditional data augmentation method, geometric transformation was performed on images through specific operations such as rotation, flipping, and cropping. The shape of the pothole target was mainly polygonal, with randomness in all directions. Four operations of rotating 90° clockwise, 180°, 90° anticlockwise, and random crop were used on the basis of the above features.

The rotation of pothole images refers to forming a new image by rotating the central point of an image at a certain angle as a reference point, which can effectively expand the amount of data.

The random crop method could produce a better effect on learning the main features of potholes while increasing model stability, which was realized by randomly cropping the corner of the image. Specifically, it was assumed that the main feature of potholes (P) is F, and the collected image contains background noise (B). That is, the pothole's main feature in the captured image is defined as C, which is expressed as (F, B). It is expected to learn F, but possible to learn (F, B), thus resulting in overfitting. Potholes are generally in the middle of the image. Their main features (F) are highly unlikely to be cut, while B is the contrary. This is equivalent to the weight distribution of the two during training. Considering the extreme case (that is, when the weight allocation is only 0 or 1), the information gain of F is large, and the learner is more likely to learn F while ignoring B, as shown in Equation (4). X_c represents several cases of learning feature information in extreme cases. For example, $X_c^{(1)} = (1 \cdot F, 0 \cdot B)$ is the case when F weighs 1, while B is 0.

$$X_c^{(1)} = (1 \cdot F, 0 \cdot B), X_c^{(2)} = (1 \cdot F, 1 \cdot B), X_c^{(3)} = (0 \cdot F, 1 \cdot B), \ldots \tag{4}$$

The two color adjustment methods and four data augmentation methods are listed in Table 3, illustrated with examples.

Table 3. Examples of data augmentation of pavement potholes.

Pre-Processing Steps	Original Image	Color Transformation (Contrast Adjustment)		Geometric Transformation (Data Augmentation)			
		Contrast	Sharpness	Rotating 90° Anticlockwise	Rotating 90° Clockwise	Rotating 180°	Random Crop
Potholes with water							
Potholes without water							

2.3. Object Detection

2.3.1. YOLOv3–ResNet101 Structure

YOLOv3 is a state-of-the-art object detection algorithm with fast speed and high accuracy. An end-to-end training and prediction method is adopted in YOLOv3, which is applicable for practical engineering applications [26]. In this paper, YOLOv3 was used as the basic framework for pothole detection. The ResNet101 network [40] was used instead of the traditional DarkNet-53 network as the feature extraction network. The proposed network structure is illustrated in Figure 2.

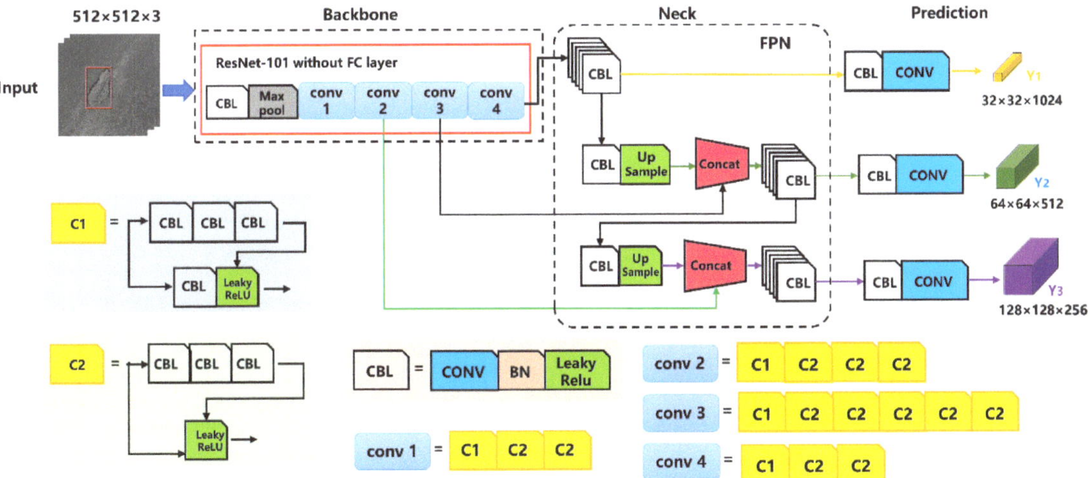

Figure 2. YOLOv3–ResNet101 network structure.

First, the pothole image was passed through a ResNet101 network without a fully connected layer to perform feature extraction. Then, upsampling and tensor concatenating were performed on the feature map. Thus, outputs at three different scales (Y1, Y2, Y3), as illustrated in Figure 2, were obtained. The multiscale method was employed to detect pothole objects of various sizes. The created detector had well-balanced performance regardless of the pothole sizes.

For a stacked layer, x is the input, and $H(x)$ is the learned feature. Thus, what is learned in the residual network is the residual $F(x) = H(x) - x$, while the original learned feature is $H(x) = F(x) + x$. This way of learning makes it easier to learn from raw features compared to the direct way. When $F(x) = 0$, the stacking layer performs identity mapping, and the performance of this network does not degrade. As a residual network, the ResNet18 network [40] can extract deeper-level features, thus having a more accurate understanding of the image. However, the ResNet18 has more network parameters than the deep residual network, resulting in a large number of training calculations. To decrease the computational cost, ResNet101 employs a bottleneck design structure. It has a convolution of inputs $7 \times 7 \times 64$, then goes through 33 (3 + 4 + 23 + 3) building blocks, each of which has three layers, and finally an FC layer (for classification), for a total of $1 + 33 \times 3 + 1 = 101$ layers. Subsequently, five feature maps of different sizes, C1, C2, C3, C4, and C5, were generated. They were different in size, and each pixel on the feature map had a different receptive field, corresponding to the original image, which was equivalent to dividing the original image into grids of different sizes. The ResNet101 network was selected. Figure 3 shows the improvement process.

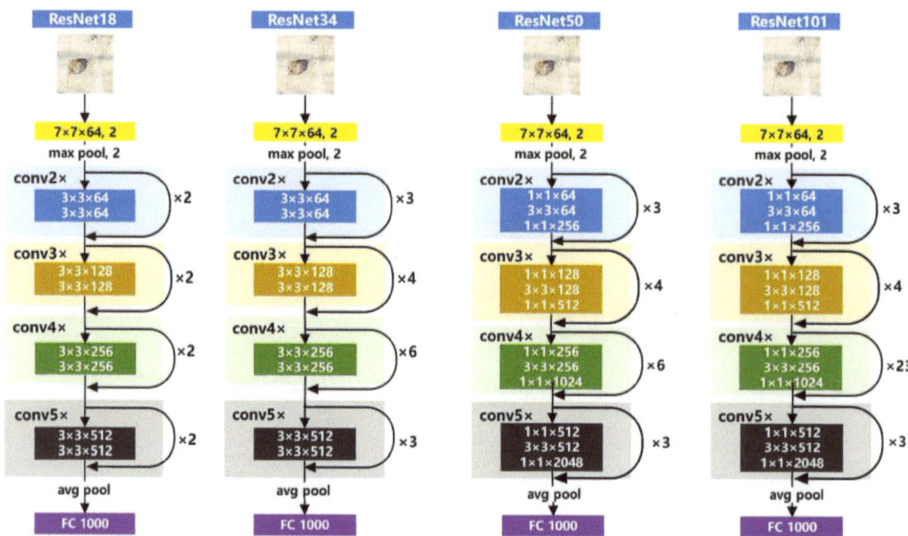

Figure 3. Comparison of ResNet101 network structure with other residual networks (ResNet18, ResNet34, and ResNet50).

To avoid gradient disappearance and speed up model convergence, ResNet101 adopts batch normalization (BN) [41]. BN converts the input distribution into a standard normal distribution with a mean of 0 and a variance of 1.

In addition, the detailed network parameters of YOLOv3–ResNet101 are given in Table 4.

Table 4. Network parameters of YOLOv3–ResNet101.

Network	Layer Name	Output Size	Parameters
ResNet101	conv1	$512 \times 512 \times 3$	conv, $7 \times 7 \times 64$, stride 2
	conv2	$256 \times 256 \times 64$	max pool, 3×3, stride 2 bottleneck: 1×1 $\begin{bmatrix} 1 \times 1 \times 64 \\ 3 \times 3 \times 64 \\ 1 \times 1 \times 256 \end{bmatrix} \times 3$
	conv3	$128 \times 128 \times 256$	bottleneck: 1×1 $\begin{bmatrix} 1 \times 1 \times 128 \\ 3 \times 3 \times 128 \\ 1 \times 1 \times 512 \end{bmatrix} \times 4$
	conv4	$64 \times 64 \times 512$	bottleneck: 1×1 $\begin{bmatrix} 1 \times 1 \times 256 \\ 3 \times 3 \times 256 \\ 1 \times 1 \times 1024 \end{bmatrix} \times 23$
	conv5	$32 \times 32 \times 1024$	bottleneck: 1×1 $\begin{bmatrix} 1 \times 1 \times 512 \\ 3 \times 3 \times 512 \\ 1 \times 1 \times 2048 \end{bmatrix} \times 3$
YOLOv3	Y_1	$32 \times 32 \times 1024$	3 anchors $\begin{bmatrix} 3 \times 3 \times 1024 \\ 1 \times 1 \times 1024 \end{bmatrix} \times 2$
	Y_2	$64 \times 64 \times 512$	3 anchors $\begin{bmatrix} 3 \times 3 \times 1024 \\ 1 \times 1 \times 1024 \end{bmatrix} \times 2$
	Y_3	$128 \times 128 \times 256$	3 anchors $\begin{bmatrix} 3 \times 3 \times 1024 \\ 1 \times 1 \times 1024 \end{bmatrix} \times 2$

2.3.2. Loss Function

The loss function of YOLOv3 consists of category loss, confidence loss, and position loss. The calculation formula is shown in Equation (5).

$$Loss = Loss_{class} + Loss_{conf} + Loss_{loc}. \tag{5}$$

Figure 4 shows the calculations of intersection over union (IoU) and other loss functions [42], where P represents the prediction box, P^{gt} represents the ground truth, and IoU is defined as the ratio of intersection and union between P, P^{gt}, which reflects the degree of overlap between the two. However, there are two problems with this calculation: first, when the prediction box is not overlapped with the real box, the IoU is 0, and the gradient return cannot be carried out; second, the calculation result is only related to the overlap area and cannot measure the way of intersection between two boxes. Therefore, the IoU is not a comprehensive and accurate measure of the degree of overlap.

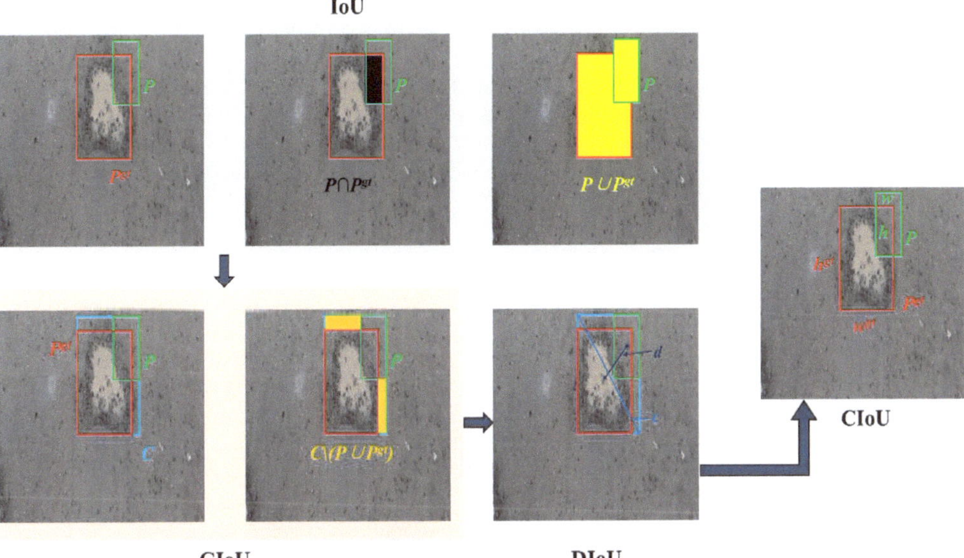

Figure 4. Calculation formula of IoU and its improvement process.

As a result, improvements to IoU have constantly been proposed. Generalized intersection over union (GIoU) considers the nonoverlapping regions; however, it is scale-invariant and is not conducive to multiscale pothole detection. The distance between P and P^{gt} is directly minimized to speed up convergence in distance IoU (DIoU); nevertheless, the aspect ratio of the two is not considered.

Complete IoU (CIoU) was selected as the loss function in this paper. Three important geometric factors of bounding-box regression loss are taken into account in CIoU: overlap area, center point distance, and aspect ratio, which allows a more stable and accurate convergence. The CIoU is calculated as shown in Equation (6).

$$\begin{cases} Loss_{CIoU} = 1 - IoU + \dfrac{\rho^2(P, P^{gt})}{c^2} \\ v = \dfrac{4}{\pi^2}\left(\arctan\dfrac{w^{gt}}{h^{gt}} - \arctan\dfrac{w}{h}\right)^2, \\ \alpha = \dfrac{v}{(1 - IoU) + v} \end{cases} \tag{6}$$

where v represents the consistency of the pothole's aspect ratio, a is the positive tradeoff parameter, w^{gt} and h^{gt} are the width and length of P^{gt}, respectively, and w and h are the width and length of P, respectively.

2.3.3. Anchor Size

An anchor is mainly used to solve the issue that the scale and aspect ratio vary too much in object detection. Through the anchor mechanism, the multiscale and aspect ratio are divided into several subspaces to reduce the difficulty of pothole detection, making the model easier to learn. In the original YOLOv3 structure, the scale and aspect ratio of the anchor are determined using the COCO dataset. However, the application object of the dataset is the pavement pothole; thus, the original anchor size is not suitable for the research scene of this paper. There is an urgent need to perform anchor clustering on the pothole dataset. The K-Means algorithm needs to determine the initial cluster center manually [26]; thus, the K-Means++ was used to cluster the real annotation boxes more accurately. The K-Means++ algorithm firstly selected a random point from the dataset as the center point and calculated the distance between the sample and each known center point. Then, M points were found far from the known center point. Lastly, a random point from M samples was chosen as the center. The clustering results were more concentrated, and the aspect ratio was more in line with the characteristics of the pothole dataset. The results are illustrated in Figure 5, and the size distribution of the anchor is shown in Table 5. The obtained anchor sizes were employed to replace the original parameters for training and testing, which reduced the difficulty of model training.

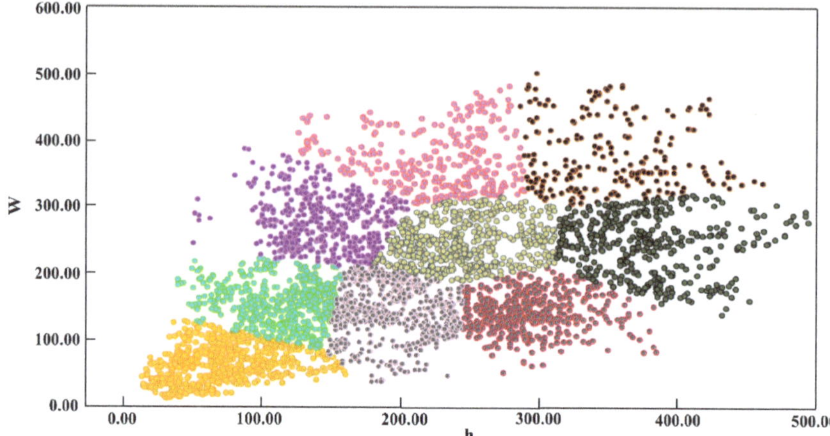

Figure 5. Visualization of training dataset anchor clustering distribution.

Table 5. Multiscale anchor size of potholes based on K-Means++.

Detection Scale	Anchor Size		
Scale 1: 32 × 32	(176, 312)	(335, 247)	(285, 362)
Scale 2: 64 × 64	(126, 236)	(285.5, 132)	(213, 201)
Scale 3: 128 × 128	(34, 35)	(72, 79)	(149, 126)

2.3.4. Robustness Verification

There are many shielding objects around potholes in actual scenes, such as leaves, road repair, and moving vehicles. Meanwhile, the same pothole can present different characteristics under different lighting conditions due to changes over time. Therefore, it is necessary to verify the proposed model's robustness in such a complex background. The existing analysis methods on robustness are mainly divided into traditional methods and adversarial attacks because of the different research fields and research angles [43]. The traditional method involves applying this model to another scene for verification and evaluating its robustness through the testing results. Adversarial attacks make the model produce false results by continuously adding tiny perturbations (adversarial examples) to the image.

Given the above problems, and considering that the potholes were interfered by shielding objects, the random occlusion (RO) method [44] was adopted to set the pixel value in a square area of the input image to zero. Details of random occlusion are shown in Figure 6, where x_e, y_e are random coefficients, and Z_e is the random occlusion area. The pixel color in the random occlusion area was set to zero.

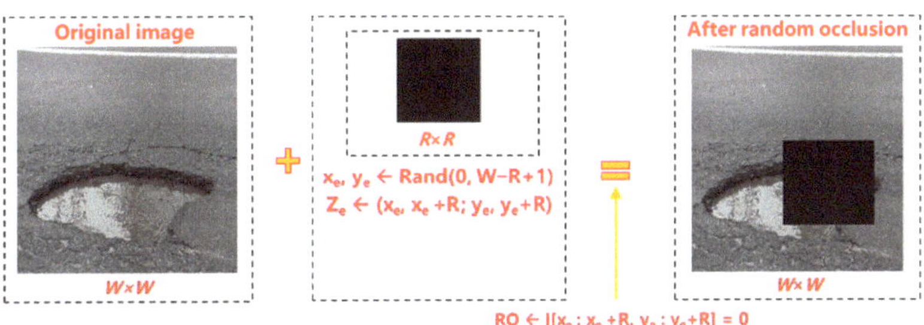

Figure 6. Generation of random occlusion.

Salt-and-pepper noise [45] and Gaussian noise [46] were also added to simulate the noise problem during the acquisition process. Figure 7 shows the noise-attacked image, in which the ratio of salt-and-pepper noise was 0.10, and the mean value and variance of Gaussian noise were 0 and 0.1, respectively. The noise had a great effect on the pothole.

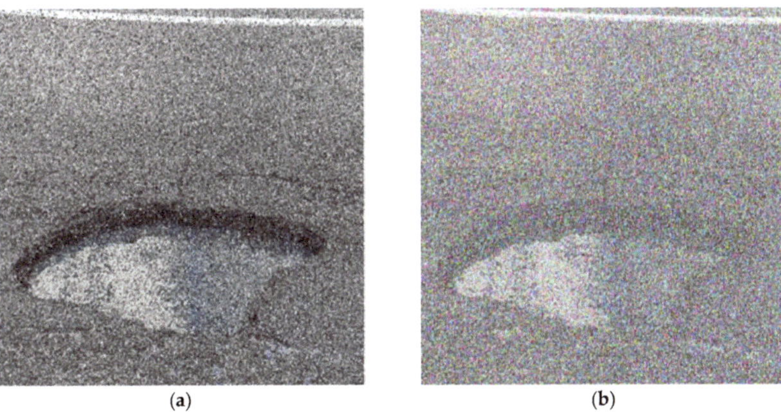

Figure 7. Pothole image after adding noise: (**a**) salt-and-pepper noise; (**b**) Gaussian noise.

2.3.5. Evaluation Index

To evaluate the accuracy and performance of the object detection model, precision (P), recall rate (R), mean average precision (mAP), and F1-score were selected. TP represents that the potholes are detected correctly when there exist potholes. FP represents that a pothole is detected when there is no pothole. FN represents that no pothole is detected when there exist potholes. TN represents that no pothole is detected when there is no pothole. The detailed calculation formulas are described in Equations (7)–(10).

$$P = \frac{TP}{TP + FP}. \tag{7}$$

$$R = \frac{TP}{TP + FN}. \tag{8}$$

$$mAP = \frac{\left(\int_0^1 P(r)dR\right)_{P1} + \left(\int_0^1 P(r)dR\right)_{P2}}{2}. \tag{9}$$

$$F1 = \frac{2 \times P \times R}{P + R}. \tag{10}$$

Furthermore, the calibration error index (ECE) [47] was selected to demonstrate the improved overfitting results after the data augmentation strategy. The calculation formula of the ECE index is described as follows [48]:

$$ECE = \sum_{b=1}^{5} \frac{n_b}{N} |A(b) - confidence(b)|. \tag{11}$$

The confidence interval [0, 1] was divided into five bins, where the detection results of each pothole image were saved. In Equation (11), b represents the b-th bin, and n_b refers to the number of samples in the b-th bin. N is the sum of samples. $A(b)$ is the average value of the ground truth of the samples in the b-th bin. Moreover, $confidence(b)$ represents the average value of the model's predicted probabilities in the b-th bin. A smaller difference between $A(b)$ and $confidence(b)$ indicates a higher model confidence.

3. Results and Discussion

The effect of the data augmentation on the detection result was studied by comparing the detection effect of the object detection models on the testing set before and after augmentation. To evaluate the proposed model from the perspectives of accuracy and performance, two control groups (the original YOLOv3 model and the YOLOv3–ResNet101 model with the new backbone) and the experimental group (YOLOv3–ResNet101 model with modified anchors) were set. Furthermore, several mainstream object detection models were also used to detect the pothole dataset in this study, including faster R-CNN, Cascade R-CNN, and SSD.

Model training and testing were conducted on a Windows 10 system using Python 3.7. The CPU was an Intel (R) Core (TM) i7-7700 CPU. The model reached convergence after 100 epochs. Detailed training parameters are presented in Table 6.

Table 6. Values of the training parameters.

Hyperparameters	Value
Batch size	2
Epochs	100
Learning rate	0.00125
Weight decay	L2
Optimizer	Momentum
Momentum	0.9

3.1. Results of Data Augmentation

Evaluation indices of the YOLOv3 model and proposed model before and after data augmentation are listed in Table 7, where "(aug)" denotes the detection result obtained by the augmented dataset. It can be seen that the P value, R value, mAP, and F1 score of the two object detection models were improved after data augmentation, with a ratio of 2–9%. Data augmentation expanded the number of samples, increased the diversities of labels, ensured the relative balance between the two labels (P1 and P2), and reduced the probability of misidentification. Therefore, the proposed augmentation method could effectively improve the detection effect of the object detection model.

Table 7. Evaluation index results of YOLOv3 model and proposed model before and after data augmentation.

Evaluation Indices	P			R			mAP			F1		
	P1	P2	Total	P1	P2	Total	P1	P2	Total	P1	P2	Total
YOLOv3	0.734	0.677	0.705	0.742	0.693	0.718	0.749	0.704	0.727	0.737	0.685	0.711
YOLOv3 (aug)	0.764	0.720	0.742	0.775	0.720	0.747	0.771	0.749	0.760	0.769	0.720	0.744
Proposed model	0.802	0.784	0.793	0.830	0.773	0.801	0.853	0.838	0.845	0.816	0.780	0.798
Proposed model (aug)	0.872	0.866	0.869	0.882	0.840	0.861	0.895	0.892	0.893	0.877	0.852	0.865

Data augmentation had different effects on different models. Figure 8 shows the enhancement effects of the two models. For the original YOLOv3 model, the improvement degree of the P value, R value, *mAP*, and F1 score was 2–5%, and the *mAP* of P2 had the highest improvement of 4.55%. Similarly, for the proposed model, the improvement degree of all evaluation indices was 4–9%, and the P value of P2 had the highest increasement of 8.22%. Generally, the proposed model had a more significant improvement effect on the evaluation indicators based on the enhanced dataset.

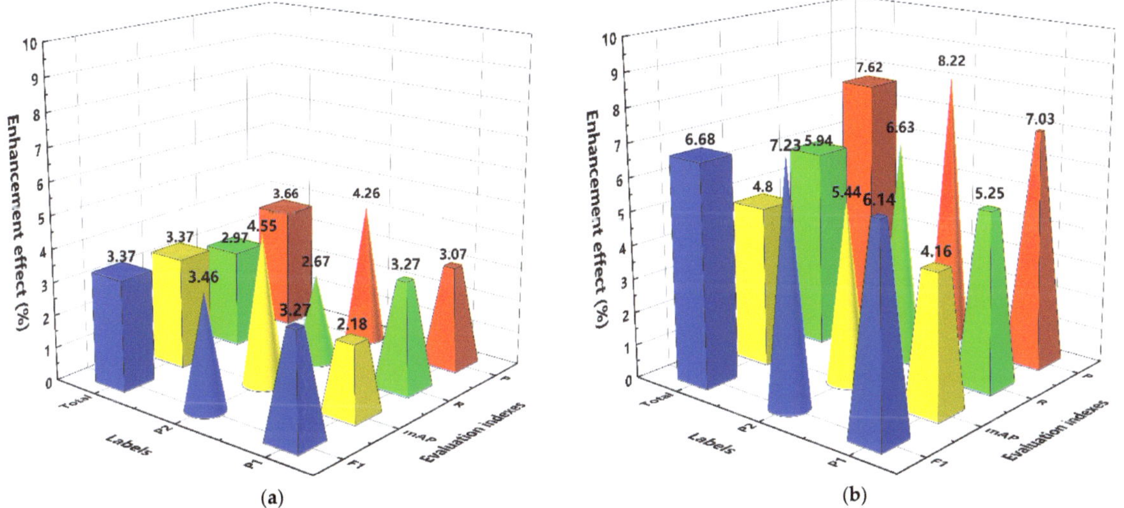

Figure 8. Data augmentation testing result comparisons: (**a**) YOLOv3 model; (**b**) proposed model.

Figure 9 shows the effect of calibrating the model using the ECE index. ECE values of the YOLOv3 model and the proposed model decreased significantly after data augmentation, by 62.8% and 69.6%, respectively. The ECE value of the proposed model dropped to 0.088 (a very small value), which shows that the data augmentation strategy is effective in promoting the confidence of the model and improving the overfitting problem.

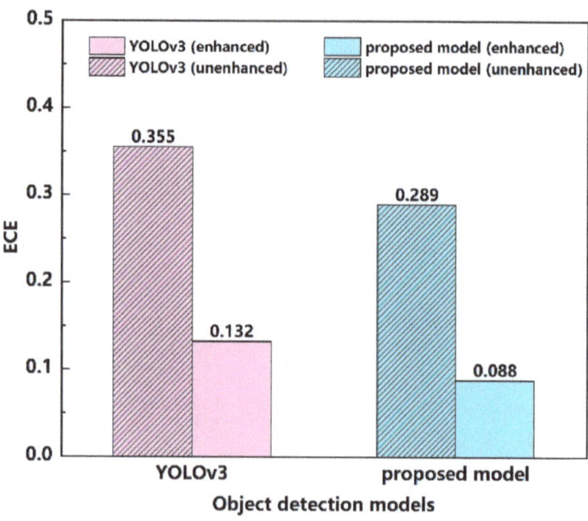

Figure 9. ECE indices of YOLOv3 model and proposed model before and after data augmentation.

3.2. Evaluation of Different DL Models

The performance of the improved model's detection effect is shown in Figure 10.

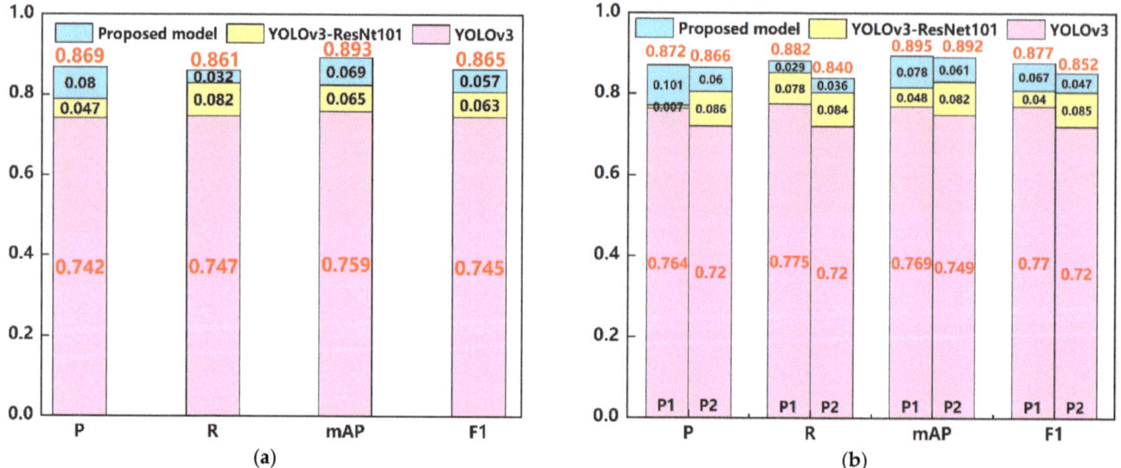

Figure 10. Evaluation indices of YOLOv3 model and other improved models. (**a**) overall evaluation indices; (**b**) evaluation indices of P1 and P2 labels.

As shown in Figure 10a, YOLOv3–ResNet101 had a significant improvement in all indices compared with the original YOLOv3. Specifically, the *mAP* and *F*1 score increased by 6.5% and 6.3%, respectively. The improvement of the P value was 4.7%, which demonstrates that the YOLOv3–ResNet101 model showed a better effect on pothole detection. It was verified that the proposed ResNet101 network was effective, allowing a deeper understanding of the deeper semantic information of the potholes. Furthermore, the proposed model had a greater improvement. The *mAP* and *F*1 score were increased by 6.9% and 5.7% compared to the YOLOv3–ResNet101 model, highlighting the great contribution of the modified anchor sizes to the detection performance.

Moreover, Figure 10b shows the detection effects of the YOLOv3 model and the improved model on P1 and P2, respectively. The indices of P1 and P2 labels were all improved in the YOLOv3–ResNet101 model; this improvement was further enlarged in the proposed model. In the original YOLOv3 model, the four indices of P1 were higher than those of P2 because the potholes with water showed different colored surfaces inside and outside the damaged area, making them easier to detect. As for YOLOv3–ResNet101, the differences between the four indices of P1 and P2 were narrowed. Furthermore, this phenomenon was obvious in the proposed model. In terms of *mAP* values, P2 was fairly close to P1. This indicates that the modified strategies in this study had a greater improvement in P2 than P1.

In addition, the detection results of different object detection models are listed in Table 8. It can be seen that the faster R-CNN model and cascade R-CNN model performed well in various indices because of their advantages, and the accuracy remained nearly above 80%. The accuracy of the SSD model and the original YOLOv3 model was slightly lower than that of the two-stage algorithms, which was caused by sacrificing accuracy for faster training and detection. The improvement effect of the proposed model was obvious; the overall P value, F1 score, and *mAP* value reached 86.9%, 86.5%, and 89.3%, respectively, which were higher than those of the two-stage methods, realizing the accurate detection of potholes. The rationality and feasibility of our modified strategies were proven in the proposed model.

Table 8. Comparison of detection results of different object detection models for potholes.

Evaluation Indices		Faster R-CNN [49]	Cascade R-CNN [50]	SSD [51]	YOLOv3–DarkNet53 [26]	YOLOv3–ResNet101	Proposed Model
P	P1	0.807	0.800	0.747	0.764	0.771	0.872
	P2	0.817	0.804	0.751	0.720	0.806	0.866
	Total	0.812	0.802	0.749	0.742	0.789	0.869
R	P1	0.793	0.794	0.733	0.775	0.853	0.882
	P2	0.822	0.810	0.738	0.720	0.804	0.840
	Total	0.807	0.802	0.736	0.747	0.829	0.861
mAP	P1	0.825	0.797	0.798	0.769	0.817	0.895
	P2	0.814	0.805	0.751	0.749	0.831	0.892
	Total	0.819	0.801	0.775	0.759	0.824	0.893
F1	P1	0.798	0.831	0.742	0.770	0.810	0.877
	P2	0.800	0.803	0.744	0.720	0.805	0.852
	Total	0.799	0.817	0.743	0.745	0.808	0.865
	AP50	0.819	0.801	0.775	0.759	0.824	0.893
	AP75	0.744	0.763	0.685	0.625	0.750	0.841
	AP90	0.782	0.783	0.730	0.692	0.784	0.867

To verify the localization accuracy of the proposed model, the detection results of faster R-CNN, cascade R-CNN, SSD, YOLOv3–Darknet53, YOLOv3–Resnet101, and the proposed model were compared. AP values with different IoU thresholds (0.50, 0.75, and 0.90) were calculated, and the results are shown in Table 8. It can be seen that the localization accuracy of the proposed model was higher that of the faster R-CNN model, with a significance difference of 7.4% between the two models. The localization accuracy of the original YOLOv3 model was 6.0% lower than that of faster R-CNN. The proposed model could effectively improve the localization accuracy of pothole detection.

3.3. Robustness Analysis

Table 9 lists the evaluation indices on the attacked testing dataset. It can be seen that, when the IoU threshold was 0.5, the overall mAP value reached 0.812, which decreased by 9.98% compared with that before attack. To more intuitively compare the performance of the proposed model on the attacked testing dataset, Table 10 describes four detection examples of typical pothole distress.

Table 9. Detection indexes of the attacked testing dataset.

Evaluation Indices	IoU_{50}			IoU_{75}			IoU_{90}		
	P1	P2	Total	P1	P2	Total	P1	P2	Total
P	0.788	0.793	0.790	0.733	0.738	0.736	0.275	0.277	0.276
R	0.763	0.802	0.783	0.718	0.746	0.732	0.266	0.248	0.257
mAP	0.810	0.814	0.812	0.764	0.773	0.769	0.211	0.325	0.268
F1	0.775	0.797	0.786	0.737	0.743	0.740	0.240	0.238	0.239

Table 10. Examples of detection results of the attacked pothole images.

Labels	Original Images	Proposed Model	Proposed Model (Random Occlusion)	Proposed Model (Salt-and-Pepper Noise Attacked)	Proposed Model (Gaussian Noise Attacked)
P1 (pothole with water)					
P2 (pothole without water)					

As shown in the table, the accuracy of the proposed model on the adversarial attack samples decreased because the random occlusion layer and the added noise destroyed the semantic information around the potholes, which hindered identification of the object. Random occlusion had great influence on the detection results. When ~50% occlusion (50% occlusion is similar to car occlusion) was set, the occluded pothole area could be correctly detected. However, in P2 (pothole without water), the occluded area was incorrectly detected as a pothole, which reduced the overall accuracy. Salt-and-pepper noise brought great damage to pothole targets, and P1 in two pictures was even wrongly detected as P2, which may be because salt-and-pepper noise destroyed the semantic characteristics of water on the surface, making it difficult to detect the existence of water. Gaussian noise was less destructive than salt-and-pepper noise, but the detection accuracy also decreased. Generally, the degree of influence of the three perturbation methods on the detection effect was as follows in this experiment: salt-and-pepper noise > random occlusion > Gaussian noise. Noise destruction is difficult to avoid in the process of image acquisition; therefore, it is of great necessity to study a reasonable image denoising method.

3.4. Generalization Performance Analysis

Generalization performance analysis was conducted. Table 11 describes the typical detection examples on a publicly available dataset [52]. The detection effect of the proposed model on the publicly available dataset was generally good. Although the detection effect was worse than that for the constructed dataset in this paper, the degree of reduction was small. It can be seen from Table 11 that the detection accuracy of the P2 label was higher, but there was a phenomenon of misdetection (row 2 and column 3 of Table 11). It was easy for the proposed model to misidentify manhole covers as P1 labels (row 2 and column 4 in Table 11), which indicated that the proposed model does not have strong detection ability for manhole covers and is not suitable for areas with many manhole covers. Additionally, the proposed model misidentified large pools of water in the public dataset as P1 labels.

Table 11. Detection examples of the open-source dataset using the proposed model.

Original image					
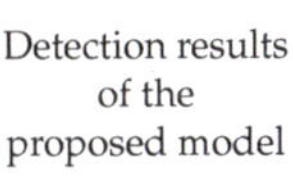Detection results of the proposed model					

4. Conclusions

A modified YOLOv3 model was proposed to detect potholes on the pavement surface under wet and dry conditions in this study. Through data augmentation, modified anchor sizes, improved feature network, and loss function, a model with higher accuracy was proposed, and its robustness was verified. The main conclusions can be drawn as follows:

(1) An effective image preprocessing strategy for improving and expanding the pothole dataset was proposed using the methods of color adjustment and geometric transformation, which ensured the detection stability of the proposed model.

(2) The potholes were further subdivided according to whether there was water. The detection results could preliminarily judge the surface state of the pothole and weather conditions.

(3) The ResNet101 network was adopted to extract features in the YOLOv3 model, which obtained abundant information on potholes. The modified anchor sizes based on the K-Means++ method were more in line with the shapes and sizes of the pothole, which improved the accuracy of the identification and location. The loss function defined by CIoU was of great help for accurate pothole detection.

(4) The robustness of the proposed model was verified by generating adversarial attack samples through random occlusion and adding noise. Results showed that the overall robustness was good. Specifically, the proposed model was more robust to Gaussian noise under the interference intensity.

It should be noted that perturbations were added to our pothole testing dataset during robustness verification. Multiple methods of perturbation, such as other types of noise and occlusion ratio, need to be evaluated in the future. Despite the limited perturbation methods, the proposed model could provide a reference for pothole detection in the actual scene.

Author Contributions: Study conceptualization and design, D.W.; data collection, Z.L. and X.G.; analysis and interpretation of results, D.W. and Z.L.; draft manuscript preparation, D.W., Y.C., W.W. and L.W. All authors have read and agreed to the published version of the manuscript.

Funding: This study was sponsored by the Key Science and Technology Research Project of Jinhua, China under Grant 2021-3-176, to which the authors are very grateful.

Data Availability Statement: Restrictions apply to the availability of these data. Data were obtained from the Jinhua Highway and Transportation Management Center, China and are available from the authors with the permission of the Jinhua Highway and Transportation Management Center, China.

Conflicts of Interest: The authors declare no conflict of interest.

References

1. Liu, S.; Tu, X.; Xu, C.; Chen, L.; Lin, S.; Li, R. An Optimized Deep Neural Network for Overhead Contact System Recognition from LiDAR Point Clouds. *Remote Sens.* **2021**, *13*, 4110. [CrossRef]
2. Liu, Z.; Gu, X.; Wu, C.; Ren, H.; Zhou, Z.; Tang, S. Studies on the validity of strain sensors for pavement monitoring: A case study for a fiber Bragg grating sensor and resistive sensor. *Constr. Build. Mater.* **2022**, *321*, 126085. [CrossRef]
3. Luo, Q.; Ge, B.; Tian, Q. A fast adaptive crack detection algorithm based on a double-edge extraction operator of FSM. *Constr. Build. Mater.* **2019**, *204*, 244–254. [CrossRef]
4. Chen, Y.; Liang, J.; Gu, X.; Zhang, Q.; Deng, H.; Li, S. An improved minimal path selection approach with new strategies for pavement crack segmentation. *Measurement* **2021**, *184*, 109877. [CrossRef]
5. Liang, J.; Gu, X.; Chen, Y. Fast and robust pavement crack distress segmentation utilizing steerable filtering and local order energy. *Constr. Build. Mater.* **2020**, *262*, 120084. [CrossRef]
6. Wang, L.; Gu, X.; Liu, Z.; Wu, W.; Wang, D. Automatic detection of asphalt pavement thickness: A method combining GPR images and improved Canny algorithm. *Measurement* **2022**, *196*, 111248. [CrossRef]
7. Liu, Z.; Chen, Y.; Gu, X.; Yeoh, J.K.; Zhang, Q. Visibility classification and influencing-factors analysis of airport: A deep learning approach. *Atmos. Environ.* **2022**, *278*, 119085. [CrossRef]
8. Krizhevsky, A.; Sutskever, I.; Hinton, G.E. Imagenet classification with deep convolutional neural networks. *Commun. ACM* **2017**, *60*, 84–90. [CrossRef]
9. Cha, Y.-J.; Choi, W.; Büyüköztürk, O. Deep Learning-Based Crack Damage Detection Using Convolutional Neural Networks. *Comput. Aided Civ. Infrastruct. Eng.* **2017**, *32*, 361–378. [CrossRef]
10. Xiong, Y.; Zhou, Y.; Wang, F.; Wang, S.; Wang, Z.; Ji, J.; Wang, J.; Zou, W.; You, D.; Qin, G. A Novel Intelligent Method Based on the Gaussian Heatmap Sampling Technique and Convolutional Neural Network for Landslide Susceptibility Mapping. *Remote Sens.* **2022**, *14*, 2866. [CrossRef]
11. Puttagunta, M.; Ravi, S. Medical image analysis based on deep learning approach. *Multimedia Tools Appl.* **2021**, *80*, 24365–24398. [CrossRef] [PubMed]
12. Liu, Z.; Wu, W.; Gu, X.; Li, S.; Wang, L.; Zhang, T. Application of Combining YOLO Models and 3D GPR Images in Road Detection and Maintenance. *Remote Sens.* **2021**, *13*, 1081. [CrossRef]
13. Xu, J.; Zhang, J.; Sun, W. Recognition of the Typical Distress in Concrete Pavement Based on GPR and 1D-CNN. *Remote Sens.* **2021**, *13*, 2375. [CrossRef]
14. Deng, T.; Zhou, Z.; Fang, F.; Wang, L. Research on Improved YOLOv3 Traffic Sign Detection Method. *Comput. Eng. Appl.* **2020**, *56*, 28–35.
15. Liu, Z.; Gu, X.; Dong, Q.; Tu, S.; Li, S. 3D Visualization of Airport Pavement Quality Based on BIM and WebGL Integration. *J. Transp. Eng. Part B Pavements* **2021**, *147*, 04021024. [CrossRef]
16. Zhou, W.N.; Sun, L.H. A Real-time Detection Method for Multi-scale Pedestrians in Complex Environment. *J. Electron. Inf. Technol.* **2021**, *43*, 2063–2070.
17. Liu, T.; Wang, Y.; Niu, X.; Chang, L.; Zhang, T.; Liu, J. LiDAR Odometry by Deep Learning-Based Feature Points with Two-Step Pose Estimation. *Remote Sens.* **2022**, *14*, 2764. [CrossRef]
18. Miao, P.; Srimahachota, T. Cost-effective system for detection and quantification of concrete surface cracks by combination of convolutional neural network and image processing techniques. *Constr. Build. Mater.* **2021**, *293*, 123549. [CrossRef]
19. Girshick, R.; Donahue, J.; Darrell, T.; Malik, J. Rich Feature Hierarchies for Accurate Object Detection and Semantic Segmentation. In Proceedings of the IEEE Conference on Computer Vision and Pattern Recognition, Columbus, OH, USA, 23–28 June 2014; pp. 580–587. [CrossRef]
20. Dai, J.; Li, Y.; He, K.; Sun, J. R-fcn: Object detection via region-based fully convolutional networks. In Proceedings of the Advances in Neural Information Processing Systems 29 (NIPS 2016), Barcelona, Spain, 5–10 December 2016.

21. He, K.; Zhang, X.; Ren, S.; Sun, J. Spatial Pyramid Pooling in Deep Convolutional Networks for Visual Recognition. *IEEE Trans. Pattern Anal. Mach. Intell.* **2015**, *37*, 1904–1916. [CrossRef]
22. Nie, M.; Wang, K. Pavement Distress Detection Based on Transfer Learning. In Proceedings of the 2018 5th International Conference on Systems and Informatics (ICSAI), Nanjing, China, 10–12 November 2018; pp. 435–439.
23. Pei, L.; Shi, L.; Sun, Z.; Li, W.; Gao, Y.; Chen, Y. Detecting potholes in asphalt pavement under small-sample conditions based on improved faster region-based convolution neural networks. *Can. J. Civ. Eng.* **2022**, *49*, 265–273. [CrossRef]
24. Song, L.; Wang, X. Faster region convolutional neural network for automated pavement distress detection. *Road Mater. Pavement Des.* **2021**, *22*, 23–41. [CrossRef]
25. Cao, X.G.; Gu, Y.F.; Bai, X.Z. Detecting of foreign object debris on airfield pavement using convolution neural network. In Proceedings of the LIDAR Imaging Detection and Target Recognition 2017, Changchun, China, 23–25 July 2017.
26. Redmon, J.; Farhadi, A. YOLOv3: An Incremental Improvement. *arXiv* **2018**, arXiv:1804.02767.
27. Zhu, J.; Zhong, J.; Ma, T.; Huang, X.; Zhang, W.; Zhou, Y. Pavement distress detection using convolutional neural networks with images captured via UAV. *Autom. Constr.* **2022**, *133*, 103991. [CrossRef]
28. Liu, Z.; Gu, X.; Yang, H.; Wang, L.; Chen, Y.; Wang, D. Novel YOLOv3 Model With Structure and Hyperparameter Optimization for Detection of Pavement Concealed Cracks in GPR Images. *IEEE Trans. Intell. Transp. Syst.* **2022**, 1–11. [CrossRef]
29. Liu, Z.; Yuan, L.; Zhu, M.; Ma, S.; Chen, L. YOLOv3 Traffic sign Detection based on SPP and Improved FPN. *Comput. Eng. Appl.* **2021**, *57*, 164–170.
30. Bochkovskiy, A.; Wang, C.-Y.; Liao, H.-Y.M. YOLOv4: Optimal Speed and Accuracy of Object Detection. *arXiv* **2020**, arXiv:2004.10934.
31. Tan, Y.; Cai, R.; Li, J.; Chen, P.; Wang, M. Automatic detection of sewer defects based on improved you only look once algorithm. *Autom. Constr.* **2021**, *131*, 103912. [CrossRef]
32. Cao, Y.; Zhou, Y. Multi-Channel Fusion Leakage Detection. *J. Cyber Secur.* **2020**, *5*, 40–52.
33. Guo, T.W.; Lu, K.; Chai, X.; Zhong, Y. Wool and Cashmere Images Identification Based on Deep Learning. In Proceedings of the Textile Bioengi-neering and Informatics Symposium (TBIS), Manchester, UK, 25–28 July 2018; pp. 950–956.
34. Tzutalin. LabelImg. Git Code. 2015. Available online: https://github.com/tzutalin/labelImg (accessed on 5 October 2015).
35. Xue, J.; Xu, H.; Yang, H.; Wang, B.; Wu, P.; Choi, J.; Cai, L.; Wu, Y. Multi-Feature Enhanced Building Change Detection Based on Semantic Information Guidance. *Remote Sens.* **2021**, *13*, 4171. [CrossRef]
36. Du, Z.; Yuan, J.; Xiao, F.; Hettiarachchi, C. Application of image technology on pavement distress detection: A review. *Measurement* **2021**, *184*, 109900. [CrossRef]
37. Liu, Z.; Gu, X.; Wu, W.; Zou, X.; Dong, Q.; Wang, L. GPR-based detection of internal cracks in asphalt pavement: A combination method of DeepAugment data and object detection. *Measurement* **2022**, *197*, 111281. [CrossRef]
38. Lae, S.; Narasimhadhan, A.V.; Kumar, R. Automatic Method for Contrast Enhancement of Natural Color Images. *J. Electr. Eng. Technol.* **2015**, *10*, 1233–1243. [CrossRef]
39. Xie, Y.; Wu, Y.; Wang, Y.; Zhao, X.; Wang, A. Light field all-in-focus image fusion based on wavelet domain sharpness evaluation. *J. Beijing Univ. Aeronaut. Astronaut.* **2019**, *45*, 1848–1854.
40. He, K.; Zhang, X.; Ren, S.; Sun, J. Deep Residual Learning for Image Recognition. In Proceedings of the 2016 IEEE Conference on Computer Vision and Pattern Recognition (CVPR), Las Vegas, NV, USA, 27–30 June 2016; pp. 770–778.
41. Ioffe, S.; Szegedy, C. Batch Normalization: Accelerating Deep Network Training by Reducing Internal Covariate Shift. *arXiv* **2015**, arXiv:1502.03167.
42. Huang, Z.; Zhao, H.; Zhan, J.; Li, H. A multivariate intersection over union of SiamRPN network for visual tracking. *Vis. Comput.* **2021**, *38*, 2739–2750. [CrossRef]
43. Ji, S.; Du, T.; Deng, S.; Cheng, P.; Shi, J.; Yang, M.; Li, B. Robustness Certification Research on Deep Learning Models: A Survey. *Chin. J. Comput.* **2022**, *45*, 190–206.
44. Hou, J.; Zeng, H.; Cai, L.; Zhu, J.; Chen, J. Random occlusion assisted deep representation learning for vehicle re-identification. *Control. Theory Appl.* **2018**, *35*, 1725–1730.
45. Wang, W.; Gao, S.; Zhou, J.; Yan, Y. Research on Denoising Algorithm for Salt and Pepper Noise. *J. Data Acquis. Processing* **2015**, *30*, 1091–1098.
46. Kindler, G.; Kirshner, N.; O'Donnell, R. Gaussian noise sensitivity and Fourier tails. *Isr. J. Math.* **2018**, *225*, 71–109. [CrossRef]
47. Tong, Z.; Xu, P.; Denœux, T. Evidential fully convolutional network for semantic segmentation. *Appl. Intell.* **2021**, *51*, 6376–6399. [CrossRef]
48. Guo, C.; Pleiss, G.; Sun, Y.; Weinberger, K.Q. On Calibration of Modern Neural Networks. *arXiv* **2017**, arXiv:1706.04599.
49. Ren, S.; He, K.; Girshick, R.; Sun, J. Faster R-CNN: Towards Real-Time Object Detection with Region Proposal Networks. *IEEE Trans. Pattern Anal. Mach. Intell.* **2017**, *39*, 1137–1149. [CrossRef]
50. Cai, Z.; Vasconcelos, N. Cascade R-CNN: Delving into High Quality Object Detection. In Proceedings of the 2018 IEEE/CVF Conference on Computer Vision and Pattern Recognition, Salt Lake City, UT, USA, 18–23 June 2018; pp. 6154–6162.
51. Liu, W.; Anguelov, D.; Erhan, D.; Szegedy, C.; Reed, S.; Fu, C.; Berg, A.C. SSD: Single Shot MultiBox Detector. *arXiv* **2015**, arXiv:1512.02325.
52. Maeda, H.; Kashiyama, T.; Sekimoto, Y.; Seto, T.; Omata, H. Generative adversarial network for road damage detection. *Comput. Aided Civ. Infrastruct. Eng.* **2021**, *36*, 47–60. [CrossRef]

remote sensing

MDPI

Article

ASFF-YOLOv5: Multielement Detection Method for Road Traffic in UAV Images Based on Multiscale Feature Fusion

Mulan Qiu [1], Liang Huang [1,2,*] and Bo-Hui Tang [1]

[1] Faculty of Land Resources Engineering, Kunming University of Science and Technology, Kunming 650093, China; qiumulan@stu.kust.edu.cn (M.Q.); tangbh@kust.edu.cn (B.-H.T.)
[2] Surveying and Mapping Geo-Informatics Technology Research Center on Plateau Mountains of Yunnan Higher Education, Kunming 650093, China
* Correspondence: kmhuangliang@kust.edu.cn

Abstract: Road traffic elements are important components of roads and the main elements of structuring basic traffic geographic information databases. However, the following problems still exist in the detection and recognition of road traffic elements: dense elements, poor detection effect of multi-scale objects, and small objects being easily affected by occlusion factors. Therefore, an adaptive spatial feature fusion (ASFF) YOLOv5 network (ASFF-YOLOv5) was proposed for the automatic recognition and detection of multiple multiscale road traffic elements. First, the K-means++ algorithm was used to make clustering statistics on the range of multiscale road traffic elements, and the size of the candidate box suitable for the dataset was obtained. Then, a spatial pyramid pooling fast (SPPF) structure was used to improve the classification accuracy and speed while achieving richer feature information extraction. An ASFF strategy based on a receptive field block (RFB) was proposed to improve the feature scale invariance and enhance the detection effect of small objects. Finally, the experimental effect was evaluated by calculating the mean average precision (mAP). Experimental results showed that the mAP value of the proposed method was 93.1%, which is 19.2% higher than that of the original YOLOv5 model.

Keywords: object detection; road traffic multiple elements; adaptively spatial feature fusion; spatial pyramid pooling fast; basic traffic geographic information database

Citation: Qiu, M.; Huang, L.; Tang, B.-H. ASFF-YOLOv5: Multielement Detection Method for Road Traffic in UAV Images Based on Multiscale Feature Fusion. *Remote Sens.* **2022**, *14*, 3498. https://doi.org/10.3390/rs14143498

Academic Editors: Valerio Baiocchi, Alessandro Mei and Xianfeng Zhang

Received: 1 July 2022
Accepted: 20 July 2022
Published: 21 July 2022

Publisher's Note: MDPI stays neutral with regard to jurisdictional claims in published maps and institutional affiliations.

1. Introduction

Road traffic elements are important components of roads, which are the main contents for basic traffic geographic information database construction and are especially important for the development of basic traffic geographic information. Road traffic element information includes road centerlines, road intersections, zebra crossings, bus stations, roadside parking spaces, and other information [1]. Their accurate recognition and detection can provide important data support for automatic driving, improving intelligent transportation systems, promoting smart cities, and updating basic traffic geographic information databases. Currently, most research has focused on the detection and recognition of single-element traffic signs [2–4], extraction of road network information [5–8], real-time monitoring of road conditions [9,10], etc. while there are few studies on the extraction of road traffic multielement information, and there are mis-detection and missing detection of small traffic elements. At the same time, the difficulty of detection of juxtaposed dense traffic elements are also one of the problems restricting the update of traffic geographic information. This has caused some trouble to the automatic detection and recognition of road traffic elements. Therefore, it is important to improve the detection accuracy and precision of small traffic elements and juxtaposed dense traffic elements for the automatic detection and recognition of road traffic elements.

Small object detection is one of the most challenging problems in computer vision. Taking the definition of an object in COCO datasets [11] as an example, small objects refer to objects with fewer than 32×32 pixel points. Small objects are easy to ignore in detection due to the small number of pixels in the image and the few available features, resulting in the phenomenon problem of missed detection. However, many scholars have improved the detection of small objects through strategies such as data enhancement, multiscale learning, contextual feature learning, and generative adversarial learning. For example, Yu et al. [12] proposed a vehicle detection method for aerial vehicles based on a deep neural network and traditional method. This method uses a deep segmentation network to mine the symbiotic relationship between roads and vehicles and then detects small vehicles based on a visual attention mechanism of spatiotemporal constraint information. Xiao et al. [13] proposed a semi-supervised fully convolutional neural network for extracting fine-grained road scene information using UAV images, which solved the problem of the high cost of labeled samples and implemented a fine-grained road scene resolution scheme for UAV remote sensing images. Wang et al. [14] proposed a multitask generative adversarial small object detection network which introduced artificial texture loss and center mask into the generator, making it easier for the generator to generate super resolution images and thus easier for small object detection. Xu et al. [15] proposed a feature enhancement network named FE-YOLO for remote sensing object detection to achieve real-time detection. FE-YOLO can accommodate remote sensing object detection under different backgrounds and can effectively improve the accuracy of remote sensing small object detection. Qing et al. [16] proposed a RepVGG-YOLO network that can be used for remote sensing image detection from arbitrary angles. Hu et al. [17] proposed a PAG-YOLO network by considering the global and local relationships in the input image. This network can adaptively redistribute the weight distribution of different scale features from spatial dimension and channel dimension through an attention-guided feature optimization module for the detection of small vessels. Kim et al. [18] proposed an ECAP-YOLO network, the model in which was based on the YOLOv5 network to discover small objects by adding a detection layer, reducing the computational power used for small target detection, improving its detection rate, and optimizing the detection of small objects. Liu et al. [19] proposed a method based on YOLOv3 for UAV perspective target detection. The method improves the whole network structure based on YOLOv3 by optimizing the resblock and adding convolution operations to enrich the spatial information and expand the UAV reception range for small target detection. Because the size of small objects only accounts for a small part of an image, their features are susceptible to the influence of weather illumination and occlusion, resulting in the phenomenon of detection error. Therefore, the detection of small objects is more difficult and challenging.

With the development of UAV technology, UAVs have been applied in all walks of life, such as the military, agriculture, electricity inspection, and transportation fields. UAVs play an important role in the transportation field because of their high efficiency, clear image acquisition, and high flexibility. UAVs can be used for cruising, vehicle tracking and identification, road traffic condition detection, etc. Lee et al. [20] proposed a set of comprehensive road monitoring systems using UAVs, aerial mapping cameras, and deep learning algorithms. The system can be used for road maintenance and management as well as autonomous vehicle roadmap planning. Some scholars use images collected by UAVs to monitor road damage, which greatly reduces the cost of manual visual interpretation. For example, Hong et al. [21] used UAVs to collect images of roads to detect and identify pavement cracks using an improved recognition method based on U-Net. The algorithm was made possible by a convolution block attention module, an improved encoder, and a strategy of fusing long and short skip connections. It enables the ability to predict road cracks with high accuracy on UAV images. Scholars have also implemented the extraction of road regions from UAV images. For example, Sultonov et al. [22] proposed an improved algorithm for automatic road extraction from UAV images, which is based on a lightweight model with depth separable convolution and ConvMixer initial blocks for extracting road regions from UAV images.

The main research of UAVs in the field of traffic and some strategies and methods for improving small target detection were introduced above. Similar to the above scholars, for this study, which collected road traffic element information by UAV, the problem that needs to be solved is the mis-detection and missed detection of small and juxtaposed dense road traffic elements. In our previous studies [1], we initially realized the automatic detection and recognition of road traffic element information. However, in the study, we also found that there are problems of incorrect and missed detection when detecting small and juxtaposed dense road traffic elements. Therefore, to solve the problems of poor detection of multiscale road traffic elements and difficult detection of small road traffic elements, an ASFF-YOLOv5 network is proposed for the automatic detection and recognition of road traffic multielements. The network improves the detection accuracy, reduces the problems of small object incorrect and missed detection, and realizes the automatic detection and recognition of road traffic multielements. It provides a new method for updating the basic traffic geographical information databases.

The main contributions of this paper are as follows:

(1) The K-means++ [23] clustering method was used for data processing to obtain the optimal candidate box size of the object so that the detection anchor box is more consistent with a multielement road traffic dataset.

(2) To address the problems of low detection accuracy, serious error detection and missed detections of road elements, and difficult recognition of small dense objects, an ASFF-YOLOV5 algorithm was proposed. In this algorithm, the classification accuracy and speed of multiple elements of road traffic are improved using the SPPF [24] structure. Moreover, by integrating the ASFF [25] structure of the RFB module [26], the receptive field is improved and the feature information of detection objects at different scales is improved to achieve richer feature information extraction, especially the detection and recognition ability of small objects.

(3) The proposed ASFF-YOLOv5 algorithm was proven to be superior in detecting multiple elements of road traffic through comparative experiments and ablation experiments and provides a new solution for updating basic traffic geographical information databases.

The rest of this paper is organized as follows. Section 2 describes the datasets used in this study. Section 3 details the proposed method, followed by the experiments and results in Section 4. A discussion is presented in Section 5. Finally, our conclusions are outlined in Section 6.

2. Datasets and Scale Statistics

In this study, the multielement road traffic datasets produced in previous studies [1] were used as the experimental data, and the sample data are shown in Figure 1. The experimental data were captured by Hava MEGA-V8 and DJI FC6310 with a spatial resolution of 0.05 m and 0.1 m, respectively. In this study, representative multielement road traffic data were selected as the research object. The selected elements are zebra crossings, roadside parking spaces, and bus stations named zebra_crossing, parking_space, and bus_station. In the dataset division, the training data accounted for 90%, and the rest were the test data.

The default size of the anchor box in the YOLOv5 network is obtained by clustering according to the COCO dataset using the K-means algorithm [27], which is not applicable to the road traffic multielement datasets proposed in this paper. Therefore, to be more suitable for the object scale range of multielement road traffic datasets, the K-means++ clustering method was used to conduct scale statistics on UAV multielement road traffic remote sensing images in this study.

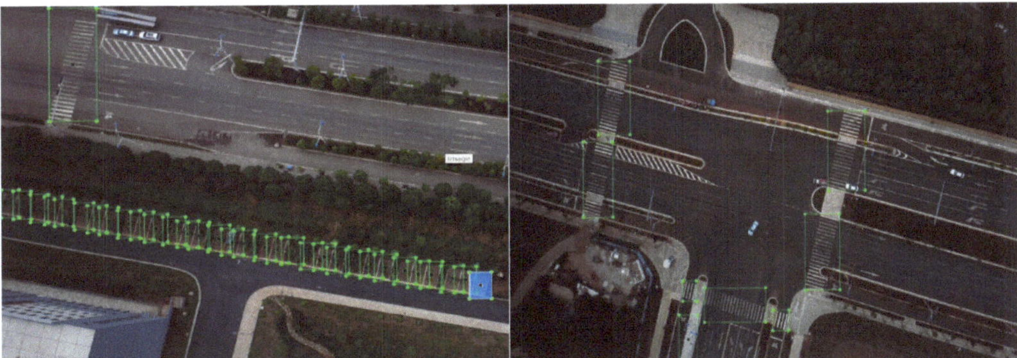

Figure 1. Sample dataset.

The K-means clustering algorithm is a typical iterative solution clustering algorithm. The core idea of the K-means clustering algorithm is to divide an object set into n clusters and randomly select a point as the cluster center. Iteration of the cluster center of each cluster continues until all the points in each cluster no longer change. However, the K-means clustering algorithm has some defects, and convergence depends heavily on the initialization of the cluster center. Compared with the K-means clustering algorithm, the K-means++ [23] clustering algorithm is improved in the initial random selection of the cluster center. Therefore, the K-means++ clustering algorithm was selected in this research for calculation. The K-means++ clustering algorithm selects cluster centers one by one instead of randomly when initializing the cluster centers, and the sample points farther away from other cluster centers are more likely to be selected as the next cluster center. The initial cluster center was obtained using the K-means++ cluster center calculation, and then the clustering result was obtained by the K-means clustering algorithm. In this study, the road traffic element scale was defined as nine clusters, and the object candidate box size was calculated by the K-means++ clustering algorithm. The results after K-means++ clustering from the test results of different K-means in Section 4.3 are shown in Table 1. Compared with the nonuse of the K-means++ algorithm, the mean average accuracy increase was 18.3%. Compared with K-means++ clustering, the mean average accuracy increase was 3%. The result shows that the effect of K-means++ clustering conformed to the datasets of road traffic elements.

Table 1. The results of K-means++ clustering.

Serial No.	1	2	3	4	5	6	7	8	9
x	10	15	16	19	23	25	33	52	125
y	17	29	19	38	22	26	70	34	123

3. Research Method

The proposed multielement road traffic detection method based on multiscale feature fusion includes data preprocessing, feature extraction, and feature fusion. First, the K-means++ clustering algorithm is used for preprocessing, and the size of the candidate box matching the multielement datasets of road traffic is obtained. Then, the main feature extraction network integrated with an SPPF structure is used to extract the features of the input feature map to achieve speed improvement and complete richer feature information extraction. Second, through a fusion of an ASFF module after the feature extraction is completely strengthened, the features of the pyramid through an RFB module expand the receptive field. To further extract the features of different scale road traffic elements, the features of the pyramid in the implementation of multiscale road traffic element feature fusion makes full use of the features of road traffic elements of different maps and the details of the semantic information. Finally, the detection results of multielement road traffic are output through the detection head, and their accuracy is evaluated. The specific flow of the proposed method is shown in Figure 2.

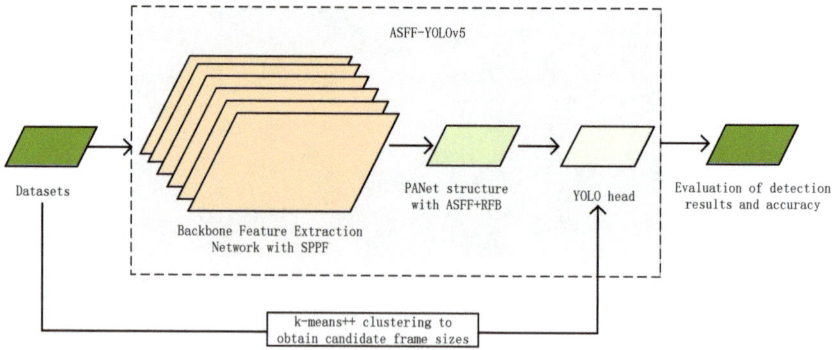

Figure 2. Flow chart of the proposed method.

3.1. ASFF-YOLOv5

The pyramid module in the original YOLOv5 network [24] was borrowed from PANet [28], and the YOLOv5 network structure diagram is shown in Figure 3. PANet achieves multiscale feature fusion of the network by aggregating features from different backbone layers and between detection layers. However, in PANet feature fusion, feature information is extracted from deep and shallow feature maps continuously, and then the feature map scale is transformed into the same scale. Then, simple addition is performed to obtain feature fusion, which cannot make full use of feature information of different scales. Therefore, the proposed ASFF-YOLOv5 algorithm makes full use of the feature information of different scales to improve the detection effect of different scale objects.

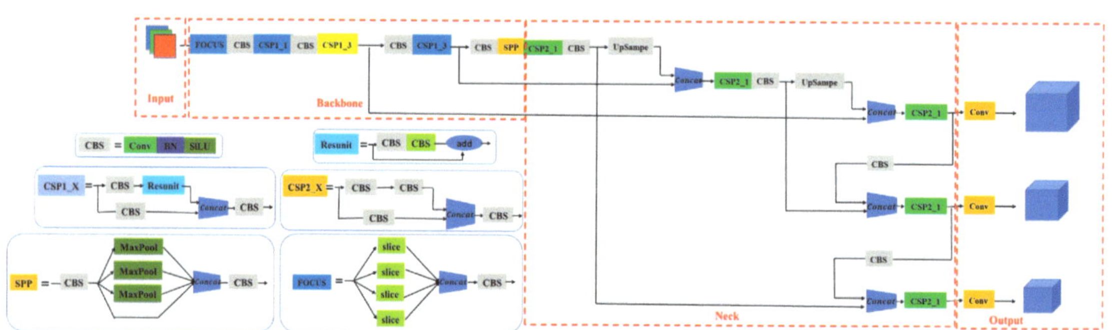

Figure 3. Network structure diagram of YOLOv5.

The ASFF-YOLOv5 network consists of three parts: a backbone feature extraction network, a feature map pyramid network, and a classifier and regressor. In the ASFF-YOLOv5 network, the backbone feature extraction network is used to extract the input feature map. The feature map pyramid network achieves the feature fusion of multiscale road traffic elements. The classifier and regressor obtain the detection results. Figure 4 shows a schematic diagram of the proposed method, in which the deepened color part is the proposed SPPF module and ASFF + RFB module. Assuming that the size of the input feature map in the network is $640 \times 640 \times 3$, the feature extraction process of the ASFF-YOLOv5 network is as follows:

(1) The height and width are compressed in the feature layer by focusing the structure, and the number of expansion channels is quadrupled to obtain a $320 \times 320 \times 12$ feature map.

(2) A $320 \times 320 \times 64$ feature map is obtained through a series of operations, such as convolution, normalization, and activation functions.

(3) In the backbone feature extraction network, three effective feature layers are obtained by stacking residual extraction, and the SPPF structure is introduced in the last effective feature layer. The SPPF structure improves the classification accuracy and speed of the feature map by using the maximum pooling of the same pooling kernel for feature extraction; at this time, the three effective feature layers obtained in the backbone feature extraction network are $80 \times 80 \times 256$, $40 \times 40 \times 512$, and $20 \times 20 \times 1024$.

(4) The obtained effective feature layer is transferred to the PANet structure, and the feature extraction is further enhanced through upsampling and downsampling. In this stage, the proposed ASFF + RFB module is integrated and used for the extraction of multiscale road traffic element information. It can achieve the enhancement of the perceptual field, realize the extraction of feature information of detectors at different scales, and complete the extraction of richer feature information.

(5) Three enhanced effective feature layers are obtained, and the prediction and regression results are obtained by the classifier and regressor.

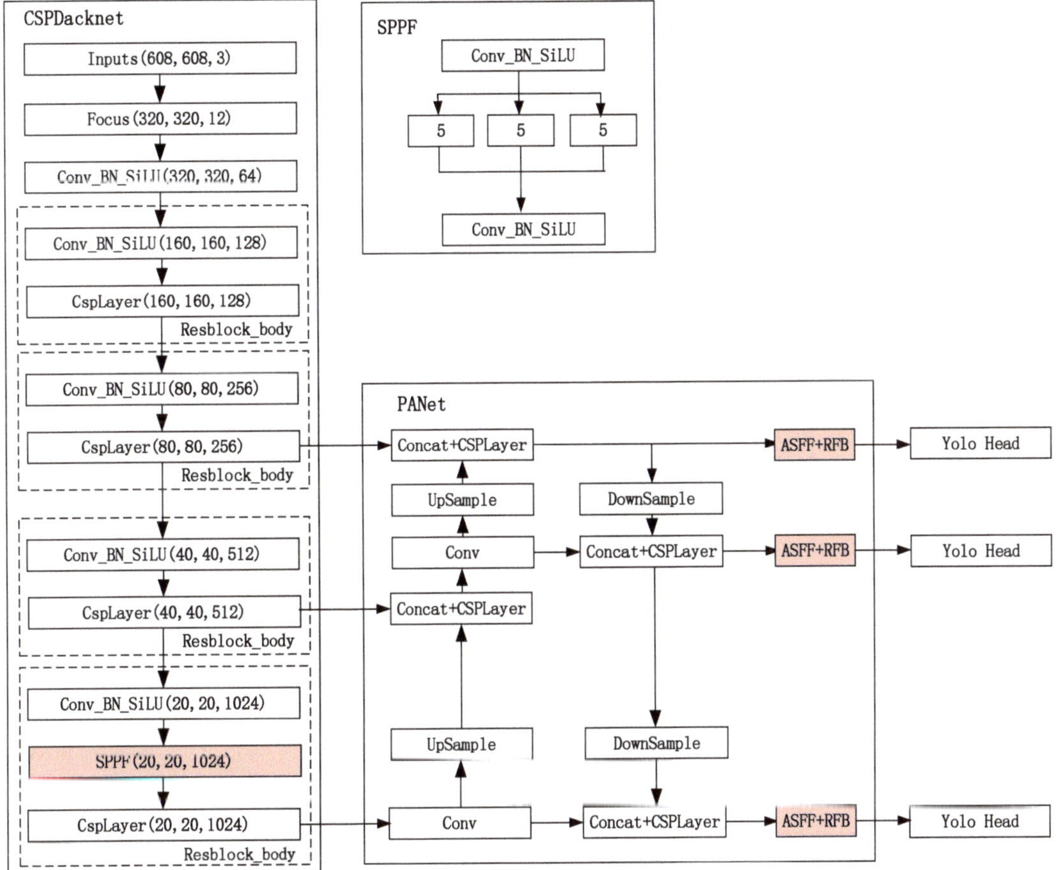

Figure 4. ASFF-YOLOv5 structure diagram.

3.2. ASFF + RFB Module

The ASFF-YOLOv5 algorithm enables improved feature scale invariance and object detection by fusing the ASFF structure into the PANet structure. In the PANet structure of the ASFF-YOLOv5 algorithm, the ASFF algorithm is introduced in each layer of the FPN structure for weighted fusion after first enhancing the semantic feature extraction top-down for the FPN structure [29]. The weight parameters are derived from the output of the convolutional feature layer, and the weight parameters become learnable after gradient backpropagation so that they can be adaptive when performing weighted fusion. Meanwhile, an RFB module layer is introduced after the ASFF algorithm. This module fuses the null convolution on top of Inception [30], thus effectively increasing the perceptual field and realizing the feature extraction capability of the network. The PANet structure of the proposed fused ASFF algorithm is shown in Figure 5.

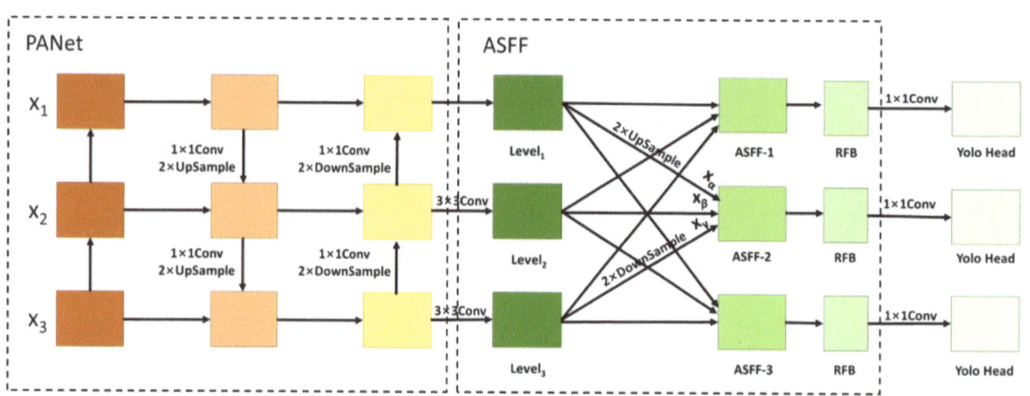

Figure 5. Integration of the PANet structure into the ASFF module.

Taking ASFF-2 computational fusion as an example, X_1, X_2, and X_3 are the feature maps extracted from the YOLOv5 backbone network. First, the feature maps $Level_1$, $Level_2$, and $Level_3$ are obtained from the PANet structure. Then, ASFF-2 is obtained by fusing them with the ASFF algorithm. The $Level_1$ feature map is convolved to obtain the same number of channels as the $Level_2$ feature map, and then the feature map with the same dimension as $Level_2$ is upsampled to obtain $X^{1 \rightarrow 2}$. For the $Level_3$ feature map, the number of channels and dimensions are adjusted using convolution and downsampling operations to keep the same number of channels and dimensions as $Level_2$, and $X^{3 \rightarrow 2}$ is obtained. The $Level_2$ feature map is adjusted by the number of channels after the convolution operation to obtain $X^{2 \rightarrow 2}$. After processing the three feature maps using the softmax function, the weight coefficients α, β, and γ of $X^{1 \rightarrow 2}$, $X^{2 \rightarrow 2}$ and $X^{3 \rightarrow 2}$ are obtained, respectively, and then the ASFF fusion calculation is performed with the following formula.

$$y_{ij}^l = \alpha_{ij}^l \cdot X_{ij}^{1 \rightarrow l} + \beta_{ij}^l \cdot X_{ij}^{2 \rightarrow l} + \gamma_{ij}^l \cdot X_{ij}^{3 \rightarrow l} \tag{1}$$

where y_{ij}^l is the new feature map obtained using the ASFF module. α_{ij}^l, β_{ij}^l, and γ_{ij}^l are the weight coefficients of the three feature maps, and α_{ij}^l, β_{ij}^l, and γ_{ij}^l after softmax function processing satisfy $\alpha_{ij}^l + \beta_{ij}^l + \gamma_{ij}^l = 1$, α_{ij}^l, β_{ij}^l, and $\gamma_{ij}^l \in [0, 1]$. $X_{ij}^{n \rightarrow l}$ denotes the feature vector of the feature map from layer n to layer l.

Through the ASFF algorithm, the multiscale feature fusion of the model is fully realized by adjusting the feature fusion with the weight parameters. In addition, this study introduces the RFB module along with the ASFF algorithm. Through multiple branch convolution and dilated convolution, the RFB module can increase the receptive field more effectively, improve the utilization of feature information, and improve the model's ability to recognize and detect small objects. In the multiple branch structures of the RFB module, the first layer of each branch is composed of convolution kernels of a specific size, and the size of the first layer convolution kernels is 1×1, 3×3, and 5×5. Figure 6a shows the effect diagram of the RFB module, and Figure 6b shows the structure diagram of the RFB module. The rate represents the expansion coefficients of different dilated convolution layers. The RFB module includes a dilated convolution layer to enhance the receptive field. The final output of the RFB module concatenates the output feature maps of different sizes and receptive fields to achieve the purpose of fusing different features.

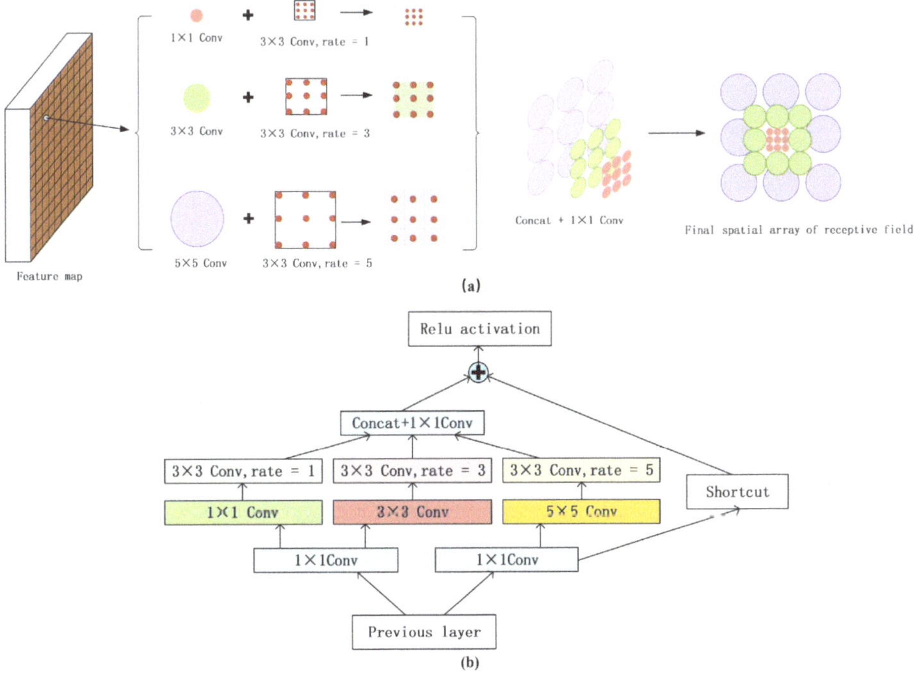

Figure 6. Schematic diagrams of (**a**) the RFB module effect and (**b**) the RFB structure.

3.3. Spatial Pyramid Pooling Fast

The role of the SPP structure [31] in the YOLOv5 network is to implement a fixed size feature vector as a fully connected layer output for images with different size inputs. The SPP structure uses three convolution kernels of different sizes, 3, 5, and 9, to extract features through the maximum pooling operation, enhance the feature expression ability of the feature graph, and improve the network receptive field. The SPP structure is shown in Figure 7a, which first performs 1×1, 3×3, 5×5, and 9×9 maximum pooling operations on the data transferred from the convolutional normalization activation function (Convolution + Banch Normalization + SiLU, CBS) in parallel and then connects them to the CBS structure by concatenation splicing. The CBS structure achieves feature fusion and completes the feature extraction operation. However, the SPP structure increases the computation of the program by parallel pooling operations with different sized convolutional kernels, which degrade the performance. Therefore, the SPPF structure is used for pooling to improve the performance of pooling while reducing the amount of program computation. The SPPF structure replaces the parallel maximum pooling operation of three convolutional kernels of different sizes in the original SPP with a serial operation of three convolutional kernels of the same size. As shown in Figure 7b, the SPPF structure is similar to the SPP structure. The operation of the SPPF layer first performs a 5×5 maximum pooling operation on the data serially transferred from the CBS structure. Then, the data are passed into the CBS structure by concatenation splicing, which achieves a speed-up while completing a richer feature information extraction.

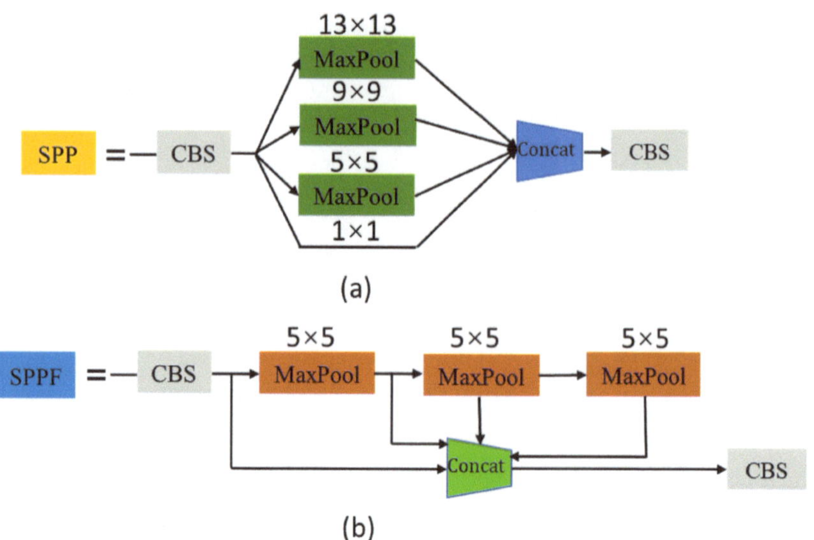

Figure 7. (**a**) SPP structure diagram and (**b**) SPPF structure diagram.

3.4. EIoU Loss

Through the above method, the object detection accuracy and overall recognition efficiency of multielement road traffic are improved. However, the overall detection accuracy is reduced if there is a parallel and dense phenomenon in road traffic elements, such as roadside parking spaces. Because the anchor boxes of the parallel and dense object detection overlap and block each other, part of the overlap is removed in the nonmaximal suppression process, leading to the phenomenon of missed detection. In previous studies [1], the CIoU loss function [32] was used to solve this problem. The CIoU loss function calculation considered the location information between the object box and the detection box, such as the aspect ratio of the distance overlap area. When there is no overlap between the object box and the detection box, backpropagation can still be carried out. However, in the process of prediction box regression, if the prediction box and the true box increase or decrease proportionally, the penalty term [32] defined by the CIoU loss function no longer takes effect. Therefore, the EIoU loss function [33] was adopted in this study to solve this problem. Compared with the CIoU loss function, EIoU [33] directly calculates penalty terms for the width and height, avoiding the problem that the penalty terms do not work when the prediction box and true box increase or decrease proportionally. At the same time, the regression accuracy is improved using the EIoU so that more attention is given to high-quality anchor boxes in the regression process.

The EIoU loss consists of three components, which are the overlap loss of the predicted and true boxes, the center distance loss of the predicted and true boxes, and the edge length loss of the height and width of the predicted and true boxes. Therefore, the EIoU loss function is defined as:

$$EIoU_{Loss} = 1 - IoU + \frac{\rho^2(b, b^{gt})}{c^2} + \frac{\rho^2(w, w^{gt})}{C_w^2} + \frac{\rho^2(h, h^{gt})}{C_h^2} \tag{2}$$

where IoU [34] is the intersection ratio of the area of the prediction box and the real box. $\rho^2(b, b^{gt})$ denotes the Euclidean distance between the prediction box and the center point of the true box. $\rho^2(w, w^{gt})$ denotes the Euclidean distance between the predicted box and the true box width. $\rho^2(h, h^{gt})$ denotes the Euclidean distance between the predicted box and the true box height. c denotes the diagonal distance between the prediction box and the minimum outer rectangle of the true box. C_w denotes the closure width between the prediction box and the true box. C_h denotes the closure height between the prediction box and the true box.

4. Experimental Results and Analysis

4.1. Experimental Environment

The computer hardware configuration used in this study is an Ubuntu 20.04 system with an Intel i7-8700 CPU and a GTX1070 graphics card configuration with 8 GB video memory. The training was performed using PyCharm software version 2020.1 (downloadable from https://www.jetbrains.com/pycharm/download/ (accessed on 1 November 2021), Prague, Czech Republic). In the experimental training, the weight decay coefficient of training was set to 0.0005, the initial learning rate was 0.001, the confidence level was 0.5, the IoU threshold was set to 0.65, and a total of 100 epochs were trained with 4000 iterations.

4.2. Evaluation Indicators

The experiments measured the accuracy of model detection by calculating the mAP, average precision (AP), precision, and recall as model quantitative evaluation metrics, which were defined as shown in Table 2.

Table 2. Evaluation indicators.

Evaluation Indicators	Calculation Formula	Definition
Precision	$Precision = \frac{TP}{TP+FP}$	*TP* represents the positive samples detected correctly, indicating the number of road traffic elements detected correctly. *FP* represents the negative samples detected incorrectly, indicating the number of objects detected as road traffic element classes but actually other classes. *FN* represents the positive samples detected incorrectly, indicating the number of other classes detected as actually road traffic element classes.
Recall	$Recall = \frac{TP}{TP+FN}$	
AP	$AP - \int_0^1 p(r)dr$	The value of AP is the size of the area enclosed by p as a function of r in the range $[0, 1]$, where p is the precision and r is the recall.
mAP	$mAP = \frac{\sum_{i=1}^N AP_i}{N}$	N represents the number of all categories in the test set, i is the ith category, and AP_i is the average precision rate of the ith category.

4.3. Comparison Experiments

To verify the effectiveness of the proposed method, the classical algorithm network of object detection was compared in this study. The SSD [35], Retinanet [36], Faster R-CNN [37], YOLOv3 [38], YOLOv4 [39], YOLOv5 [24] networks as well as the previous research were selected for comparison experiments with the proposed method. In the experiment, the values of the AP, precision, recall, and mAP evaluation indexes were calculated and compared when the multielement datasets of road traffic were trained. As shown in Table 3, the recognition accuracy of multielement road traffic under different network models was counted. Among them, the increased mAP values were calculated by comparing each network and the proposed method.

Table 3. Detection results of different networks.

Network Model	Transport Elements	AP/%	Precision/%	Recall/%	mAP/%	Rise Points
Faster R-CNN	zebra crossings	64.3	59.0	71.4	56.9	36.2
	bus stations	71.5	73.3	71.0		
	roadside parking spaces	35.0	31.5	48.8		
Retinanet	zebra crossings	70.0	87.8	61.2	57.3	35.8
	bus stations	67.3	86.4	61.3		
	roadside parking spaces	34.5	74.3	24.4		
SSD	zebra crossings	52.9	76.8	32.6	53.9	39.2
	bus stations	75.1	**100.0**	54.8		
	roadside parking spaces	33.8	73.4	14.7		
YOLOv3	zebra crossings	84.3	87.4	80.4	81.5	11.6
	bus stations	83.8	88.9	77.4		
	roadside parking spaces	76.5	76.1	75.9		
YOLOv4	zebra crossings	81.8	90.0	79.2	74.7	18.4
	bus stations	76.8	90.5	61.3		
	roadside parking spaces	65.4	70.2	71.0		
YOLOv5	zebra crossings	85.3	75.2	82.0	73.9	19.2
	bus stations	65.2	50.9	58.1		
	roadside parking spaces	78.9	74.2	77.5		
YOLOv4 + ECA	zebra crossings	**94.3**	**90.1**	93.9	90.5	2.6
	bus stations	**99.6**	91.3	**100.0**		
	roadside parking spaces	77.4	81.0	78.1		
ASFF-YOLOv5 (YOLOv5 + ASFF + RFB)	zebra crossings	94.0	85.9	94.4	93.1	
	bus stations	96.2	93.1	96.3		
	roadside parking spaces	**83.7**	**86.9**	**88.6**		

Note: the best results are in bold font and the second best results are underlined.

To verify the accuracy of anchor positioning obtained by K-means++ clustering in this study, the experimental results of the original YOLOv5 network, YOLOv5 + K-means and YOLOv5 + K-means++ were compared. Table 4 shows the detection results of different K-means.

Table 4. Detection results of different K-means.

Network Model	Transport Elements	AP/%	Precision/%	Recall/%	mAP/%	Rise Points
YOLOv5	zebra crossings	85.3	75.2	82.0	73.9	18.3
	bus stations	65.2	50.9	58.1		
	roadside parking spaces	78.9	74.2	77.5		
YOLOv5 + K-means	zebra crossings	93.4	84.0	92.5	89.2	3
	bus stations	95.3	74.0	83.4		
	roadside parking spaces	84.7	84.9	88.7		
YOLOv5 + K-means++	zebra crossings	**94.6**	**84.8**	**93.8**	**92.2**	
	bus stations	**95.5**	**78.2**	**93.9**		
	roadside parking spaces	**86.0**	**86.9**	**88.9**		

Note: the best results are in bold font and the second best results are underlined.

To verify the accuracy of the K-means++ clustering effect on anchor boxes, images were selected for detection, and the detection effect under different K-means is shown in Figure 8. From the viewpoint of detection accuracy and the size and location of the anchor box, YOLOv5 had the lowest detection accuracy for the detectors, with an accuracy of only 70.6%. The location of the candidate box does not include all the detectors to be detected. The detection accuracy of YOLOv5 + K-means for detected objects ranked second, with an accuracy of 87.9%. Although the positioning accuracy of the candidate boxes was improved, there was still a slight gap. YOLOv5 + K-means++ had the highest detection accuracy of 95.7% for detected objects, the positioning of candidate boxes was the most accurate, there was no missing part, and the size of candidate boxes was appropriate. The results showed that the K-means++ clustering algorithm used could make the position of candidate boxes more accurate and the detection effect better.

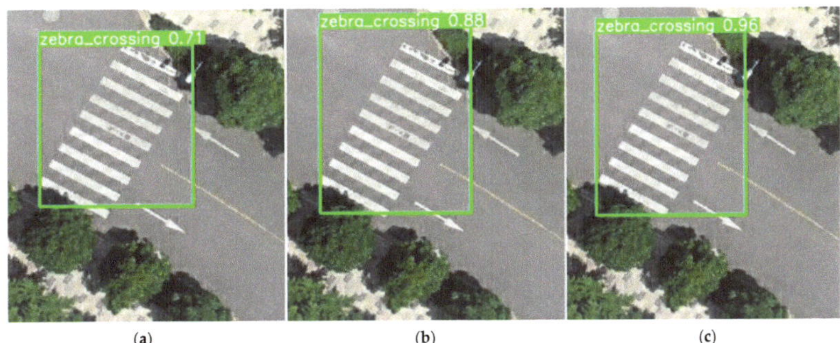

Figure 8. Detection effects under different K-means: (**a**) YOLOv5; (**b**)YOLOv5 + K-means; and (**c**) YOLOv5 + K-means++.

To verify the effectiveness of the proposed algorithm, an ablation experiment was conducted on the multielement road traffic datasets, and the experimental results after fusion of K-means++, SPPF, and ASFF were compared. The AP, precision, recall, and mAP values were calculated and compared. The ablation test results are shown in Table 5.

Table 5. Results of ablation tests.

Network Model	Transport Elements	AP/%	Precision/%	Recall/%	mAP/%	Rise Points
	zebra crossings	85.3	75.2	82.0		
YOLOv5	bus stations	65.2	50.9	58.1	73.9	19.2
	roadside parking spaces	78.9	74.2	77.5		
	zebra crossings	**94.9**	90.3	94.2		
YOLOv5 + SPPF	bus stations	**100.0**	83	88.8	90.7	2.4
	roadside parking spaces	<u>85.6</u>	86.2	**89.1**		
	zebra crossings	<u>94.6</u>	84.8	93.8		
YOLOv5 + K-means++	bus stations	95.5	78.2	93.9	92.2	0.9
	roadside parking spaces	**86.0**	86.9	88.9		
	zebra crossings	91.3	**92.1**	94.1		
YOLOv5 + ASFF	bus stations	**100.0**	91.6	<u>95.1</u>	92.8	0.3
	roadside parking spaces	77.1	**91.4**	<u>89.0</u>		
	zebra crossings	94.0	85.9	**94.4**		
ASFF-YOLOv5 (YOLOv5 + ASFF + RFB)	bus stations	<u>96.2</u>	**93.1**	**96.3**	**93.1**	
	roadside parking spaces	83.7	<u>86.9</u>	88.6		

Note: the best results are in bold font and the second best results are underlined.

To verify the practicality and effectiveness of the proposed method, prediction experiments were conducted in different scenes separately. Small scene image maps taken by an UAV and large scene image maps generated by commercial software processing were selected for detection. The small scene images were normal road scenes and small object element scenes. Among them, the normal road scenes were intersection sections imaged by the UAV, including 1 bus station and 10 zebra crossings. The small object element scenes were roadside parking spaces captured by the UAV, divided into scene 1 and scene 2. Scene 1 was an unobstructed roadside parking space, including nine juxtaposed and dense roadside parking spaces. Scene 2 was an obscured object to be detected, including 45 roadside parking spaces and 2 zebra crossings. The parking spaces were juxtaposed, dense, and partially obscured by trees. The prediction effect under the normal road scene is shown in Figure 9. Figure 10 shows scene 1 under small target elements. Figure 11 shows scene 2 under small target elements. Table 6 shows the detection results under small scenes.

The prediction effect can be seen from the results shown in Figure 9 and Table 6. In the normal road scenario, the ablation experiments were able to correctly detect the multielement road traffic with no detection error or missed detections. The ablation experiments all had good ability to detect multiple elements of road traffic. However, the detection accuracy obtained by the proposed method was the highest.

Figure 10 shows the unobstructed detection in the small target element scenario, where the object detection was nine side-by-side and dense roadside parking spaces. Figure 10 and Table 6 show that the ablation experiments detected all roadside parking spaces in the small object scenario without obstruction and there were no detection errors or missed detections. Ablation experiments have a good ability to detect multiple elements of road traffic in unobstructed environments. However, the detection accuracy of the method in this paper reached 91.2% when detecting small objects without occlusion, which is the highest compared with other methods.

Figure 11 shows the detection with occlusion in the small object element scene, and the detection objects were 45 parallel and dense roadside parking spaces and 2 zebra crossings. Figure 11 and Table 6 show that the ablation experiments achieved 99.5% and 100% detection of zebra crossings, respectively. It is proven that the above ablation experiments could achieve good results for zebra crossing detection and that there were no detection errors or missed detections. However, the above ablation experiments were missed in the detection of street parking spaces, and the number of missed street parking spaces ranged from 3 to 11. From the overall results, although the YOLOv5 + SPPF method resulted in the highest average detection accuracy, the number of missed spots was the most serious at 11. Compared with the serious missed detections in the ablation experiment, the proposed method only missed three detections, which was the minimum number of missed detections in the ablation experiment. The detection accuracy of the proposed method was 93.5% for obscured roadside parking spaces, which is only different by 0.28 from the optimal detection accuracy, but the missed detections were greatly reduced. This proves that the proposed method ensured the correctness of detection while maintaining the detection accuracy, especially in the detection of obscured small objects and juxtaposed dense small objects with high improvement.

Figure 12 shows the prediction result map under an orthophoto image. The image is an orthophoto map generated by a remote sensing image map taken by UAV and processed using commercial software, and the total area of this image area is 302,813 m^2. The image size is 37,548 × 21,438 pixels. The small and medium objects in this image are 140 × 82 pixels and 514 × 254 pixels, respectively. The image contains 18 zebra crossings, 5 bus stations, and 58 roadside parking spaces. The specific prediction results are shown in Table 7.

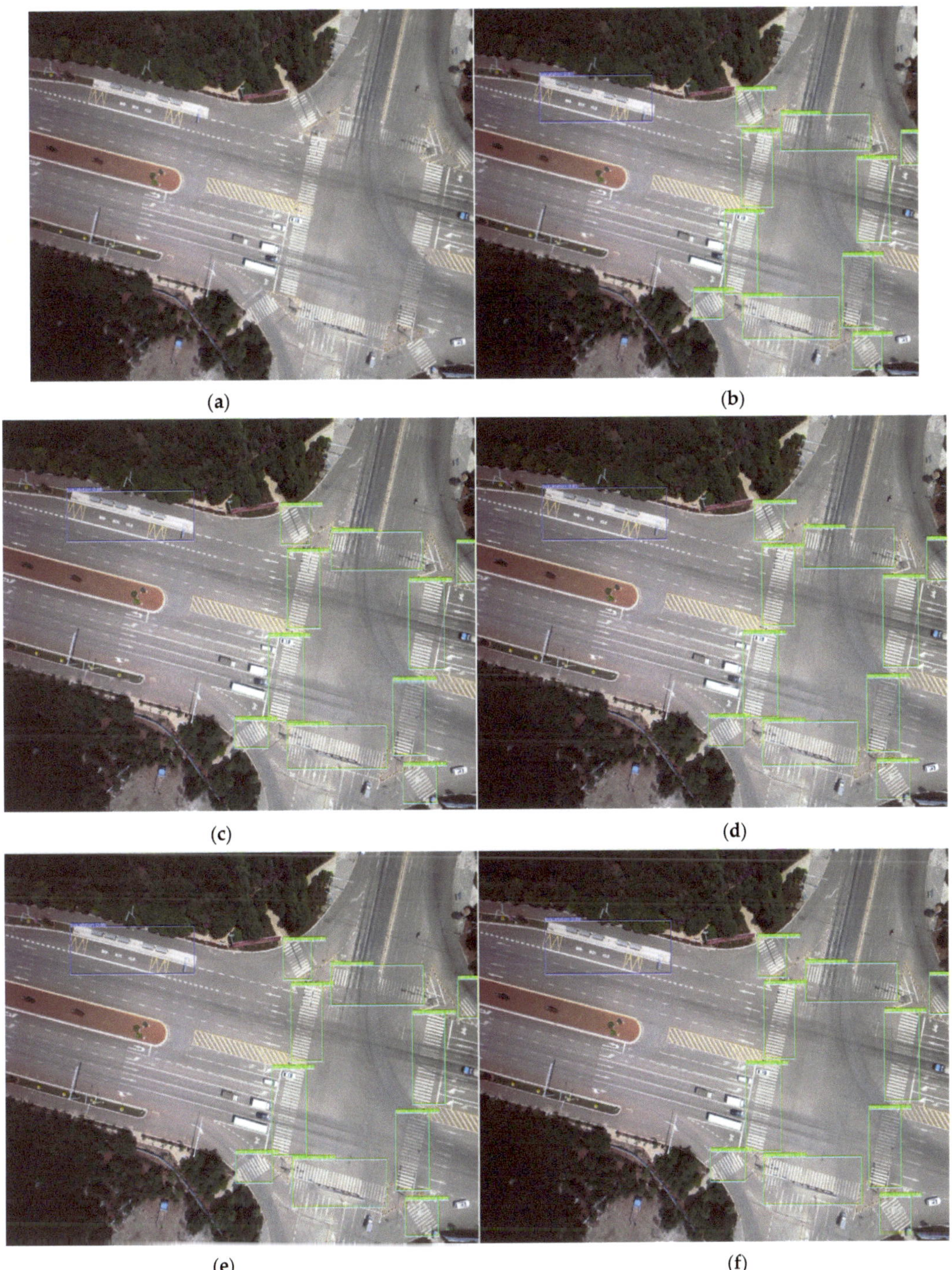

(a)

(b)

(c)

(d)

(e)

(f)

Figure 9. Detection results in normal road scenarios: (**a**) detection of the original image; (**b**)YOLOv5; (**c**) YOLOv5 with SPPF; (**d**) YOLOv5 with K-means++; (**e**) YOLOv5 with ASFF; and (**f**) ASFF-YOLOv5.

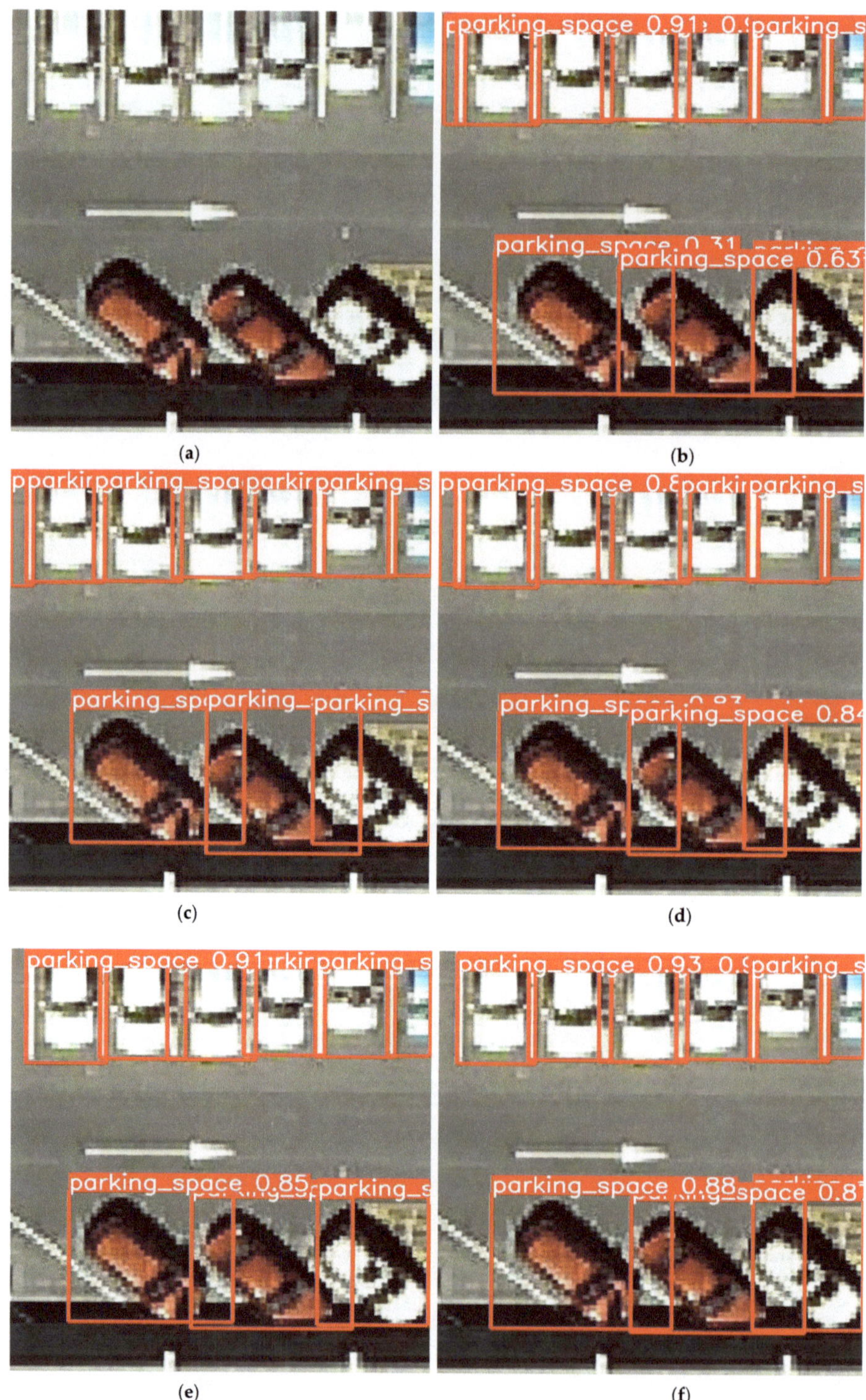

Figure 10. Unobstructed detection in small object scenes: (**a**) detection of the original image; (**b**)YOLOv5; (**c**) YOLOv5 with SPPF; (**d**) YOLOv5 with K-means++; (**e**) YOLOv5 with ASFF; and (**f**) ASFF-YOLOv5.

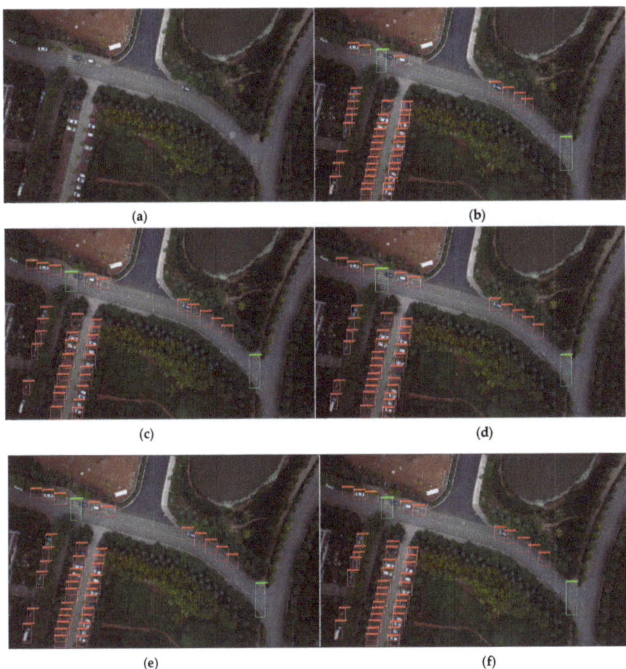

Figure 11. Effect of detection with occlusion in small target scenes: (**a**) detection of the original image; (**b**) YOLOv5; (**c**) YOLOv5 with SPPF; (**d**) YOLOv5 with K-means++; (**e**) YOLOv5 with ASFF; and (**f**) ASFF-YOLOv5.

Table 6. Predicted results of ablation experiments in a small scenario.

Network Model	Transport Elements	Normal Road Scene		Small Target Scenes Scene 1 (Unobstructed)		Small Target Scenes Scene 2 (Obstructed)	
		Number	AP/%	Number	AP/%	Number	AP/%
YOLOv5	zebra crossings	10	87.0	-	-	2	87
	bus stations	1	81.7	-	-	-	-
	roadside parking spaces	-	-	9	72.9	42	81.6
YOLOv5 + SPPF	zebra crossings	10	85.3	-	-	2	99.5
	bus stations	1	87.8	-	-	-	-
	roadside parking spaces	-	-	9	76.6	34	**96.3**
YOLOv5 + K-means++	zebra crossings	10	87.1	-	-	2	**100**
	bus stations	1	84.8	-	-	-	-
	roadside parking spaces	-	-	9	82.9	38	93.3
YOLOv5 + ASFF	zebra crossings	10	86.7	-	-	2	**100**
	bus stations	1	85.7	-	-	-	-
	roadside parking spaces	-	-	9	89.7	40	93.3
ASFF-YOLOv5 (YOLOv5 + ASFF + RFB)	zebra crossings	10	**88.4**	-	-	2	99.5
	bus stations	1	**89.1**	-	-	-	-
	roadside parking spaces	-	-	9	**91.2**	42	93.5

Note: "-" in the table indicates that the images measured do not contain this category, and the bolded font is the best result for each. Number is the number of road traffic elements that were correctly detected.

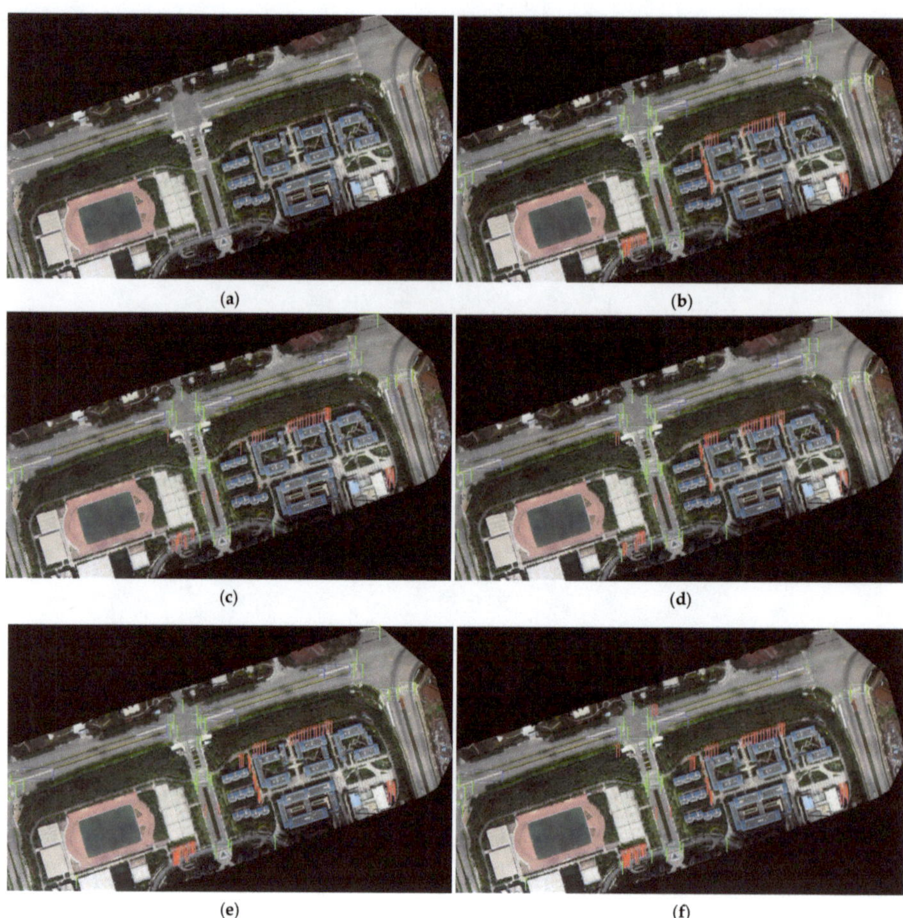

Figure 12. Predicted results of the ablation experiment for a large and complex scenario: (**a**) detection of the original image; (**b**) YOLOv5; (**c**) YOLOv5 with SPPF; (**d**) YOLOv5 with K-means++; (**e**) YOLOv5 with ASFF; and (**f**) ASFF-YOLOv5.

Figure 12 and Table 7 show that several of the above algorithms showed different degrees of missed detections when applied to complex large scenes. However, the proposed method had the lowest number of missed detections. When detecting bus stations and zebra crossings, the missed detections in the ablation experiment were not obvious, and all methods could correctly detect the objects. However, in the detection of roadside parking spaces, the missed detections were serious, and the number of missed detections ranged from 8 to 29. The reason for this is that for large images, roadside parking is a small target detection and most of the parking spaces are covered by greenery, so the feature information is not obvious, thus causing missed detections. The proposed method is optimized for multiscale feature extraction, which improved the detection accuracy of small objects to a certain extent. The proposed method achieved an average detection accuracy of 80.3% for roadside parking spaces in complex and large scenes. Compared with the ablation experiments, the number of missed detections was the lowest, and only eight roadside parking spaces obscured by greenery were missed. The average detection accuracy for zebra crossings and bus stations in complex large scenes reached 89.5% and 89.7%, respectively, which were the best results of the ablation experiment.

Table 7. Predicted results of ablation experiments for a large and complex scenario.

Network Model	Transport Elements	Number 1	Number 2	AP/%
YOLOv5	zebra crossings	18	0	87.2
	bus stations	4	1	75.5
	roadside parking spaces	40	18	62.3
YOLOv5 + SPPF	zebra crossings	17	1	89.1
	bus stations	5	0	80.3
	roadside parking spaces	29	29	68.7
YOLOv5 + K-means++	zebra crossings	18	0	88.4
	bus stations	4	1	81.7
	roadside parking spaces	37	21	73.3
YOLOv5 + ASFF	zebra crossings	18	0	88.6
	bus stations	4	1	84.0
	roadside parking spaces	44	14	71.8
ASFF-YOLOv5 (YOLOv5 + ASFF + RFB)	zebra crossings	18	0	**89.5**
	bus stations	5	0	**89.7**
	roadside parking spaces	50	8	**80.3**

Note: The bolded font is the best result for each. Number 1 is the number of road traffic elements that were correctly detected. Number 2 is the number of road traffic elements that were missed.

5. Discussion

Compared with other methods, the mAP of the proposed method had a great improvement, with an increase ranging from 0.3% to 39.2%, which verified that the proposed ASFF-YOLOV5 algorithm can effectively improve the detection accuracy of multielement road traffic. Different from other scholars' research [40–45], this paper considered multiscale object detection, especially in small object detection whose detection accuracy was substantially improved. Compared with the results of different network models, the detection accuracy of the proposed method was improved both in terms of the overall accuracy and individual objects for detection, which proves the superiority of the proposed method for multiscale object detection.

To verify the practicability and correctness of the proposed method, ablation experiments were conducted for different scenarios. The prediction results showed that the proposed method achieved the optimal detection accuracy compared with other methods. At the same time, the proposed method had the lowest number of missed detections, the highest detection accuracy, and the best detection effect compared with other methods for large complex scenes. This proves the practicality and superiority of the proposed method. It can be directly applied to image maps of large scenes and provides a more intelligent and convenient method for updating geographic information databases. However, in complex large scenes, there is inevitably the case of missed roadside parking space detections with occlusion since the semantic information of these parking spaces is more inconspicuous and the features are difficult to extract, which is also one of the challenges to be further addressed in subsequent work.

6. Conclusions

To address the problems of little data extraction of traffic element information, low automation but high demand, small element scale, and detection susceptibility to environmental interference, an ASFF-YOLOv5 algorithm for UAV remote sensing images was proposed in this paper. In this algorithm, an adaptive spatial feature fusion method based on a receptive field module is adopted to make full use of different scale information, improve the invariance of the feature scale, and improve the detection effect of small objects.

When detecting multiple road traffic elements, the mAP of the proposed method reached 93.1%, which is 19.2% higher than that of the original YOLOv5 network. A comparison experiment and ablation experiment proved that the proposed method could solve the problem of detection errors and missed detections for multielement road traffic, improve the detection accuracy of small and medium objects and intensive objects of multifactor road elements, and provide a new solution for the construction of basic geographic traffic information databases.

However, there are shortcomings of the proposed method. The experiments only focused on three types of elements, zebra crossings, bus stations, and roadside parking spaces. Subsequent work will expand the dataset to complete the automatic recognition and detection of more elements. At the same time, there are also missed detections of obscured small objects, and subsequent work will further improve and enhance the detection accuracy of small objects.

Author Contributions: M.Q.: algorithm proposal and testing, data processing, manuscript writing, and research conceptualization. L.H.: funding acquisition, directing, and manuscript review. B.-H.T.: project administration and supervision. All authors have read and agreed to the published version of the manuscript.

Funding: This research was funded by National Natural Science Foundation of China (41961039), Yunnan Fundamental Research Projects (grant NO. 202201AT070164 and 202101AT070102).

Data Availability Statement: The original data have not been made publicly available, but it can be used for scientific research. Other researchers can send emails to the first author if needed.

Conflicts of Interest: The authors declare no conflict of interest.

References

1. Huang, L.; Qiu, M.L.; Xu, A.Z.; Sun, Y.; Zhu, J.J. UAV imagery for automatic multi-element recognition and detection of road traffic elements. *Aerospace* **2022**, *9*, 198. [CrossRef]
2. Dewi, C.; Chen, R.C.; Liu, Y.T.; Jiang, X.Y.; Hartomo, K.D. YOLO v4 for advanced traffic sign recognition with synthetic training data generated by various GAN. *IEEE Access* **2021**, *9*, 97228–97242. [CrossRef]
3. Haque, W.A.; Arefin, S.; Shihavuddin, A.S.M.; Hasan, M.A. DeepThin: A novel lightweight CNN architecture for traffic sign recognition without GPU requirements. *Expert Syst. Appl.* **2021**, *168*, 114481. [CrossRef]
4. Zhou, K.; Zhan, Y.F.; Fu, D.M. Learning region-based attention network for traffic sign recognition. *Sensors* **2021**, *21*, 686. [CrossRef]
5. Cui, Y.L.; Yu, Y.; Cai, Z.Y.; Wang, D.H. Optimizing road network density considering automobile traffic efficiency: Theoretical approach. *J. Urban Plan. Dev.* **2022**, *148*, 04021062. [CrossRef]
6. Zhou, G.D.A.; Chen, W.T.; Gui, Q.S.; Li, X.J.; Wang, L.Z. Split depth-wise separable graph-convolution network for road extraction in complex environments from high-resolution remote-sensing images. *IEEE Trans. Geosci. Remote Sens.* **2022**, *60*, 1–15. [CrossRef]
7. Liu, C.F.; Yin, H.; Sun, Y.X.; Wang, L.; Guo, X.D. A grade identification method of critical node in urban road network based on multi-attribute evaluation correction. *Appl. Sci.* **2022**, *12*, 813. [CrossRef]
8. Dai, J.G.; Ma, R.C.; Ai, H.B. Semi-automatic extraction of rural roads from high-resolution remote sensing images based on a multifeature combination. *IEEE Geosci. Remote Sens. Lett.* **2020**, *19*, 1–5. [CrossRef]
9. Hu, C.H.; Fan, W.C.; Zeng, E.L.; Hang, Z.; Wang, F.; Qi, L.Y.; Bhuiyan, M.Z.A. Digital twin-assisted real-time traffic data prediction method for 5G-enabled internet of vehicles. *IEEE Trans. Ind. Inform.* **2021**, *18*, 2811–2819. [CrossRef]
10. Li, Z.S.; Xiong, G.; Tian, Y.L.; Lv, Y.S.; Chen, Y.Y.; Hui, P.; Su, X. A multi-stream feature fusion approach for traffic prediction. *IEEE Trans. Intell. Transp. Syst.* **2022**, *23*, 1456–1466. [CrossRef]
11. Lin, T.Y.; Maire, M.; Belongie, S.; Hays, J.; Perona, P.; Ramanan, D.; Dollar, P.; Zitnick, C.L. Microsoft COCO: Common objects in context. In Proceedings of the 13th European Conference on Computer Vision, Zurich, Switzerland, 6–12 September 2014.
12. Yu, R.N.; Li, H.G.; Jiang, Y.L.; Zhang, B.C.; Wang, Y.F. Tiny vehicle detection for mid-to-high altitude UAV images based on visual attention and spatial-temporal information. *Sensors* **2022**, *22*, 2354. [CrossRef]
13. Xiao, R.; Wang, Y.Z.; Tao, C. Fine-grained road scene understanding from aerial images based on semisupervised semantic segmentation networks. *IEEE Geosci. Remote Sens. Lett.* **2022**, *19*, 1–5.
14. Wang, H.F.; Wang, J.Z.; Bai, K.M.; Sun, Y. Centered multi-task generative adversarial network for small object detection. *Sensors* **2021**, *21*, 5194. [CrossRef]
15. Xu, D.Q.; Wu, Y.Q. FE-YOLO: A feature enhancement network for remote sensing target detection. *Remote Sens.* **2021**, *13*, 1311. [CrossRef]

16. Qing, Y.H.; Liu, W.Y.; Feng, L.Y.; Gao, W.J. Improved Yolo network for free-angle remote sensing target detection. *Remote Sens.* **2021**, *13*, 2171. [CrossRef]
17. Hu, J.M.; Zhi, X.Y.; Shi, T.J.; Zhang, W.; Cui, Y.; Zhao, S.G. PAG-YOLO: A portable attention-guided YOLO network for small ship detection. *Remote Sens.* **2021**, *13*, 3059. [CrossRef]
18. Kim, M.; Jeong, J.; Kim, S. ECAP-YOLO: Efficient Channel Attention Pyramid YOLO for Small Object Detection in Aerial Image. *Remote Sens.* **2021**, *13*, 4851. [CrossRef]
19. Liu, M.J.; Wang, X.H.; Zhou, A.J.; Fu, X.Y.; Ma, Y.W.; Piao, C.H. Uav-yolo: Small object detection on unmanned aerial vehicle perspective. *Sensors* **2020**, *20*, 2238. [CrossRef]
20. Lee, S.; Kim, S.; Moon, S. Development of a car-free street mapping model using an integrated system with unmanned aerial vehicles, aerial mapping cameras, and a deep learning algorithm. *J. Comput. Civil. Eng.* **2022**, *36*, 04022003. [CrossRef]
21. Hong, Z.H.; Yang, F.; Pan, H.Y.; Zhou, R.Y.; Zhang, Y.; Han, Y.L.; Wang, J.; Yang, S.H.; Chen, P.; Tong, X.H.; et al. Highway crack segmentation from unmanned aerial vehicle images using deep learning. *IEEE Geosci. Remote Sens. Lett.* **2021**, *19*, 1–5. [CrossRef]
22. Sultonov, F.; Park, J.H.; Yun, S.; Lim, D.W.; Kang, J.M. Mixer U-Net: An improved automatic road extraction from UAV imagery. *Appl. Sci.* **2022**, *12*, 1953. [CrossRef]
23. Yoder, J.; Carey, E. Semi-supervised k-means++. *J. Stat. Comput. Simul.* **2017**, *87*, 2597–2608. [CrossRef]
24. Ultralytics. YOLOv5. 2020. Available online: https://github.com/ultralytics/yolov5 (accessed on 1 November 2021).
25. Liu, S.T.; Huang, D.; Wang, Y.H. Receptive field block net for accurate and fast object detection. In Proceedings of the 15th European Conference on Computer Vision, Munich, Germany, 8–14 September 2018.
26. Liu, S.T.; Di, H.; Wang, Y.H. Learning spatial fusion for single-shot object detection. In Proceedings of the 15th European Conference on Computer Vision, Munich, Germany, 8–14 September 2018.
27. Pelleg, D.; Moore, A. X-means: Extending k-means with efficient estimation of the number of clusters. In Proceedings of the Seventeenth International Conference on Machine Learning, San Francisco, CA, USA, 29 June–2 July 2000.
28. Liu, S.; Qi, L.; Qin, H.F.; Shi, J.P.; Jia, J.Y. Path aggregation network for instance segmentation. In Proceedings of the 31st IEEE Conference on Computer Vision and Pattern Recognition, Salt Lake City, UT, USA, 18–23 June 2018.
29. Lin, T.Y.; Dollar, P.; Girshick, R.; He, K.M.; Hariharan, B.; Belongie, S. Feature pyramid networks for object detection. In Proceedings of the 30th IEEE/CVF Conference on Computer Vision and Pattern Recognition, Honolulu, HI, USA, 21–26 July 2017.
30. Szegedy, C.; Ioffe, S.; Vanhoucke, V.; Alemi, A.A. Inception-v4, inception-resnet and the impact of residual connections on learning. In Proceedings of the 31st AAAI Conference on Artificial Intelligence, San Francisco, CA, USA, 4–9 February 2017.
31. He, K.M.; Zhang, X.Y.; Ren, S.Q.; Sun, J. Spatial pyramid pooling in deep convolutional networks for visual recognition. In Proceedings of the 13th European Conference on Computer Vision, Zurich, Switzerland, 6–12 September 2014.
32. Zheng, Z.H.; Wang, P.; Liu, W.; Li, J.Z.; Ye, R.G.; Ren, D.W. Distance-IoU loss: Faster and better learning for bounding box regression. In Proceedings of the 34th AAAI Conference on Artificial Intelligence, New York, NY, USA, 7–12 February 2020.
33. Zhang, Y.F.; Ren, W.Q.; Zhang, Z.; Jia, Z.; Wang, L.; Tan, T.N. Focal and efficient IoU loss for accurate bounding box regression. *arXiv* **2021**, arXiv:2101.08158.
34. Jiang, B.R.; Luo, R.X.; Mao, J.Y.; Xiao, T.T.; Jiang, Y.N. Acquisition of localization confidence for accurate object detection. In Proceedings of the 15th European Conference on Computer Vision, Munich, Germany, 8–14 September 2018.
35. Liu, W.; Anguelov, D.; Erhan, D.; Szegedy, C.; Reed, S.; Fu, C.Y.; Berg, A.C. SSD: Single shot multibox detector. In Proceedings of the 14th European Conference on Computer Vision, Amsterdam, The Netherlands, 8–16 October 2016.
36. Lin, T.Y.; Goyal, P.; Girshick, R.; He, K.M.; Dollar, P. Focal loss for dense object detection. *IEEE Trans. Pattern Anal. Mach. Intell.* **2020**, *42*, 318–327. [CrossRef]
37. Ren, S.Q.; He, K.M.; Girshick, R.; Sun, J. Faster R-CNN: Towards real-time object detection with region proposal networks. *IEEE Trans. Pattern Anal. Mach. Intell.* **2017**, *39*, 1137–1149. [CrossRef]
38. Redmon, J.; Farhadi, A. YOLOv3: An Incremental Improvement. Available online: https://arxiv.org/abs/1804.02767 (accessed on 8 April 2018).
39. Bochkovskiy, A.; Wang, C.Y.; Liao, H.Y.M. YOLOv4: Optimal Speed and Accuracy of Object Detection. Available online: https://arxiv.org/abs/2004.10934 (accessed on 23 April 2020).
40. Tian, Y.; Gelernter, J.; Wang, X.; Li, J.Y.; Yu, Y.Z. Traffic sign detection using a multi-scale recurrent attention network. *IEEE Trans. Intell. Transp. Syst.* **2020**, *20*, 4466–4475. [CrossRef]
41. Chen, P.D.; Huang, L.; Xia, Y.; Yu, X.N.; Gao, X.X. Detection and recognition of road traffic signs in UAV images based on Mask R-CNN. *Remote Sens. Land Resour.* **2020**, *32*, 61–67.
42. Lin, C.J.; Jhang, J.Y. Intelligent traffic-monitoring system based on YOLO and convolutional fuzzy neural networks. *IEEE Access* **2022**, *10*, 14120–14133. [CrossRef]
43. Wang, M.Y.; Liu, R.F.; Yang, J.B.; Lu, X.S.; Yu, J.Y.; Ren, H.W. Traffic Sign Three Dimensional Reconstruction Based on Point Clouds and Panoramic Images. *Photogramm. Rec.* **2022**, *37*, 87–110. [CrossRef]
44. Liang, T.J.; Bao, H.; Pan, W.G.; Pan, F. Traffic sign detection via improved sparse R-CNN for autonomous vehicles. *J. Adv. Transp.* **2022**, *2022*, 3825532. [CrossRef]
45. Mishra, J.; Goyal, S. An effective automatic traffic sign classification and recognition deep convolutional networks. *Multimed. Tools Appl.* **2022**, *81*, 18915–18934. [CrossRef]

remote sensing

Article

A Recurrent Adaptive Network: Balanced Learning for Road Crack Segmentation with High-Resolution Images

Yi Zhang [1,2], Junfu Fan [1,3,*], Mengzhen Zhang [1], Zongwen Shi [1], Rufei Liu [4] and Bing Guo [1]

[1] School of Civil and Architectural Engineering, Shandong University of Technology, Zibo 255000, China; zhangyi_csu@csu.edu.cn (Y.Z.); zhangmz@lreis.ac.cn (M.Z.); 21407010766@stumail.sdut.edu.cn (Z.S.); guobing@sdut.edu.cn (B.G.)
[2] School of Geosciences and Info-Physics, Central South University, Changsha 410083, China
[3] State Key Laboratory of Resources and Environmental Information System, Chinese Academy of Sciences, Beijing 100101, China
[4] College of Geodesy and Geomatics, Shandong University of Science and Technology, Qingdao 266590, China; liurufei@sdust.edu.cn
* Correspondence: fanjf@sdut.edu.cn

Citation: Zhang, Y.; Fan, J.; Zhang, M.; Shi, Z.; Liu, R.; Guo, B. A Recurrent Adaptive Network: Balanced Learning for Road Crack Segmentation with High-Resolution Images. *Remote Sens.* **2022**, *14*, 3275. https://doi.org/10.3390/rs14143275

Academic Editors: Valerio Baiocchi, Alessandro Mei and Xianfeng Zhang

Received: 25 May 2022
Accepted: 5 July 2022
Published: 7 July 2022

Publisher's Note: MDPI stays neutral with regard to jurisdictional claims in published maps and institutional affiliations.

Abstract: Road crack segmentation based on high-resolution images is an important task in road service maintenance. The undamaged road surface area is much larger than the damaged area on a highway. This imbalanced situation yields poor road crack segmentation performance for convolutional neural networks. In this paper, we first evaluate the mainstream convolutional neural network structure in the road crack segmentation task. Second, inspired by the second law of thermodynamics, an improved method called a recurrent adaptive network for a pixelwise road crack segmentation task is proposed to solve the extreme imbalance between positive and negative samples. We achieved a flow between precision and recall, similar to the conduction of temperature repetition. During the training process, the recurrent adaptive network (1) dynamically evaluates the degree of imbalance, (2) determines the positive and negative sampling rates, and (3) adjusts the loss weights of positive and negative features. By following these steps, we established a channel between precision and recall and kept them balanced as they flow to each other. A dataset of high-resolution road crack images with annotations (named HRRC) was built from a real road inspection scene. The images in HRRC were collected on a mobile vehicle measurement platform by high-resolution industrial cameras and were carefully labeled at the pixel level. Therefore, this dataset has sufficient data complexity to objectively evaluate the real performance of convolutional neural networks in highway patrol scenes. Our main contribution is a new method of solving the data imbalance problem, and the method of guiding model training by analyzing precision and recall is experimentally demonstrated to be effective. The recurrent adaptive network achieves state-of-the-art performance on this dataset.

Keywords: road crack segmentation; convolutional neural networks; balanced deep learning; adaptive loss

1. Introduction

Road crack segmentation has important application value in road maintenance, especially for highways. It is challenging for road management departments to quickly determine the large-scale technical conditions of asphalt road pavement because of the rapid increase in road network mileage [1]. Scholars proposed various traditional computer vision-based road crack segmentation methods before convolutional neural networks became mainstream. Traditional digital image processing (DIP) methods, such as the Canny edge detector [2], wavelet transform [3], and crack index [4], have been carried out for crack segmentation. However, these algorithms focus on a small group of pixels and lack an understanding of the image content. Notably, DIP-based methods are very sensitive to noise [5], such as shadows and stains, and are unable to meet the needs of large-scale road maintenance. In the past five years, methods for extracting road cracks via convolutional neural networks have gradually replaced traditional methods and have become mainstream approaches in scientific research. Deep learning methods can better preserve the geometry of detected crack patterns, and their predictions (in terms of the pixels belonging to a crack) have ultimately become more accurate than those of the threshold method [6]. Therefore, the DIP method is not discussed in the following sections.

With the development of deep learning technology, convolutional neural networks such as ResNet [7] and DeepLabV3 [8] have achieved great success in computer vision and have been naturally introduced to the road crack segmentation task. Due to their powerful ability to learn high-level feature representations, road crack segmentation approaches using supervised convolutional neural networks have achieved good results. However, most studies have only applied advanced computer vision methods to the field of road damage segmentation and have not deeply considered the differences between road cracks and natural images. Compared with the semantic segmentation task involving natural images, the road crack segmentation task has unique characteristics:

- Small targets: According to the latest "Highway Technical Condition Evaluation Standard JTG 5210-2018", cracks with average widths greater than 1 mm should be extracted. Many road crack instances are only two or three pixels wide, making them much smaller than natural images in datasets such as PASCAL Visual Object Class (VOC), and Microsoft Common Objects in Context (MS-COCO), even in very high-resolution images.
- Imbalance at the sample level: The positive and negative data are imbalanced at the sample level. For example, in the DeepCrack dataset [9], the ratio of positive and negative samples is approximately 1:33, and in the HRRC dataset, the ratio greatly increases to 1:705 at the pixel level. The randomness and sparsity of the positive class are even more apparent in the engineered HRRC dataset.
- Imbalance at the feature level: Driven from samples, imbalance also occurs at the feature level. Deep high-level features in backbones have more semantic meanings, while shallow low-level features are more descriptive in terms of content [10]. The gradient generated by a small number of positive features, especially at a deep level, may be submerged by the gradient generated by a large number of negative features, and the utilized model will have difficulty achieving enough positive characteristics.

Due to the problems above, extracting road cracks from real road inspection images is still a challenging task. A neural network method based on supervised learning cannot achieve good results in real road inspection scenarios, as shown in Table 1.

In the case of supervised learning, the small target and imbalance problems greatly reduce the ability of convolutional neural networks to perform road crack segmentation. We are in urgent need of a simple, universal, and effective method to solve such problems and meet the needs of engineering applications. Two questions need to be rethought: what network structure is best suited for small target recognition? How can the extreme imbalance problem be solved with a stable, universal, and adaptive solution?

To solve the first question, we tested the current dominant computer vision-based convolutional neural network architecture in real road inspection scenarios to determine a powerful baseline network. To solve the second problem, we follow a simple line of thought: a poor positive sample recognition ability will result in high precision and low recall. Can we obtain better results if we maintain a relative balance between precision and recall? The second law of thermodynamics states that heat can be spontaneously transferred from a hotter body to a cooler body but cannot be spontaneously transferred from a cooler body to a hotter body. Inspired by this, we put forward a main point of view: is it possible to establish a channel between precision and recall in which indicators flow dynamically like temperature? Therefore, we propose a precision–recall–flow (PRF) index to adaptively control the strength and direction of the flow and take measures at the sample level and feature level to drive the flow recurrently.

In this paper, our main contributions are as follows. (1) We review and analyze several kinds of convolutional neural networks. The experiments were performed to prove that the HRNet [11] architecture with rich high-resolution representation has advantages in road crack extraction tasks. By replacing batch normalization with group normalization (HRNet) and group normalization (HRNet-GN), the network performance was improved. (2) We propose a recurrent adaptive network for equilibrium learning. The network treats precision and recall as temperatures in different regions of the model and forces energy to flow from high- to low-temperature regions by automatically adjusting sampling and loss weights. The proposed method balances precision and recall and improves the ability of the model to handle unbalanced data. (3) A new high-resolution road crack database named HRRC is established. Its road images were continuously captured by high-resolution cameras during the driving process. The road crack areas in these images were carefully labeled by LabelMe [12] and were sliced into 24,704 images, each with dimensions of 512×512 pixels. The dataset has abundant image noise, such as shadows, stains, filling joints, and cars, and was used to verify the capability of the proposed method. This paper presents a new way of addressing data imbalance in deep learning tasks by analyzing precision versus recall and represents the first study that attempts to measure data imbalance in terms of the relationship between precision and recall rather than the data ratio.

The remainder of this paper is organized as follows. In Section 2, we introduce some existing convolutional neural network structures for segmentation and methods for solving the imbalance problem. Section 3 describes the proposed method. The experimental details and results are reported in Sections 4 and 5, our findings and future works are discussed in Section 6, and this paper is concluded in Section 7.

2. Related Work

Road crack segmentation methods can be divided into two categories: traditional digital image processing and machine learning-based approaches. In the past five years, machine learning-based crack segmentation methods have repeatedly used advanced state-of-the-art techniques [9,13] and have replaced traditional methods. In this section, we review works involving convolutional neural networks for solving network architecture and imbalance problems. The road crack segmentation dataset used in the mainstream study is presented.

2.1. Network Architectures

The fully convolutional network (FCN) [14] architecture transforms fully connected layers into convolution layers, enabling a classification network to output a heatmap. A convolutional layer is used after the pooling layer to convert the fully connected layers in VGG16 [15] and up-sample the predictions back to pixels in a single step. The FCN architecture builds "fully convolutional" networks that take inputs of an arbitrary size and produce correspondingly sized outputs with efficient inference and learning processes.

With an FCN architecture, Chun [16] proposed a fully convolutional neural network-based road surface damage detection approach with semisupervised learning. The model is updated by using pseudolabeled images derived from semisupervised learning methods to improve the performance of road surface damage detection techniques.

The U-Net [17] architecture achieves very good performance in very different biomedical segmentation applications and has become one of the most commonly used network structures. The architecture consists of a contracting path to capture context and a symmetric expanding path that enables precise localization. Using skip connections between the two parts, U-Net can mix shallow, low-level feature maps obtained from the encoder and deep, semantic features derived from the decoder. Benefiting from skip connections, the network can better retain low-dimensional features with high resolution to achieve good results in medical segmentation tasks with rich details, such as boundaries and seams.

With the U-Net architecture, Rodrigo [18] proposed a novel synthetic dataset and a weakly supervised learning method to overcome the inaccurate annotation problem and improved the results by up to 12%. Hong [19] proposed an improved identification technique based on the U-Net architecture that was enhanced with a convolutional block attention module, an improved encoder, and the strategy of fusing long- and short-skip connections. This method could effectively predict highway cracks in unmanned aerial vehicle (UAV) images.

UNet++ [20] alleviates the unknown network depth problem with an efficient ensemble of U-Nets with varying depths, which partially share an encoder and simultaneously perform co-learning using deep supervision. Furthermore, the skip connections are redesigned to aggregate features of varying semantic scales at the decoder. These network structures effectively preserve shallow features and are widely used in medical image segmentation.

With the UNet++ architecture, Yang [21] proposed a method composed of two stages, crack recognition and crack semantic segmentation, making it easy to meet the needs of efficient and reliable detection for large-scale collected images. The fine-tuned VGG16 model in the first stage can accurately identify crack images and avoid the computing costs incurred by the further processing of non-crack images, and the UNet++ model in the second stage provides pixel-level semantic segmentation for crack images.

The pyramid scene parsing network (PSPNet) [22] was proposed to embed difficult scenery context features in an FCN-based pixel prediction framework. The pyramid pooling module fuses features under four different pyramid scales, aggregates multilevel context to global context information, and provides global-scene-level information to the developed model. This architecture exploits the capability of global context information derived from different-region-based context aggregation through a pyramid pooling module and a suitable strategy to utilize global scene category clues.

A feature pyramid network (FPN) [23] naturally leverages the pyramidal shape of a ConvNet's feature hierarchy while creating a feature pyramid that has strong semantics at all scales. An FPN relies on an architecture that combines low-resolution, semantically strong features with high-resolution, semantically weak features via a top-down pathway and lateral connections, thereby developing a top-down architecture to build high-level semantic feature maps at all scales.

With a similar network structure, DeepCrack [9] resizes, concatenates, and fuses these multiscale feature maps at all scales, and a guide filter [24] is used to perform denoising at the feature level before prediction. APLCNet [25] combines a channel attention mechanism and a spatial attention mechanism during FPN feature fusion, highlights the attribute information and location information of cracks, and achieves good results on road crack instance segmentation tasks.

DeepLabV3+ [26] improves the encoder–decoder structure of other approaches. The encoder module encodes multiscale contextual information by applying atrous convolution on multiple scales, while the simple and efficient decoder module refines the segmentation results along the target boundary. The Xception model is further explored for the segmentation task. Deep separable convolution is applied to the ASPP and decoder modules to produce faster and stronger encoder–decoder networks.

A high-resolution network (HRNet) [11] maintains high-resolution representations by connecting high-to-low resolution convolutions in parallel and strengthens high-resolution representations by repeatedly performing multiscale fusions across parallel convolutions. HRNet has achieved great success in key point detection, attitude estimation, and multiperson attitude estimation tasks and exhibits enormous potential for scientific research and applications.

Based on HRNet, Chen [27] proposed an enhanced version called HRNete by removing the down-sampling operation in the initial stage, reducing the number of high-resolution representation layers, using dilated convolution, and introducing hierarchical feature integration. Using HRNet as the backbone, Bai [28] proposed the robust mask R-CNN, an end-to-end deep neural network for crack detection and segmentation on structures or their components that may be damaged during extreme events, such as earthquakes.

2.2. Imbalance Problem

Libra R-CNN [29] utilizes a simple but effective framework for balanced learning. It integrates IOU-balanced sampling by splitting the sampling interval into K bins according to their IOU measures and uniformly selects samples from these bins. It obtains a balanced feature pyramid by resizing the multilevel features in an FPN, averaging them to obtain an integrate feature, making refinements with convolutions, and resizing the features to their original sizes. A balanced L1 loss is set by separating inliers from outliers and clipping the large gradients produced by outliers with a maximum value of 1.0.

A two-stage convolutional neural network [13] was proposed for road crack detection and segmentation in images at the pixel level. The first stage serves to remove noise or artifacts and isolate the potential cracks in a small area via a classification network that is composed of five-layer convolutional neural networks and two fully connected (FC) layers. The second stage is a U-Net-structured segmentation network that can learn the context of cracks in the detected area. In this two-step framework, the first network filters out pure negative sample images, and images containing positive samples are allowed to proceed to the second stage; this approach can reduce the proportion of negative samples.

Focal loss [30] addresses the class imbalance problem by reshaping the standard cross-entropy loss such that it down-weights the losses assigned to well-classified examples. Focal loss focuses training on a sparse set of hard examples and prevents the vast number of easy negatives from overwhelming the detector during training. Two preset hyperparameters are employed; one is used to reduce the loss contributions provided by easy examples and extend their range, and the other addresses class imbalance.

Dice loss [31] considers that predictions are strongly biased toward the background if the area of interest occupies only a very small region of an image, and a loss function based on the Dice coefficient (related to precision and recall) can ameliorate this situation. The F1 score considers both the precision and recall of classification models and can comprehensively describe the performance of a model. The Dice loss is defined as 1 minus the F1 score, which involves maximizing the value of the F1 score as the optimization condition.

RAO-UNet [32] proposed a foreground perception optimization method. This method calculates the ratio of the sum of the probabilities of all pixels predicted as the background to the sum of the probabilities of all pixels predicted as the foreground in the predicted probability map. This ratio is used as an adaptive parameter for weighted binary cross entropy (BCE) loss in each batch to improve the performance damage caused by data imbalance.

Generative adversarial networks (GANs) [33] can extend the data by generating virtual data to solve the data imbalance problem. This approach has been used to deal with the problem of unbalanced landslide remote sensing data [34]. In Ref. [35], crack detection was enhanced by a generative adversarial network. CrackGAN [36] uses the generator as a segmentation network and adds a new constraint, generative adversarial loss, to regularize the objective function, which makes the network always generate a crack-GT detection result. These studies demonstrate the effectiveness of GANs in combating sample imbalance.

The technique of randomly under-sampling the majority class (RUMC) [37] involves randomly selecting examples from the majority class and removing them for the training dataset. The majority-class instances are discarded at random until a balanced class distribution in the training set is reached.

2.3. Dataset Situation of Road Crake Segmentation

At present, existing road crack segmentation datasets, such as DeepCrack [9], Crack-Forest [38], and 2StageCrack [13], are used by researchers to conduct experiments. The DeepCrack dataset contains 537 images with dimensions of 544 × 384 pixels. The Crack-Forest dataset contains 118 labeled images of 544 × 384 pixels. The 2StageCrack dataset, the newest open-source dataset, contains 1276 images for training and 354 for testing, each with dimensions of 96 × 96 pixels.

3. The Proposed Method

The main idea of our method is inspired by the second law of thermodynamics: heat can be spontaneously transferred from a hotter body to a cooler body. We hope to establish a similar mechanism during the convolutional neural network training process to regard precision and recall as body temperature and make them naturally flow from high to low. Therefore, we need to solve three problems: how to determine the direction of the flow, how to determine the strength of the flow, and how to make the precision and recall flow spontaneously.

To overcome these problems, our proposed method is as follows. (1) We define an adaptive parameter called PRF, which is associated with precision and recall, to evaluate the gap between precision and recall at a defined interval. (2) The flow direction is determined according to the positive and negative values of the PRF, and the flow intensity is determined according to the absolute value of the PRF. (3) To bridge precision and recall at the sample level and feature level, the sampling method and loss weight are repeatedly adjusted during the whole training process. Finally, spontaneous flow and a dynamic balance between precision and recall are achieved to narrow the gap between them and obtain a better trained model. The overall design of the recurrent adaptive network framework is shown in Figure 1.

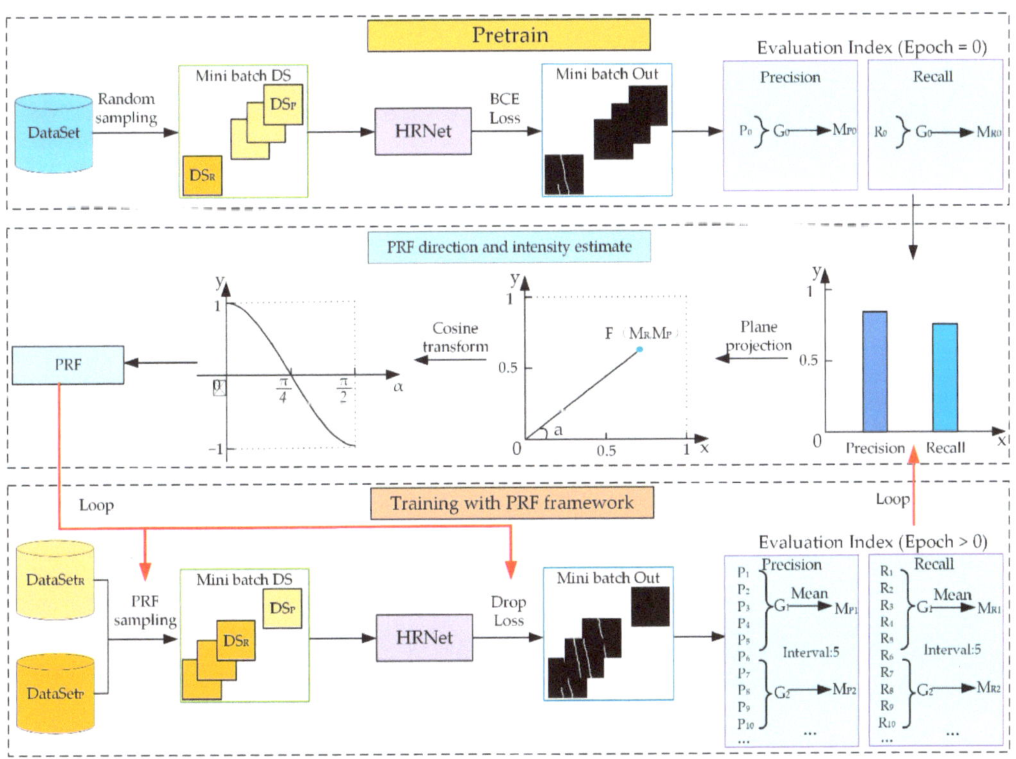

Figure 1. The overall design of the recurrent adaptive network framework.

3.1. Precision–Recall Flow Definition

During the training process, the average values of precision and recall during each epoch are calculated and added to the list L (R_i, P_i). To reduce the deviation caused by occasional shocks during the training procedures, we organize the values in L into groups at a certain interval and calculate the average precision and recall values in each group to obtain L_g (M_{Ri}, M_{Pi}). Restricted to supervised training, the offset degree can be displayed intuitively in L_g with an imbalanced dataset—there will be a gap between recall and precision. As the degree of data imbalance increases, this gap will also become larger. A rectangular coordinate system is plotted for each pair of values in L_g. Point F (M_R, M_P) is obtained with its corresponding recall and mean precision as the coordinates, and the azimuth of line OF is calculated as:

$$\alpha = \arctan\left(\frac{M_P}{M_R}\right), M_P \in [0,1], M_R \in [0,1] \tag{1}$$

where M_P is the average value of precision at interval, M_R is the average value of recall at interval, and α is the angle between OF and X-axis in Figure 1.

The PRF is defined as the evaluation index:

$$PRF = \cos(2\alpha), \alpha \in \left[0, \frac{\pi}{2}\right], PRF \in [-1, 1] \tag{2}$$

3.2. Precision–Recall Flow Direction and Intensity Estimate

In different datasets, the proportions of positive and negative samples are different. Negative samples, which form the larger imbalanced class during the supervised learning of road crack segmentation, overwhelm the BCE loss, Dice loss, and Focal loss. To avoid the lopsidedness of precision and recall, we need to evaluate the model's bias toward precision or recall, measure the degree of such an imbalance, and improve it.

The PRF parameters are obtained by cosine transformation, which can effectively meet the requirements of determining the flow direction and intensity; the direction of the flow is indicated by its positive or negative sign and the intensity of the flow is measured by its absolute value. When the positive sample recognition ability is strong, the precision is lower than the recall. In contrast, a precision higher than the recall means that the model is more advantageous in identifying negative samples. Therefore, the range of the PRF can clearly indicate the flow direction of precision and recall and can be expressed by the following equation:

$$\begin{aligned} M_P > M_R &\quad PRF \in (0,1] \\ M_P = M_R &\quad PRF = 0 \\ M_P < M_R &\quad PRF \in [-1,0) \end{aligned} \tag{3}$$

where M_P is the average value of precision at interval and M_R is the average value of recall at interval.

Furthermore, we need to estimate the strength of the flow; the gap between the precision and recall should be proportional to the flow strength. According to the calculation process, when the distance between precision and recall approaches 0, the absolute value of the PRF approaches 0, and when the distance approaches 1, the absolute value of the PRF approaches 1. Therefore, the PRF can measure the degree of deviation: the larger the absolute value of the PRF is, the more serious the deviation is and a greater flow intensity is needed. The PRF can effectively solve the problem of intensity estimation.

3.3. Precision–Recall Flow Sampling Method

When training a model, a random sampling scheme usually leads to the selected samples being dominated by negative samples in the road crack segmentation task, as shown in Table 1. To avoid this situation, we propose the PRF sampling method. The dataset is divided into two subsets according to the proportion of the positive sample pixels of each image; the images with positive sample pixels accounting for more than 0.5% of the total are put into dataset R, and the rest are placed into dataset P. We sample from the two subsets at different probabilities (controlled by the parameter, S). The initial value of S is 0.1; iterating with the PRF produces updates, as shown in the following equation:

$$S = \begin{cases} 0.01 & S < 0.01 \\ S(1 + PRF) & 0.2 < S < 0.01 \quad S \in [0.01, 0.2] \\ 0.2 & S > 0.2 \end{cases} \tag{4}$$

where S is the probability of sampling from dataset P.

The sampling strategy is as follows. Before each sampling step, take a random number $R \in [0,1]$. When R is less than S, the sample is from dataset P; otherwise, the sample is from dataset R, as shown in Figure 2. In this way, dominant classes are suppressed during sampling, and smaller classes are more likely to enter the network. S is cyclically and continuously modified to ensure that balance is maintained at each stage. As a result, the precision and recall will flow from higher to lower at the data sampling level.

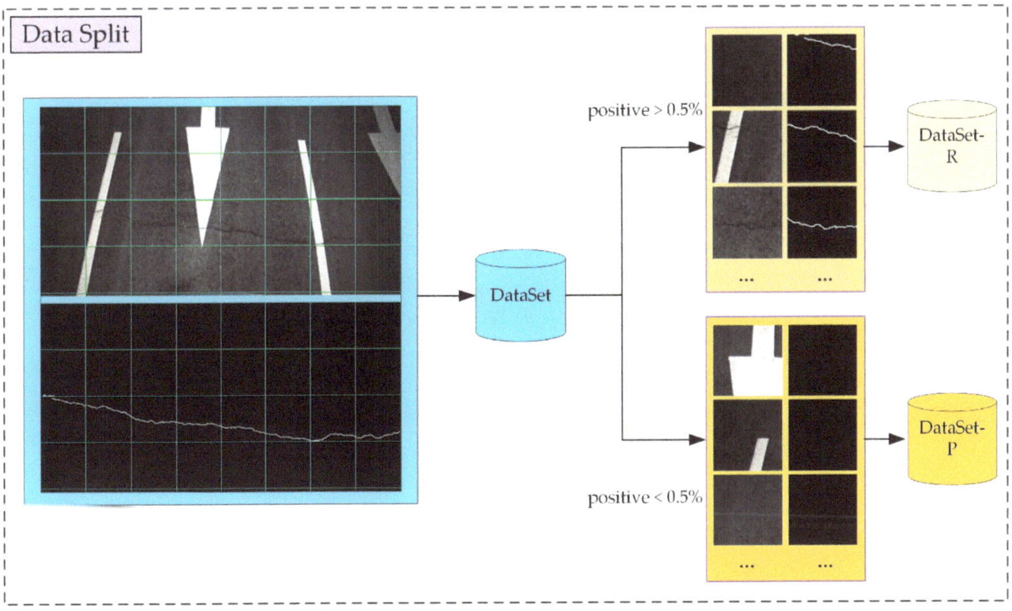

Figure 2. The dataset is divided into dataset R and dataset P.

3.4. Precision–Recall Flow Loss Method

At the feature level, researchers usually solve the imbalance problem by adjusting the structure or weight of the employed loss function. The Dice loss [31] directly maximizes the F1 score as the goal, which can ameliorate the imbalance between precision and recall but easily leads to gradient disappearance or explosion, especially when the given data are extremely unbalanced. The Focal loss [30] improves upon the BCE loss in terms of stability and robustness, but the hyperparameters need to be set according to expert experience. The hyperparameters are different for diverse networks and datasets. They cannot be universal in various scenes and cannot be adaptively modified during the training process.

To overcome the above problems, we propose an adaptive loss function named the drop loss, which is an improvement of the BCE loss. The definition of the BCE loss is:

$$L_i = [y_i \cdot \log(P_i) + (1 - y_i) \cdot \log(1 - P_i)] \tag{5}$$

where i is the pixel index, P_i is the probability that pixel i is predicted as belonging to the positive category, y_i is the true probability that pixel i belongs to the positive category, and L_i is the BCE loss value generated on pixel i:

$$BCELoss = -\frac{1}{N}\sum_{i=1}^{N} L_i \tag{6}$$

where N is the total number of pixels in the batch.

A balanced loss weight and intensity coefficient β is introduced:

$$\beta = |PRF| \quad \beta \in [0,1] \tag{7}$$

The gradient caused by positive features is dropped when the recall is stronger, and the positive features are dropped in the opposite case:

$$C_i = \begin{cases} y_i \cdot \log(P_i) \cdot \beta & PRF < 0 \\ (1 - y_i) \cdot \log(1 - P_i) \cdot \beta & PRF > 0 \end{cases} \tag{8}$$

where C_i is the loss value after features dropped on pixel i.

It is desired that the loss function has some randomness to enhance the ability of the neural network to move away from local optimal solutions. A random tensor r is introduced to improve diversity. The tensor r has the same shape as the tensor label of the input neural network and is populated by random numbers that are uniformly distributed on the interval [0,1]. Each D_i in the drop loss is calculated as:

$$D_i = L_i + C_i \cdot r_i \tag{9}$$

where D_i is the drop loss value after the gradient is dropped on pixel i, L_i is the BCE loss value generated on pixel i, and r_i is a random tensor between 0 and 1.

Finally, the expression for the drop loss is:

$$DropLoss = \begin{cases} -\frac{1}{N} \sum_{i=1}^{N} [y_i \cdot \log(P_i) \cdot (1 + \beta \cdot r_i) + (1 - y_i) \cdot \log(1 - P_i)] & PRF < 0 \\ -\frac{1}{N} \sum_{i=1}^{N} [y_i \cdot \log(P_i) + (1 - y_i) \cdot \log(1 - P_i)] & PRF = 0 \\ -\frac{1}{N} \sum_{i=1}^{N} [y_i \cdot \log(P_i) + (1 - y_i) \cdot \log(1 - P_i) \cdot (1 + \beta \cdot r_i)] & PRF > 0 \end{cases} \tag{10}$$

where i is the pixel index, P_i is the probability that pixel i is predicted as belonging to the positive category, y_i is the true probability that pixel i belongs to the positive category, N is the total number of pixels in the batch, β is the intensity coefficient in Equation (7), and r_i is a random tensor defined in Equation (9).

By discarding part of the gradient, the precision and recall are in dynamic equilibrium in unbalanced situations, as shown in Figure 3. The drop loss can balance the loss weight at all stages of the training process automatically.

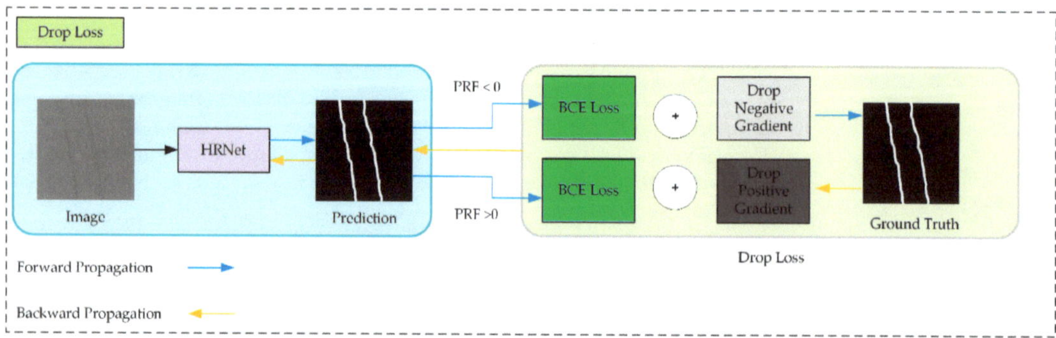

Figure 3. Schematic diagram of the drop loss.

4. Experimental Details

4.1. Dataset Preparation

Although a number of open-source road crack segmentation datasets already exist in the academic community, these datasets have certain limitations.

- Insufficient quantity: To date, most datasets used by scholars contain only a few hundred images, which is far from sufficient for deep learning. For example, the MS-COCO-2018 dataset [39] contains 118,287 images for training, and the ILSVRC-14 dataset contains 456,567 for training. By comparison, the datasets for road crack segmentation are two orders of magnitude smaller. The use of a small amount of data easily leads to the occurrence of overfitting, even early in the training process, and it is difficult to verify the capability of a network on a small dataset. Therefore, a new dataset with a large amount of data is urgently needed.
- Discontinuous sampling: Most existing datasets only include images of very small areas where the roads are damaged and ignore the large areas surrounding healthy road surfaces. When only a small number of normal road surfaces are involved in the training process, the model will misidentify normal objects such as road markings, road facilities, and automobile parts as road cracks.
- Lack of resolution: Most existing datasets are collected by mobile phone cameras or ordinary civilian cameras, whose photosensitive elements are unable to obtain sufficiently clear photos while driving on highways.
- Single shooting time: A dataset obtained entirely at one time cannot provide sufficient diversity for complex external elements such as external light, road shadows, road stains, and road materials.

In view of the above defects, existing datasets cannot be used to verify the performance of neural networks in real road damage inspection scenarios.

We propose a new dataset: the high-resolution road crack dataset (HRRC). The resolution of each HRRC image is 12 million pixels with a width of 4096, a height of 2080, and 5 fps. The dataset contains multiple disturbances, such as overexposure, road markings, road materials, road stains, and shadows. Experts have carefully labeled each image, creating semantically segmented labels. Each large image is divided into 512 × 512 slices, and a total of 24,704 images are obtained. Figure 4 shows examples of samples belonging to the HRRC dataset.

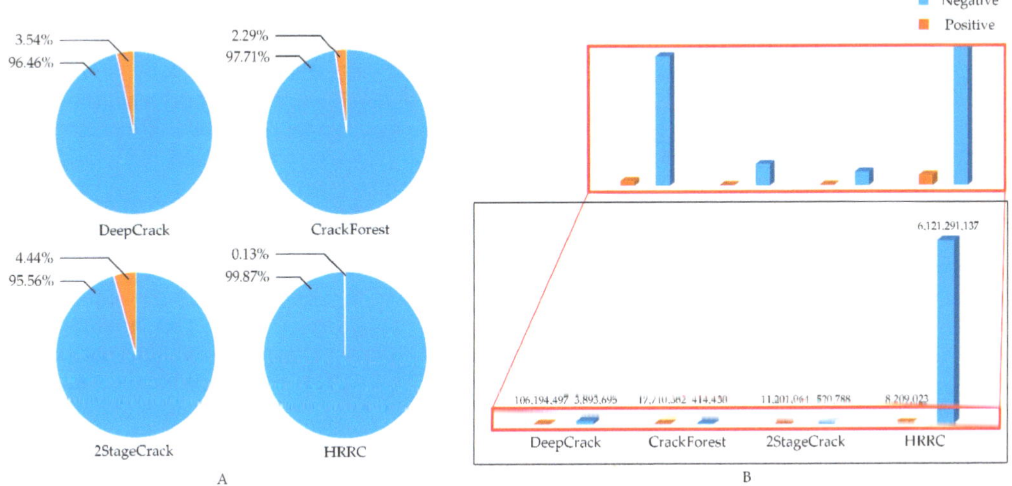

Figure 4. The total amount of data in each dataset, as well as their positive and negative sample ratios. (**A**) Positive and negative sample ratios. (**B**) Number of pixels in each dataset.

As shown in Figure 4, the HRRC dataset is the largest road crack dataset at present. A high-resolution Complementary Metal-Oxide-Semiconductor (CMOS) sensor ensures the integrity of various road elements, such as lines and vehicles. Millimeter-level cracks can be seen clearly, and multiple sections of expressway with different materials can be photographed under different weather and lighting conditions. Therefore, this dataset is a high-quality dataset that alleviates the above problems to a certain extent. However, a higher complexity and a smaller positive sample ratio make it difficult for a model to converge and impose higher requirements on the robustness of the model and the advancement of the training method.

HRRC dataset images were captured continuously during the automobile traveling process by an industrial camera, as shown in Figure 5.

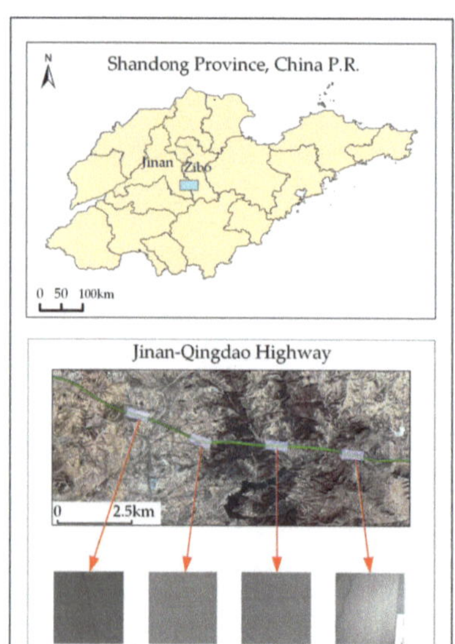

Figure 5. Geographical location distribution and collection equipment of the HRRC dataset.

4.2. Implementation Environment

All the experiments were based on the following hardware environment: an i7 9700 CPU, an Nvidia GeForce GTX2080ti GPU, and 64 GB of RAM. The software environment is based on Python 3.8, PyTorch 1.6, OpenCV 4.5, and the Windows 10 operating system. The training parameters of the model were set as follows: the learning rate is 0.001, the training batch size is 8, and the validation batch size is 1. All experiments were conducted based on the HRRC dataset only.

4.3. Implementation Details

For the first set of experiments, we aimed to find the most suitable convolutional neural network architecture for the road crack segmentation task. Before training, the entire dataset was divided into training and validation sets at a ratio of 9:1. In addition, we used 100 unsliced image-labeled data as an independent test set. The training set was randomly shuffled before breaking it up in order to ensure that each image has the same probability of being input into the network. Due to the large number of images in the dataset, data augmentation was not used to extend the amount of data. The BCE loss function and Adam optimizer were used during the training process. The experiments were conducted with the DeepLabV3+, UNet, UNet++, PSPNet, DeepCrack, FPN, and HRNet frameworks, with 100 training epochs for each network.

To verify the precision of the network architecture, MobileNet_v2 was used as the backbone under the DeepLabV3+, UNet, UNet++, PSPNet, and FPN architectures, and HRNet was also trained using HRNet-18w-small with minimal parameters. The hyperparameters of the model remained the same as in the original paper except for the backbone. The reasons for this are as follows:

(1) MobileNet_v2 has good performance and can be seamlessly integrated with the above network architectures. MMSegmentation [40], PaddleSeg [41], segmentation_models [42], and other standardized open-source code suites support this network, making it convenient for researchers to reproduce.
(2) A network with MobileNet_v2 easily converges and has a fast training speed. This significantly reduces the time required for training when faced with large dataset training tasks.
(3) Regarding the application requirements of a lightweight network, to ensure full coverage of the road surface during shooting, the shooting frequency of the vehicle camera is five shots per second, which leads to the actual application of the data volume being massive. A lightweight network can ensure that staff obtain real-time inspection results.

In addition, we note that batch normalization, which is widely used in computer tasks, facilitates network training. However, the mean and variance obtained are not representative of the global dataset because the batch size used during the training process is not sufficiently large. The variances between different kinds of samples in a small region are large, and the use of a batch size of one during validation and testing causes the batch normalization layer to be more biased to images in which the large class is located; it cannot be well normalized to the regions where the small class is located, thus reducing the recognition ability of the model for the small class. Therefore, this mechanism reduces the recognition ability of the network in the face of unbalanced datasets, so the batch normalization process in HRNet is replaced with group normalization (HRNet-GN), which is the most suitable technique for the road crack segmentation task, to mitigate this problem. The structure of HRNet is shown in Figure 6.

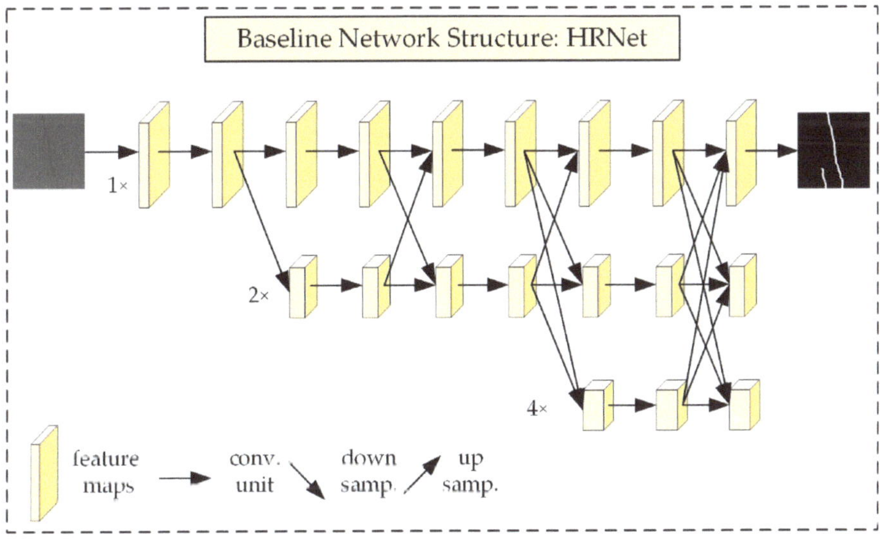

Figure 6. HRNet structure schematic (1×, 2×, 4× are sampling scales).

The goal of the second set of experiments was to verify the ability of each loss function to deal with the imbalance problem at the feature level. This set of experiments used HRNet-

GN, which is the best network in the first set of experiments, as the baseline. Comparisons were made between the BCE loss, Dice loss, Focal loss, and drop loss. The effectiveness of the drop loss was verified with the same data, network, and training parameters in the road crack segmentation task.

The goal of the third set of experiments was to verify the generality and advancement of the PRF method proposed in this paper, which includes direction determination, intensity evaluation, sampling, and loss weight adjustment. This method adaptively adjusts itself throughout the training process and can be integrated into all convolutional neural networks without adding extra parameters. The experiments were first conducted to evaluate the performance improvement yielded by this method for convolutional neural networks by conducting ablation tests on the sample and loss methods with HRNet-GN as the baseline. Second, alignment tests were also conducted on HRNet to verify the generality and universality of this method.

Under the PRF sampling method, the dataset was divided into two sub-datasets. The probabilities of images from different datasets were changed, and the sampling method was also changed from non-release sampling to release sampling. The number of images in dataset R was used as the dataset length in these experiments, resulting in a smaller number of samples per epoch. To fully train the model, we adjusted the number of epochs to 800.

4.4. Methods for Evaluation

To objectively evaluate the performance of the network, the precision, recall, and IOU metrics were used to verify the performance of the semantic segmentation model. Their definitions are as follows:

$$precision = \frac{TP}{TP + FP} \tag{11}$$

$$recall = \frac{TP}{TP + FN} \tag{12}$$

$$IOU = \frac{TP}{TP + FP + FN} \tag{13}$$

$$F1score = \frac{2TP}{2TP + FP + FN} \tag{14}$$

TP, FP, and FN are the numbers of true-positive, false-positive, and false-negative pixels, respectively. The threshold for splitting positive and negative in a predicted probability map is 0.5.

5. Results

In this section, we present and analyze the experimental results in three parts according to the experimental design.

5.1. Network Structure Experiment Results

The first set of experiments involving different network architectures is shown in Table 1. Surprisingly, the DeepLabV3+ results are the worst, and this result does not match the performance of DeepLabV3+ in natural image semantic segmentation. To explain this phenomenon, we believe that dilated convolution expands the view of the convolution but reduces the spatial resolution of the feature layers. This approach can be advantageous in dealing with large targets but cannot cope well with road cracks that are only a few pixels in width. Apart from DeepLabV3+, the precision rankings of the other networks do not differ much from their performance in natural images. HRNet performs the best and leads the other network architectures by a large margin. It is worth mentioning that the IOU of HRNet-GN improves by 4.6%, indicating that group normalization has a significant advantage over the batch normalization structure when facing imbalance problems. The experimental findings indicate that the HRNet structure had outstanding performance and that group normalization had an advantage over batch normalization in this task.

Table 1. Scores of different network structures trained with the BCE loss.

Network Structure	IOU (%)	F1 (%)	Recall (%)	Precision (%)
DeepLabV3+ [26]	21.79	30.45	24.27	86.6
U-Net [17]	23.93	33.79	25.46	89.35
PSPNet [22]	25.19	34.26	34.14	78.85
UNet++ [20]	27.69	36.74	31.08	88.75
DeepCrack [9]	34.7	48.69	60.97	46.32
FPN [23]	35.61	46.53	41.6	84.42
HRNet [11]	40.65	52.21	49.99	79.57
HRNet-GN	44.61	56.53	58.65	75.37

Table 2 shows the parameters, multiply-accumulate operations (MAdd), floating point operations (Flops), and GPU memory values of different network structures for inference purposes. Among the tested methods, PSPNet has the smallest number of parameters, computational complexity, and GPU memory needed, but its precision is 22.27-percentage points lower than the best precision achieved by HRNet-GN. The advantage of small computation complexity does not offset the excessive capacity degradation of PSPNet. The DeepCrack structure has the largest number of parameters and MAdd value, but its IOU is 7.53-percentage points lower than that of HRNet-GN, so it is not a good choice. The number of HRNet parameters is only larger than that of PSPNet, but its MAdd, Flops, and GPU memory are the second largest—only being smaller than those of DeepCrack. This is because the HRNet structure retains more feature maps with different resolutions and requires more data operations and memory allocations within the network to achieve cross-resolution fusion with the highest parameter utilization rate. Considering the precision, number of parameters, Flops, and memory requirements, we believe that HRNet-GN has the highest precision and fewest parameters and is the most suitable convolutional neural network structure for road crack semantic segmentation applications at this stage.

Table 2. Information about number of parameters, number of calculations, and GPU memory usage levels of the tested networks.

Network Structure	Parameters	MAdd	Flops	GPU Memory
DeepLabV3+ [26]	4,378,513	2.3 G	1.16 G	112.66 MB
U-Net [17]	6,628,945	5.16 G	2.59 G	124.13 MB
PSPNet [22]	94,969	305.55 M	157.82 M	59.88 MB
UNet++ [20]	6,824,721	6.85 G	3.43 G	157.67 MB
DeepCrack [9]	14,720,389	30.76 G	15.39 G	155.87 MB
FPN [23]	4,215,425	3.7 G	1.86 G	88.92 MB
HRNet [11]	3,934,629	10.62 G	5.34 G	222.56 MB
HRNet-GN	3,934,629	10.62 G	5.34 G	222.56 MB

5.2. Experimental Loss Function Comparison Results in a Case with Imbalance

Table 3 shows the comparison between the BCE loss, the Dice loss, the Focal loss, and our drop loss. The Dice loss is unable to learn enough positive features, resulting in a failure to converge. Without parameter tuning, the Focal loss with its default parameters ($\gamma = 2$, $\alpha = 0.25$) is even worse than the BCE loss for the road crack segmentation task. From the experimental results, it can be concluded that the poor performance of the Dice and Focal losses is caused by the fact that the recall is greatly below the precision in the case of extreme imbalance among the HRRC data categories. The BCE loss achieves good results and stable convergence with no need to adjust parameters, as well as low training difficulty. Supported by the recurrent adaptive network framework, the drop loss inherits these advantages and yields an exciting achievement: a 9.35% IOU increase.

Table 3. Comparison among the Dice loss, Focal loss, and drop loss.

Network Structure	IOU (%)	F1 (%)	Recall (%)	Precision (%)
DICE loss	2.73×10^{-9}	2.73×10^{-9}	2.73×10^{-9}	100
Focal loss	25.19	48.59	34.14	78.85
BCE loss	44.61	56.53	58.65	75.37
Drop loss	53.39	66.46	71.16	71.72

5.3. Ablation Experiments Involving the PRF Method

Ablation experiments are also performed on both HRNet and HRNet-GN to verify the reliability of the proposed method. According to the test results in Table 4, the following conclusions can be drawn:

- Both PRF sampling and the drop loss method can clearly increase the IOU score: 13.1% on HRNet and 9.56% on HRNet-GN.
- The recurrent adaptive network framework minimizes the performance gap between HRNet and HRNET-GN from 3.96% to 0.41%.
- The drop loss method applied to the feature yields slightly higher results than those of the sampling method applied to the data layer.

Table 4. Ablation experiment results obtained based on HRNet and HRNet-GN.

Network Structure	IOU (%)	F1 (%)	Recall (%)	Precision (%)
HRNet	40.65	52.21	49.99	79.57
HRNet + PRF sampling	52.78	66.19	70.98	68.01
HRNet + Drop loss	53.25	66.75	70.11	70.37
HRNet + PRF sampling + Drop loss	53.63	67.28	71.22	69.16
HRNet-GN	44.61	56.53	58.65	75.37
HRNet-GN + PRF sampling	53.33	66.61	70.01	71.56
HRNet-GN + Drop loss	53.39	66.46	71.16	71.72
HRNet-GN + PRF sampling + Drop loss	54.03	67.39	69.77	71.96

With the recurrent adaptive network framework, HRNet-GN obtains an IOU score of 0.5416, which is the state-of-the-art value on the HRRC dataset (The codes used in this work can be accessed via the URL provided in the Supplementary Materials section).

Figure 7 shows the variations in precision and recall with the IOU and PRF values during the training processes of the HRNet-GN and HRNet-GN+ the recurrent adaptive network framework. By observation, we find the following:

- Convergence acceleration: In Figure 7f,g, the training process starts with a precision greater than 0.98 and a recall less than 0.02, and this phenomenon lasts for five epochs before it starts to improve. In Figure 7b,c, the model starts to converge well at the second epoch, and the IOU value is greater than 0.5 in the third epoch in Figure 7a, while in Figure 7e reaches the same level after 50 epochs.
- Balancing precision and recall: In Figure 7g, the maximum value of recall is less than 0.6, and the minimum value of precision is greater than 0.75 throughout the training process in Figure 7f. This large gap continues over the entire procedure. In Figure 7d, the recurrent adaptive network framework starts to work at the second epoch, and the gap between precision and recall decreases rapidly so that it remains stable at approximately 0.1; this is an obvious reduction in the precision–recall gap.
- Adaptive adjustment throughout the whole process: At the beginning of training, the first epoch is used for pre-training to obtain the first PRF value, and from the second epoch until the end of training, the PRF value is adaptively updated after a specific interval (five epochs in these experiments). This mechanism makes the sampling and loss weights more suitable for the next training stage and adjusts the training parameters according to the model states observed throughout the process.

- Significant performance improvement: As training proceeds, the IOU and PRF curve amplitudes gradually decrease. This means that the PRF framework makes the training process stable and reduces the training difficulty. More importantly, with the optimized training of the recurrent adaptive network, the IOU value of the model greatly improves, which significantly improves the performance of the model in the face of unbalanced datasets. Figure 8 shows that PRF sample and drop loss method in recurrent adaptive network are used together to obtain the best performance.

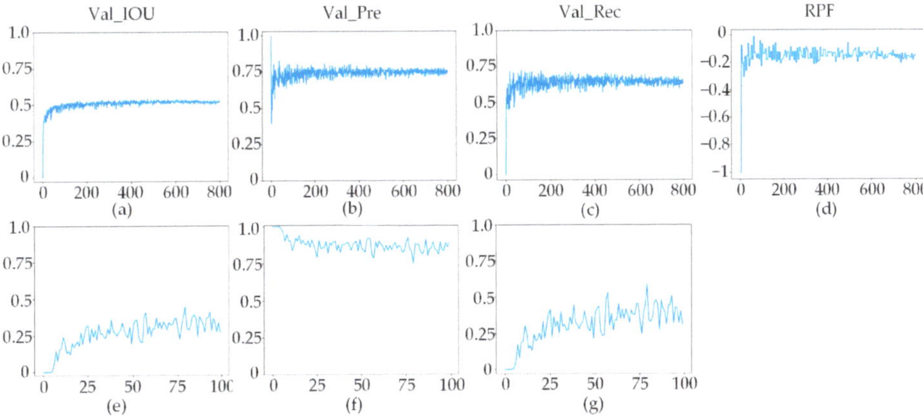

Figure 7. Variations in the precision, recall, and IOU curves during training. (**a–c**) show validation IOU, precision and recall when training whit recurrent adaptive network. (**d**) shows values and changes of PRF during training. (**e–g**) show validation IOU, precision and recall with normal training.

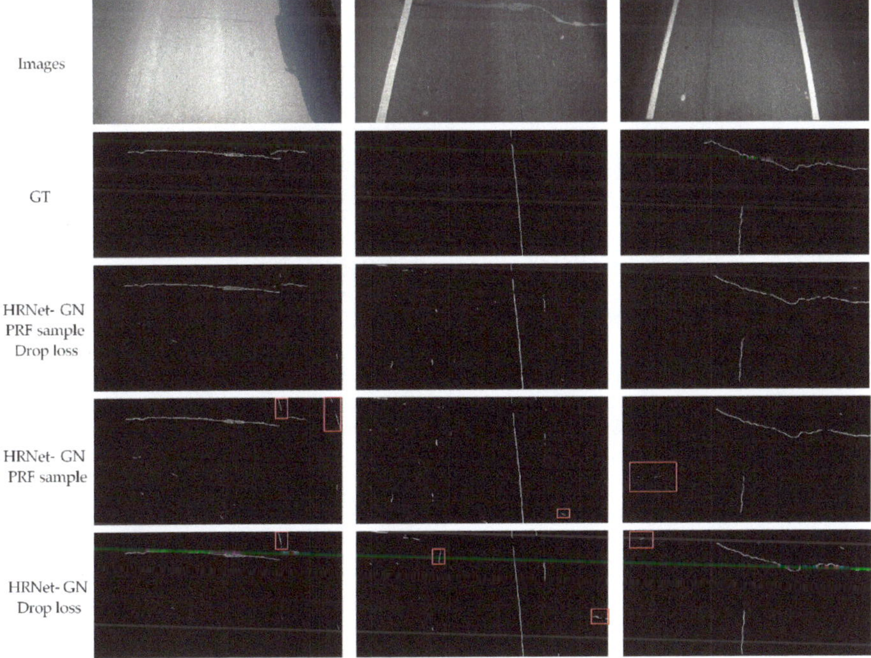

Figure 8. Results of ablation experiments (red boxes represent the regions of poor performance that model obtained from the single method training in the ablation experiment).

6. Discussion

It is well known that the performance of a model can be improved by adjusting the sampling and loss weights in a suitable way when faced with data imbalance. The most interesting finding was that we guided the direction and value of parameter tuning in sampling and loss by measuring the deviation between precision and recall, and we repeated the process at regular intervals during training. We refer to this training framework as a "recurrent adaptive network".

In this study, there are two key points: (1) The large gap between precision and recall in the road crack segmentation task is an external manifestation of positive and negative sample imbalance. Their changes need to be analyzed to guide the adaptive adjustment of the balance parameters. (2) The PRF keeps changing during the training process, which indicates that the negative impact of positive and negative sample imbalance varies at different stages of the training process. The parameters need to be adjusted periodically as the training progresses.

In previous studies, no researcher has demonstrated that narrowing the deviation between precision and recall can improve model performance. What is surprising is that the experimental results in this paper demonstrate that this approach is effective in solving the imbalance problem in the road crack segmentation task. The training converges faster and leads to an improved model performance, which is supported by the construction of a recurrent adaptive network.

Of course, there are shortcomings and limitations in our work. The most effective way to solve the imbalance problem is to increase the number of samples in small categories, which can be achieved by GAN networks, but this is not considered in this method. In the experiments described in this paper, the precision was higher than the recall throughout the training process, which shows that the imbalance is only mitigated and not eliminated. The desired state is that precision and recall curves interpolate with each other with oscillation, but it is obvious that the experiment does not meet this expectation.

In the next study, we will try to continue our research by using a GAN to generate simulated data based on the framework of this paper and using the device to collect more positive samples to alleviate the degree of imbalance between positive and negative samples. Moreover, the question of how a complete classification can be provided from the smaller frames for segmentation needs to be addressed in our next research. In addition, there is road crack incompleteness in the prediction results, and this study does not improve the completeness of the small category targets.

7. Conclusions

In this study, a new high-resolution road crack database (HRRC) was produced; it came from a real application scene, and data were collected by a vehicle-mounted industrial camera. Currently, HRRC is the largest and most complicated dataset in the field of road crack research. We analyzed semantic segmentation frameworks based on convolutional neural networks, and performance tests were conducted on HRRC. The results show that HRNet-GN is the most suitable convolutional neural network framework for road crack segmentation. A recurrent adaptive network framework based on the idea of precision–recall flow, including flow direction determination, intensity estimation, sampling, and drop loss calculation, is proposed. This framework measures the deviation between precision and recall to determine the degree of observed imbalance. Regarding the data-level and feature-level aspects, this imbalance is automatically bridged to obtain a better trained model. This framework maintains a balance between precision and recall at the data and feature levels and performs well in training tasks with extremely unbalanced datasets, providing adaptive tuning, accelerated network convergence, balanced precision and recall, and significant performance gains. As a result, the recurrent adaptive network achieves a state-of-the-art performance on this dataset, with an F1-score of 56.53, which is far better than the other methods.

Supplementary Materials: The experimental codes can be found here: https://aistudio.baidu.com/aistudio/projectdetail/4242896?contributionType=1&shared=1 (accessed on 19 June 2022).

Author Contributions: Conceptualization, Y.Z. and J.F.; methodology, Y.Z. and J.F.; software, Y.Z.; validation, M.Z., R.L. and Z.S.; formal analysis, Y.Z.; investigation, R.L.; resources, M.Z. and Z.S.; data curation, Y.Z. and R.L.; writing—original draft preparation, Y.Z. and J.F.; writing—review and editing, J.F. and B.G.; visualization, M.Z. and Z.S.; supervision, J.F.; project administration, J.F.; funding acquisition, J.F. All authors have read and agreed to the published version of the manuscript.

Funding: This research was funded by the National Natural Science Foundation of China (grant nos. 42171413, 42101306 and 42001414), a grant from the State Key Laboratory of Resources and Environmental Information System, the Shandong Provincial Natural Science Foundation (grant nos. ZR2020MD015, ZR2020MD018 and ZR2019BD033), the National Key Research and Development Program of China (grant no. 2017YFB0503500), and the Young Teacher Development Support Program of Shandong University of Technology (grant no. 4072-115016).

Conflicts of Interest: The authors declare no conflict of interest.

References

1. Pan, Y.; Zhang, X.; Cervone, G.; Yang, L. Detection of Asphalt Pavement Potholes and Cracks Based on the Unmanned Aerial Vehicle Multispectral Imagery. *IEEE J. Sel. Top. Appl. Earth Obs. Remote Sens.* **2018**, *11*, 3701–3712. [CrossRef]
2. Xu, H.; Tian, Y.; Lin, S.; Wang, S. Research of image segmentation algorithm applied to concrete bridge cracks. In Proceedings of the IEEE Third International Conference on Information Science and Technology (ICIST), Yangzhou, China, 23–25 March 2013; pp. 1637–1640. [CrossRef]
3. Zhou, X.; Xu, L.; Wang, J. Road crack edge detection based on wavelet transform. *IOP Conf. Series: Earth Environ. Sci.* **2019**, *237*, 032132. [CrossRef]
4. Abdellatif, M.; Peel, H.; Cohn, A.G.; Fuentes, R. Pavement Crack Detection from Hyperspectral Images Using a Novel Asphalt Crack Index. *Remote. Sens.* **2020**, *12*, 3084. [CrossRef]
5. Guo, B.; Wei, C.; Yu, Y.; Liu, Y.; Li, J.; Meng, C.; Cai, Y. The Dominant Influencing Factors of Desertification Changes in the Source Region of Yellow River: Climate Change or Human Activity? *Sci. Total Environ.* **2022**, *813*, 152512. [CrossRef]
6. Rezaie, A.; Achanta, R.; Godio, M.; Beyer, K. Comparison of crack segmentation using digital image correlation measurements and deep learning. *Constr. Build. Mater.* **2020**, *261*, 120474. [CrossRef]
7. He, K.; Zhang, X.; Ren, S.; Sun, J. Deep residual learning for image recognition. In Proceedings of the 2016 IEEE Conference on Computer Vision and Pattern Recognition (CVPR), Las Vegas, NV, USA, 27–30 June 2016; pp. 770–778. [CrossRef]
8. Chen, L.-C.; Papandreou, G.; Schroff, F.; Adam, H. Rethinking atrous convolution for semantic image segmentation. *arXiv* **2017**, arXiv:1706.05587. [CrossRef]
9. Liu, Y.; Yao, J.; Lu, X.; Xie, R.; Li, L. DeepCrack: A deep hierarchical feature learning architecture for crack segmentation. *Neurocomputing* **2019**, *338*, 139–153. [CrossRef]
10. Zeiler, M.D.; Fergus, R. Visualizing and understanding convolutional networks. In Proceedings of the 13th European Conference on Computer Vision (ECCV), Zurich, Switzerland, 6–12 September 2014; pp. 818–833. [CrossRef]
11. Sun, K.; Xiao, B.; Liu, D.; Wang, J. Deep High-Resolution Representation Learning for Human Pose Estimation. In Proceedings of the IEEE/CVF Conference on Computer Vision and Pattern Recognition (CVPR), Long Beach, CA, USA, 16–20 June 2019; pp. 5686–5696. [CrossRef]
12. Russell, B.C.; Torralba, A.; Murphy, K.P.; Freeman, W.T. LabelMe: A Database and Web-Based Tool for Image Annotation. *Int. J. Comput. Vis.* **2007**, *77*, 157–173. [CrossRef]
13. Nguyen, N.H.T.; Perry, S.; Bone, D.; Le, H.T.; Nguyen, T.T. Two-stage convolutional neural network for road crack detection and segmentation. *Expert Syst. Appl.* **2021**, *186*, 115718. [CrossRef]
14. Shelhamer, E.; Long, J.; Darrell, T. Fully convolutional networks for semantic segmentation. *IEEE Trans. Pattern Anal. Mach. Intell.* **2017**, *39*, 640–651. [CrossRef] [PubMed]
15. Simonyan, K.; Zisserman, A. Very deep convolutional networks for large-scale image recognition. *arXiv* **2014**, arXiv:1409.1556. [CrossRef]
16. Chun, C.; Ryu, S.-K. Road Surface Damage Detection Using Fully Convolutional Neural Networks and Semi-Supervised Learning. *Sensors* **2019**, *19*, 5501. [CrossRef]
17. Ronneberger, O.; Fischer, P.; Brox, T. U-Net: Convolutional networks for biomedical image segmentation. In *Medical Image Computing and Computer-Assisted Intervention 2015*; Navab, N., Hornegger, J., Wells, W.M., Frangi, A.F., Eds.; Springer International Publishing: Cham, Switzerland, 2015; pp. 234–241. [CrossRef]
18. Rill-García, R.; Dokladalova, E.; Dokládal, P. Pixel-accurate road crack detection in presence of inaccurate annotations. *Neurocomputing* **2022**, *480*, 1–13. [CrossRef]
19. Hong, Z.; Yang, F.; Pan, H.; Zhou, R.; Zhang, Y.; Han, Y.; Wang, J.; Yang, S.; Chen, P.; Tong, X.; et al. Highway Crack Segmentation from Unmanned Aerial Vehicle Images Using Deep Learning. *IEEE Geosci. Remote Sens. Lett.* **2021**, *19*, 6503405. [CrossRef]

20. Zhou, Z.; Siddiquee, M.M.R.; Tajbakhsh, N.; Liang, J. UNet++: Redesigning Skip Connections to Exploit Multiscale Features in Image Segmentation. *IEEE Trans. Med. Imaging* **2020**, *39*, 1856–1867. [CrossRef]

21. Yang, Q.; Ji, X. Automatic Pixel-Level Crack Detection for Civil Infrastructure Using Unet++ and Deep Transfer Learning. *IEEE Sensors J.* **2021**, *21*, 19165–19175. [CrossRef]

22. Zhao, H.; Shi, J.; Qi, X.; Wang, X.; Jia, J. Pyramid scene parsing network. In Proceedings of the 2017 IEEE Conference on Computer Vision and Pattern Recognition, Honolulu, HI, USA, 21–26 July 2017; pp. 2881–2890. [CrossRef]

23. Lin, T.Y.; Dollár, P.; Girshick, R.; He, K.; Hariharan, B.; Belongie, S. Feature pyramid networks for object detection. In Proceedings of the 30th IEEE Conference on Computer Vision and Pattern Recognition, CVPR, Honolulu, HI, USA, 21–26 July 2017; pp. 936–944. [CrossRef]

24. He, K.; Sun, J.; Tang, X. Guided Image Filtering. *IEEE Trans. Pattern Anal. Mach. Intell.* **2013**, *35*, 1397–1409. [CrossRef]

25. Zhang, Y.; Chen, B.; Wang, J.; Li, J.; Sun, X. APLCNet: Automatic Pixel-Level Crack Detection Network Based on Instance Segmentation. *IEEE Access* **2020**, *8*, 199159–199170. [CrossRef]

26. Chen, L.-C.; Zhu, Y.; Papandreou, G.; Schroff, F.; Adam, H. Encoder-Decoder with Atrous Separable Convolution for Semantic Image Segmentation. In *Proceedings of the Computer Vision—ECCV 2018*; Ferrari, V., Hebert, M., Sminchisescu, C., Weiss, Y., Eds.; Springer International Publishing: Cham, Switzerland, 2018; pp. 833–851. [CrossRef]

27. Chen, H.; Su, Y.; He, W. Automatic crack segmentation using deep high-resolution representation learning. *Appl. Opt.* **2021**, *60*, 6080. [CrossRef]

28. Bai, Y.; Sezen, H.; Yilmaz, A. End-to-end Deep Learning Methods for Automated Damage Detection in Extreme Events at Various Scales. In Proceedings of the 25th International Conference on Pattern Recognition (ICPR), Milan, Italy, 10–15 January 2021; pp. 6640–6647. [CrossRef]

29. Pang, J.; Chen, K.; Shi, J.; Feng, H.; Ouyang, W.; Lin, D. Libra R-CNN: Towards Balanced Learning for Object Detection. In Proceedings of the IEEE/CVF Conference on Computer Vision and Pattern Recognition (CVPR), Long Beach, CA, USA, 15–20 June 2019; pp. 821–830. [CrossRef]

30. Lin, T.Y.; Goyal, P.; Girshick, R.; He, K.; Dollar, P. Focal loss for dense object detection. *IEEE Trans. Pattern Anal. Mach. Intell.* **2020**, *42*, 318–327. [CrossRef]

31. Milletari, F.; Navab, N.; Ahmadi, S.-A. V-net: Fully convolutional neural networks for volumetric medical image segmentation. In Proceedings of the 2016 Fourth International Conference on 3D Vision (3DV), Stanford, CA, USA, 25–28 October 2016; pp. 565–571.

32. Fan, L.; Zhao, H.; Li, Y.; Li, S.; Zhou, R.; Chu, W. RAO-UNet: A residual attention and octave UNet for road crack detection via balance loss. *IET Intell. Transp. Syst.* **2021**, *16*, 332–343. [CrossRef]

33. Goodfellow, I.; Pouget-Abadie, J.; Mirza, M.; Xu, B.; Warde-Farley, D.; Ozair, S.; Courville, A.; Bengio, Y. Generative adversarial nets. Advances in Neural Information Processing Systems. *arXiv* **2014**, arXiv:1406.2661. [CrossRef]

34. Al-Najjar, H.A.H.; Pradhan, B.; Sarkar, R.; Beydoun, G.; Alamri, A. A New Integrated Approach for Landslide Data Balancing and Spatial Prediction Based on Generative Adversarial Networks (GAN). *Remote Sens.* **2021**, *13*, 4011. [CrossRef]

35. Duan, L.; Geng, H.; Pang, J.; Zeng, J. Unsupervised Pixel-level Crack Detection Based on Generative Adversarial Network. In Proceedings of the 5th International Conference on Multimedia Systems and Signal Processing, Chengdu, China, 8–10 May 2020; pp. 6–10. [CrossRef]

36. Zhang, K.; Zhang, Y.; Cheng, H.-D. CrackGAN: Pavement Crack Detection Using Partially Accurate Ground Truths Based on Generative Adversarial Learning. *IEEE Trans. Intell. Transp. Syst.* **2020**, *22*, 1306–1319. [CrossRef]

37. Fiorentini, N.; Losa, M. Handling Imbalanced Data in Road Crash Severity Prediction by Machine Learning Algorithms. *Infrastructures* **2020**, *5*, 61. [CrossRef]

38. Shi, Y.; Cui, L.; Qi, Z.; Meng, F.; Chen, Z. Automatic Road Crack Detection Using Random Structured Forests. *IEEE Trans. Intell. Transp. Syst.* **2016**, *17*, 3434–3445. [CrossRef]

39. Lin, T.Y.; Maire, M.; Belongie, S.; Bourdev, L.; Girshick, R.; Hays, J.; Perona, P.; Zitnick, C.L.; Dollár, P. Microsoft COCO: Common Objects in Context. In *Proceedings of the European Conference on Computer Vision*; Springer: Cham, Switzerland, 2014; pp. 740–755. [CrossRef]

40. Chen, K.; Wang, J.; Pang, J.; Cao, Y.; Xiong, Y.; Li, X.; Sun, S.; Feng, W.; Liu, Z.; Xu, J.; et al. MMDetection: Open mmlab detection toolbox and benchmark. *arXiv* **2019**, arXiv:1906.07155. [CrossRef]

41. Liu, Y.; Chu, L.; Chen, G.; Wu, Z.; Chen, Z.; Lai, B.; Hao, Y. Paddleseg: A high-efficient development toolkit for image segmentation. *arXiv* **2021**, arXiv:2101.06175. [CrossRef]

42. Yakubovskiy, P. Segmentation Models Pytorch. 2020. Available online: https://github.com/qubvel/segmentation_models (accessed on 1 March 2022).

Article

Damage Properties of the Block-Stone Embankment in the Qinghai–Tibet Highway Using Ground-Penetrating Radar Imagery

Shunshun Qi [1,2,3], Guoyu Li [1,2,3,*], Dun Chen [1,2,3,4], Mingtang Chai [5], Yu Zhou [1,2,3], Qingsong Du [1,2,3], Yapeng Cao [1,2,3], Liyun Tang [6] and Hailiang Jia [6]

[1] State Key Laboratory of Frozen Soil Engineering, Northwest Institute of Eco-Environment and Resources, Chinese Academy of Sciences, Lanzhou 730000, China; qishunshun@nieer.ac.cn (S.Q.); chendun@lzb.ac.cn (D.C.); zhouyu@lzb.ac.cn (Y.Z.); xbdqs@lzb.ac.cn (Q.D.); caoyapeng@lzb.ac.cn (Y.C.)

[2] University of Chinese Academy of Sciences, Beijing 100049, China

[3] Da Xing'anling Observation and Research Station of Frozen-Ground Engineering and Environment, Northwest Institute of Eco-Environment and Resources, Chinese Academy of Sciences, Jagedaqi 165000, China

[4] State Key Laboratory for Geomechanics and Deep Underground Engineering, China University of Mining and Technology, Xuzhou 221116, China

[5] School of Civil and Hydraulic Engineering, Ningxia University, Yinchuan 750014, China; chaimingtang@nxu.edu.cn

[6] School of Architecture and Civil Engineering, Xi'an University of Science and Technology, Xi'an 710064, China; tangly@xust.edu.cn (L.T.); hailiang.jia@xust.edu.cn (H.J.)

* Correspondence: guoyuli@lzb.ac.cn

Citation: Qi, S.; Li, G.; Chen, D.; Chai, M.; Zhou, Y.; Du, Q.; Cao, Y.; Tang, L.; Jia, H. Damage Properties of the Block-Stone Embankment in the Qinghai–Tibet Highway Using Ground-Penetrating Radar Imagery. *Remote Sens.* **2022**, *14*, 2950. https://doi.org/10.3390/rs14122950

Academic Editors: Valerio Baiocchi, Alessandro Mei and Xianfeng Zhang

Received: 3 May 2022
Accepted: 17 June 2022
Published: 20 June 2022

Publisher's Note: MDPI stays neutral with regard to jurisdictional claims in published maps and institutional affiliations.

Abstract: The block-stone embankment is a special type of embankment widely used to protect the stability of the underlying warm and ice-rich permafrost. Under the influence of multiple factors, certain damages will still occur in the block-stone embankment after a period of operation, which may weaken or destroy its cooling function, introducing more serious damages to the Qinghai–Tibet Highway (QTH). Ground-penetrating radar (GPR), a nondestructive testing technique, was adopted to investigate the damage properties of the damaged block-stone embankment. GPR imagery, together with the other data and methods (structural characteristics, field survey data, GPR parameters, etc.), indicated four categories of damage: (i) loosening of the upper sand-gravel layer; (ii) loosening of the block-stone layer; (iii) settlement of the block-stone layer; and (iv) dense filling of the block-stones layer. The first two conditions were widely distributed, whereas the settlement and dense filling of the block-stone layer were less so, and the other combined damages also occurred frequently. The close correlation between the different damages indicated a causal relationship. A preliminary discussion of these observations about the influences on the formation of the damage of the block-stone embankment is included. The findings provide some points of reference for the future construction and maintenance of block-stone embankments in permafrost regions.

Keywords: ground-penetrating radar; Qinghai–Tibet highway; block-stone embankment; permafrost; embankment ratios

1. Introduction

The QTH, built in 1954, is a major artery connecting the interior of Chinese territory and the remote Qinghai–Tibet Plateau (QTP). Of its total length of 1937 km, 528.5 km is over continuous permafrost terrain [1]. Due to global warming, the permafrost table on the QTP has fallen significantly [2–4], and the presence of the QTH has modified the water and heat exchange conditions of the original soil on the QTP. This has created a surface temperature difference between the black asphalt pavement and the natural ground surface, which has accelerated the permafrost thaw under the highway embankment [5–11]. In addition, the sunny–shady slopes effect, which means that the sunny slope (east-facing slope) of the highway embankment absorbs more solar heat and exhibits a higher surface ground temperature, has led to the asymmetry of the thaw depth in the highway embankment [12,13]. The insulation effect of snow on highway embankment slopes in winter further warms the permafrost [14]. These effects have affected the stability of the higher and ice-rich permafrost under the embankment along the QTP engineering corridor. This destabilizing effect is the main reason for embankment damages in permafrost regions, being closely related to the temperature and ice content of the permafrost [15–17]. Therefore, a large number of special highway embankments have been built in the QTH to reduce the impact of changes in the permafrost.

Two categories of special highway embankments are currently employed in the QTH: passive and proactive temperature-controlling measures [18]. The passive measures mainly rely on maintaining the current ground temperature to slow down the degradation of the permafrost, such as thermal-insulated highway embankments [19,20]. Proactive measures have been designed to actively change the thermal condition of the permafrost so that it develops in a direction that is conducive to the stability of the embankment, such as duct-ventilated embankments, block-stone embankments, and thermosyphon embankments (Figure 1) [18,21–24]. Block-stone embankments protect the underlying permafrost by changing the embankment structure and using the convective heat transfer of air in the porous block-stone layer to increase the cooling capacity of the embankment [25,26]. However, after this system has run for a certain period, various damages occur in the block-stone embankment [27], and its cooling effect is weakened or disappears. The weakening or loss of the cooling effect will produce more serious damage in this section of the highway. Studies to date have mainly focused on the design and optimization of the structure of block-stone embankments [28–32], with little research on their damages.

The QTH is used to transport materials and a large number of tourists in and out of Tibet. Therefore, its normal operation is vital to local economic development and social stability. The traditional methods, such as hammer sounding, destructive coring, and testing pits, were used to investigate the damage characteristics of the highway that would affect the normal operation of the highway [17,22]. Meanwhile, the QTP environment is harsh. The large-scale surveys are time-consuming and labor-intensive, which is a burden on the technicians and the fragile QTP environment. In the past, investigations of QTH damages mainly focused on visual inspection, supplemented by local field surveys. The highway conditions can only be generally evaluated with limited borehole data observations since it is difficult to obtain specific information about the conditions under the embankment [33].

Ground-penetrating radar (GPR) detects and characterizes underground targets using changes in the electromagnetic characteristics of the medium, and allows the continuous, large-scale, long-distance, non-destructive detection of targets. [34–36]. Therefore, GPR is widely used in geotechnical engineering, environmental protection, archeological site detection, geological research, underground pipeline detection, detection of urban public facilities, and military detection [37–46]. Based on this, GPR has more advantages in continuous long-distance highway engineering detection. For example, Solla et al. [47] reviewed the published literature to demonstrate the method, advantages, and disadvantages of GPR in highway infrastructure detection, along with up-to-date research results, and potential. Peng et al. [48] systematically expounded on the application of GPR in highway damage detection together with typical engineering examples, proved its superiority, and discussed the future development of GPR. Krysiński et al. [49] used GPR to study pavement cracks in semi-rigid highways and analyzed the characteristics of GPR signals in detail.

There are two main applications of GPR in the permafrost regions. One is to research the distribution characteristics and degradation mechanism of permafrost, and the determination of parameters related to specific strata [50–55]. The other is to research the impact of the presence and degradation of permafrost on engineering facilities (highways, railways, airports, pipelines, etc.) in these regions [56–60]. The QTH has been seriously deformed or even damaged due to changes in the properties of the underlying permafrost. Thus, the use of GPR for QTH damage research has a very broad application potential.

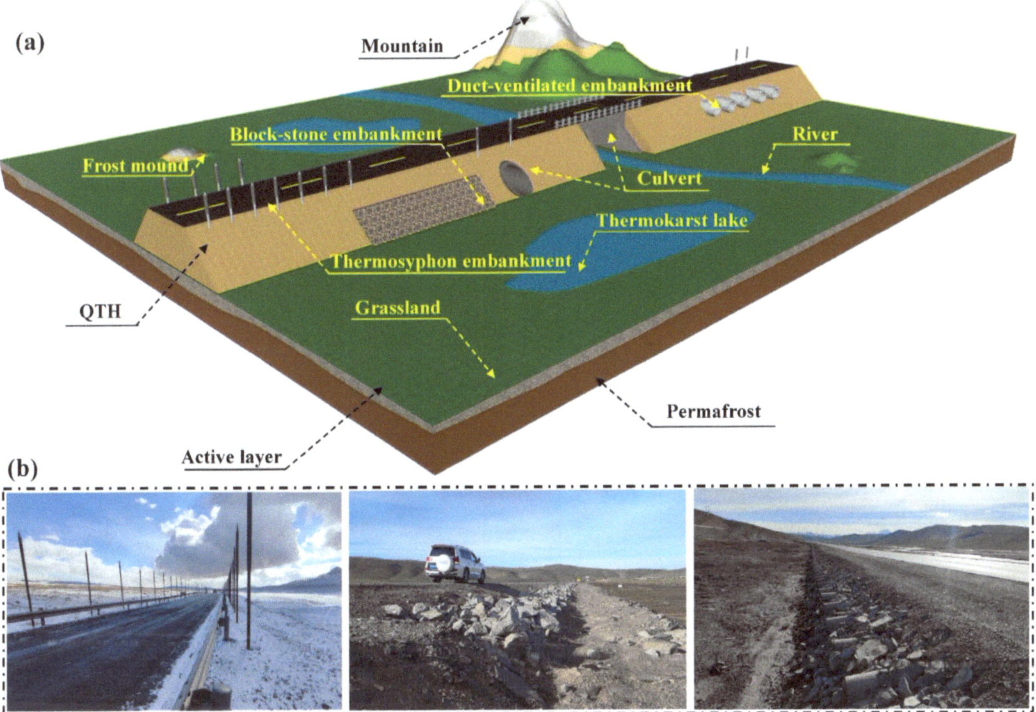

Figure 1. (**a**) Proactive temperature-controlling measures and the surrounding environment along the QTH; (**b**) field photographs, from left to right: thermosyphon embankment; block-stone embankment, and duct-ventilated embankment.

In this paper, we present a highway embankment monitoring technology based on GPR image characteristics. According to that, different damages of the block-stone embankment were identified and classified based on their GPR imagery, and the distribution and mutual relationships of different damages were investigated. Then, from the findings of the study, a preliminary discussion of the damage-causing process of the block-stone embankment is included. The research results provide references for the subsequent construction and maintenance of block-stone embankments in permafrost regions.

2. Study Area

2.1. Physical Geography of the Study Region

The permafrost on the QTP is the largest high-altitude permafrost in the world, covering an area of 1.06×10^6 km^2, and accounting for about 40.2% of the land area of the QTP [61–64]. There are many linear infrastructures such as the QTH, the Qinghai–Tibet Railway, and the Golmud–Lhasa Oil Product Pipeline. The QTH is located at an elevation of 4000–5231 m, the annual average temperature varies from −2 to −7 °C, and the annual freezing period lasts for 7–8 months. The permafrost region has poor geological conditions due to lots of periglacial phenomena (e.g., frost mounds, ice layers, and thermokarst lakes) [22,65].

The study area is the block-stone embankment section from K3024 to K3025 (Figure 2), which is located in the Kekexili Nature Reserve. The average elevation of the mountainous area is 4600–4700 m, the average annual temperature is −5.5 to −6.5 °C, the permafrost temperature is −0.5 to −1.8 °C, and the permafrost table at natural sites is 1.5–3.5 m below the ground surface [22,64]. The vegetation on both sides of the highway is sparse, and the vegetation grows at some distance from the highway.

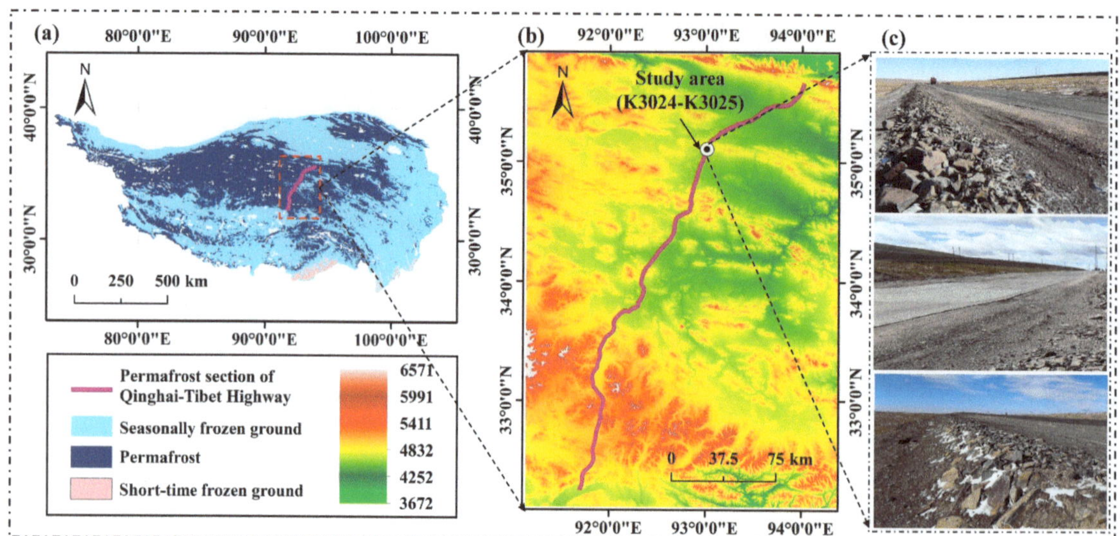

Figure 2. Study area: (**a**) spatial distribution of permafrost region in the QTP (the permafrost map is from [61,62]); (**b**) the location of k3024 to K3025 in the QTH; (**c**) field photographs of k3024 to k3025, from top to bottom: "wavy" pavement; local settlement of embankment and the block-stone embankment.

The permafrost type underlying the highway embankment in this section belongs to the icy permafrost exhibiting an ice layer with litter soil at a certain depth, classified as a sub-stable area depending on the overall stability of the entire QTH [22]. The engineering geological condition of the permafrost has been evaluated as a poor one. The overall highway in the permafrost areas is severely damaged, and the embankment has settled unevenly, with dense transverse and longitudinal cracks on the asphalt pavement. This has resulted in a "wavy" pavement as a whole and vehicle traffic ability is poor.

2.2. Block-Stone Embankment

2.2.1. The Structure and Materials of the Block-Stone Embankment

The purpose of the block-stone embankment is to reduce the temperature in the embankment by placing a layer of stone in the lower part of the highway embankment and removing the heat in the embankment in winter. Figure 3 shows the typical block-stone embankment structure. This includes the gravel layer, bottom sand-gravel layer, block-stone layer, and top sand-gravel layer from bottom to top [24,66,67].

The bottom gravel layer (general thickness 0.3–0.5 m) is in direct contact with the original ground, being densely compacted by vibration rolling. It is overlain by a layer of geotextile and sand-gravel, mainly to provide a good foundation for the block-stone layer. The overlying block-stone layer, which is the core of the block-stone embankment, is generally between 1.0 and1.5 m thick. The stone size is 150–300 mm, the porosity is not less than 25%, and the slenderness ratio is designed to be less than 3. In addition, single-layer filling and double-layer filling are selected depending on the ice content of the permafrost under the embankment. The structure shown in Figure 3 is double-layer filling.

The overlying sand-gravel layer is divided into upper and lower sand-gravel layers and is part of the auxiliary protection structure. The top sand-gravel layer (generally 0.3-m-thick) above the block-stone layer is separated by a geotextile and gravel layer. Their main function is to prevent fines from entering the block-stone layer, ensuring its convection effect. A lower 0.5-m-thick sand-gravel layer is located above the gravel layer [24,66,67].

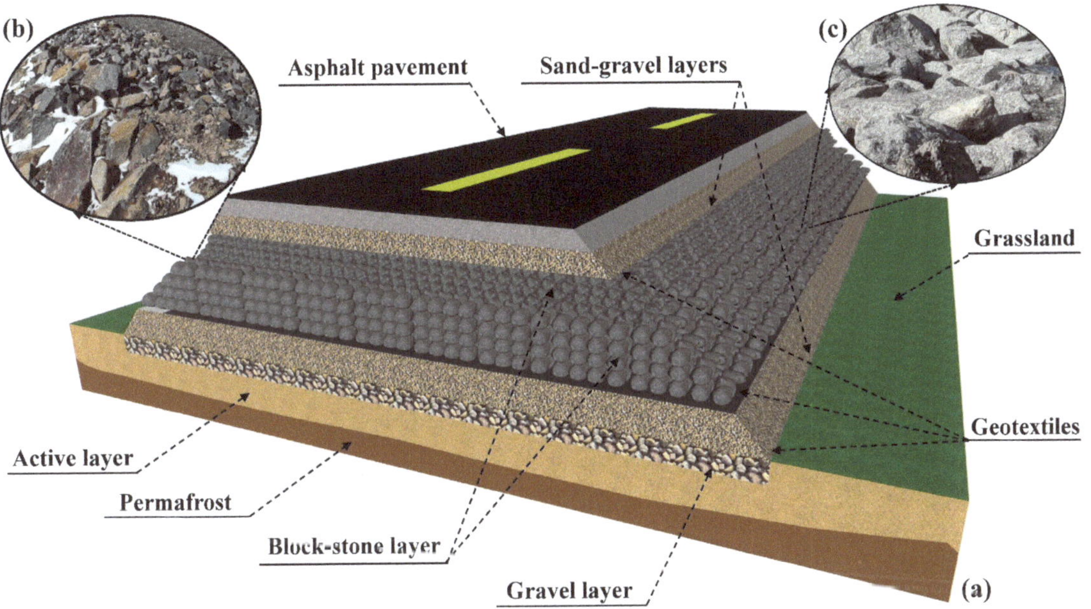

Figure 3. Schematic diagram of the block-stone embankment: (**a**) block-stone embankment structure; (**b**,**c**) field photographs of block-stone layer.

2.2.2. Working Principle of the Block-Stone Embankment

During the cold season, the cold air with a large density flows downward via the pores in the block-stone layer and the warm air rises correspondingly, causing a temperature difference by convective heat transfer, thereby removing the heat from the embankment and the active layer (Figure 4a). This cycle lowers the temperatures of the embankment and the underlying active layer and permafrost, protecting the permafrost from thaw [24,66].

During the warm season, the atmospheric heat is transferred into the embankment through the asphalt pavement and the slopes of the embankment by heat conduction, heating the air in the block-stone layer. Because the warmer air is in the upper part of the block-stone layer and the cold air is in the lower part, there is no convective heat transfer in the stone layer. The reduced heat exchange between the upper part and lower part of the block-stone layer happens only by conduction, which equivalently forms a layer of a heat insulation barrier at the bottom in summer (Figure 4b). This "thermal diode" effect takes place to cool the underlying permafrost [24,67].

2.3. Factors Leading to Block-Stone Embankment Damage

The stability of the block-stone highway embankment and the underlying permafrost is influenced by the surrounding environmental factors which may be summarized as natural or human factors.

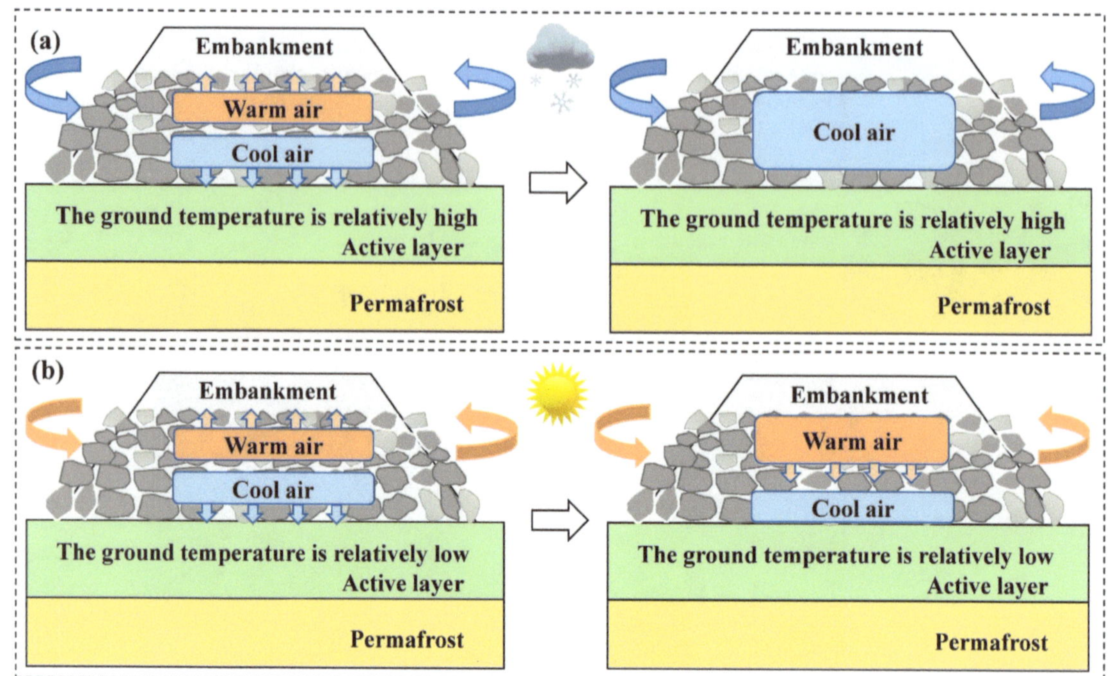

Figure 4. Schematic diagram of the working principle of a block-stone embankment: (**a**) cold season; (**b**) warm season.

2.3.1. Natural Factors

Natural factors are the main reasons for damages in block-stone embankments. They are climatic conditions and the properties of the underlying permafrost.

The natural environment on the QTP is harsh with a cold and dry climate, a freezing period for 7–8 months a year, strong solar radiation, unevenly distributed precipitation, and strong winds. This harsh environment causes severe damage to the highway, greatly reducing its service life [22,66]. In addition, the strong wind has a weathering effect on the block-stone layer, reducing the strength of the block-stones themselves and warming the permafrost under the embankment [27]. Moreover, the high ground temperature and ice content of the underlying permafrost are the main factors controlling the stability of highway embankments. Additionally, the sunny–shady effect along the QTH is another important factor leading to embankment damage [13].

2.3.2. Human Factors

Human factors, mainly including engineering construction, vehicle driving, and highway maintenance and replacement, are also important factors influencing the stability of the block-stone embankment.

Human activities destroy the thermal balance of the original ground, change the environmental and engineering geological conditions of the permafrost, and decrease the stability of the embankment [5–11]. In addition, the construction quality also affects the service life of the block-stone embankment to a great extent [24]. The traffic vehicles on the QTH are mainly heavy trucks whose long-term rolling and frequent braking lead to severe damage to the highway. Because of the harsh natural environment, highway maintenance is not immediate, which leads to gradual damage development.

3. Data and Methods

3.1. Working Principle of GPR

GPR is a geophysical method of detecting underground mediums by the reflection and refraction of electromagnetic waves in the medium [34–36]. Figure 5 shows the classical GPR process for detecting subsurface targets. The electromagnetic waves emitted by the transmitting antenna are reflected by the layers under the highway and then return to the receiving antenna (Figure 5b). The two-way travel time t of the electromagnetic wave is calculated from

$$t = \frac{\sqrt{4h^2 + x^2}}{v} \tag{1}$$

where h is the depth of the target (m) and x is the offset between the transmitting antenna and the receiving antenna (m). When the transmitting and receiving antenna are shielded, x is ignored. v is the velocity of electromagnetic wave propagation in the medium, which is given by

$$v = \frac{C}{\sqrt{\varepsilon}} \tag{2}$$

where C is the propagation velocity in free space (3×10^8 m/s) and ε is the dielectric constant of the medium. Therefore, the buried depth of the target can be calculated if the two-way travel time and dielectric constant are known. The use of this method can accurately calculate the depth of, for example, underground pipes, wires, and tombs [37,41,42], etc. The dielectric constant can be calculated if the buried depth of the target and the two-way travel time of the electromagnetic wave is known [68]. This method is mostly used in practice to estimate the dielectric constant of the mixture of actual surveys; for example, the dielectric constant of permafrost and active layer calculated by this method can inversely deduce the ice content of permafrost and the water content of the active layer [50,54,55,57].

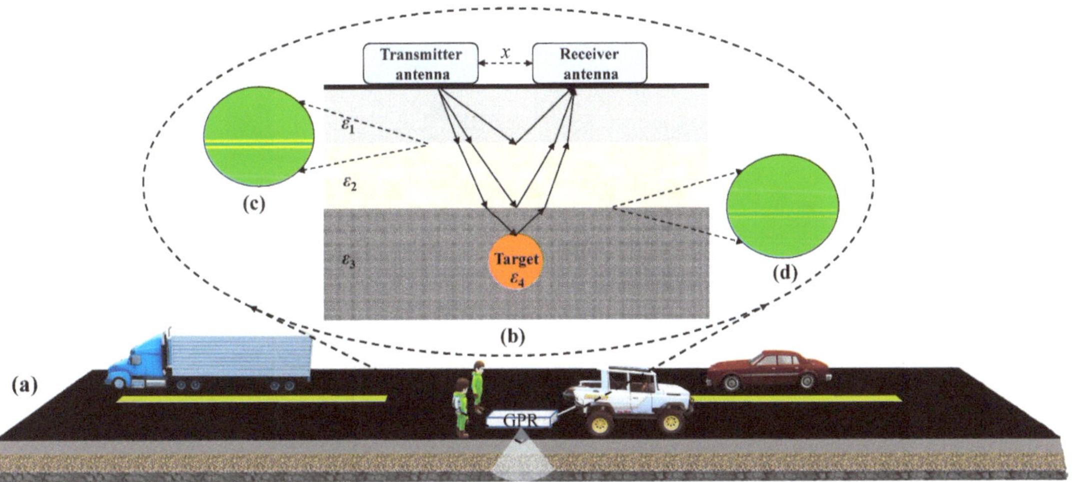

Figure 5. Schematic diagram of the GPR survey on the highway: (**a**) highway; (**b**) the electromagnetic waves emitted by the transmitting antenna are reflected by the different layers and then return to the receiving antenna; (**c**) strong electromagnetic wave reflection; (**d**) weak electromagnetic wave reflection.

In addition, the frequency for GPR detection should be selected in combination with the actual situation: a low-frequency signal has a deeper penetration but a poorer resolution, and a high-frequency antenna gives a shallower penetration depth and a better resolution [37]. At present, the GPR of many companies shield two sets of antennas with different frequencies, which can meet the requirements of depth and accuracy at the same time.

3.2. Data Acquisition

With the financial support of a Chinese research project, named the Second Tibetan Plateau Scientific and Research Expedition, XPRT Crossover CO730 two-dimensional GPR (Impulse Radar, Sweden) was used to conduct a field survey in the whole permafrost section of the QTH in October 2020 (Figure 6). This GPR system shielded antennae with frequencies centered around 70 and 300 MHz to measure different depths at different precisions. The GPR parameter settings are listed in Table 1. During the survey, the abnormal highway conditions and surrounding conditions were recorded in detail, and GPS positioning was used to locate the start and end positions of each dataset, providing references for subsequent image interpretation.

Figure 6. Diagram of the field GPR survey on QTH: (**a**) GPR survey system; (**b**) site survey.

Table 1. Parameter setting of GPR.

Parameters and Device	Setting A	Setting B
Antenna frequency (MHz)	300	70
Antenna offset (m)	0.23	0.6
Time window (ns)	93	375
Sampling rate (m)	0.05	0.05
Samples	300	300
Trigger device	Wheel	Wheel

3.3. Data Processing and Presentation

The purpose of the data processing was to suppress noise, enhance the signal, and improve the signal-to-noise ratio of the data, extracting useful information about GPR data [69].

Data processing software Reflexw (Sandmeier Geophysical Research, Karlsruhe, Germany) was used to process raw GPR data as follows (Figure 7) including 6 steps [35,70,71], i.e., input data, removing direct waves (static correction), exponential gain, 2-D filtering (average path extraction), 1-D filtering (band-pass Butterworth filtering), and 2-D filtering (moving average). The GPR data profile processing is shown in Figure 7f, and the reflection of the target horizon is clearer. In addition, the steps and methods of data processing should be adjusted according to the requirements of image interpretation to achieve the best processing effect.

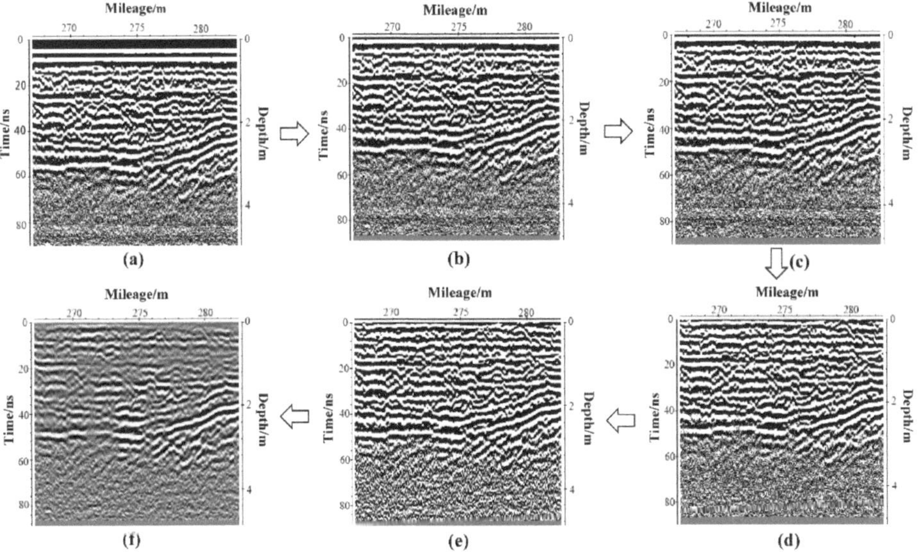

Figure 7. GPR data processing flow: (**a**) input data; (**b**) remove direct waves; (**c**) exponential gain; (**d**) average path extraction; (**e**) 1-D filtering; (**f**) 2-D filtering.

The radargrams were presented in the form of two-dimensional images. The horizontal axis was determined by highway mileage. The left-hand vertical axis shows the two-way travel time of the electromagnetic signal. The right-hand vertical axis converts the time to depth using $v = 0.11$ m/ns.

3.4. GPR Data Interpretation

The interpretation of GPR data not only needs to understand all aspects of data acquisition and processing, but also needs to combine other data and methods. In addition, due to the clear structure and materials of the block-stone embankment, it can focus more effort on the abnormal changes of different layers when interpreting the GPR data, which greatly reduces the difficulty. Therefore, the interpretation of GPR data is based on the following points: (1) field survey data, (2) effective permittivity and wave velocity, (3) reflection coefficient, (4) attenuation, and (5) GPR data analogy.

3.4.1. Field Survey Data

Detailed pavement damages, repairs, cracks, and whether the embankment settlement and block-stone layers are blocked and weathered, can provide important references for the interpretation. The condition of the embankment can directly reflect the occurrence and degree of the damage, so for the GPR data, the interpretation should consider whether or not the embankment settlement occurred in reality in the block-stone layer settlement. In addition, the environmental factors, such as vegetation conditions around the highway and thermokarst lakes, should also be fully considered.

3.4.2. Effective Permittivity and Wave Velocity

The damages to the sand-gravel and block-stone layers, whether loose or dense, will cause changes in their electrical parameters, which are largely caused by the changes in the air volume. Therefore, the Rayleigh model [72,73] introduces the calculated effective permittivity of the mixed medium, which is generally considered to be a function of the permittivity of each component and its volume content, and it can be calculated from

$$\frac{\varepsilon_{eff} - \varepsilon_1}{\varepsilon_{eff} + 2\varepsilon_1} = f\frac{\varepsilon_2 - \varepsilon_1}{\varepsilon_2 + 2\varepsilon_1} \tag{3}$$

where ε_1 and ε_2 represent the dielectric constant of the matrix medium and air in the mixture, respectively, ε_{eff} is the dielectric constant of the two-phase mixture, and f is the volume fraction of air in the mixed medium. According to the Rayleigh model, as the volume of air increases from the mixed medium, the effective permittivity will decrease, and the wave velocity will increase. Since the radargrams convert the time to the depth at a certain velocity, the increase or decrease in the wave velocity makes the same layer of the radargrams locally elongated or shortened. We need to pay attention to such changes in GPR data interpretation.

3.4.3. Reflection Coefficient

A component of the propagating GPR waves are reflected when they encounter layers composed of different mediums. The strength of the reflection from two mediums is expressed as the reflection coefficient R calculated from

$$R = \frac{\sqrt{\varepsilon_1} - \sqrt{\varepsilon_2}}{\sqrt{\varepsilon_1} + \sqrt{\varepsilon_2}} \tag{4}$$

where ε_1 and ε_2 are the dielectric constants of the medium. According to the reflection coefficient, the reflection is more obvious when the difference in the dielectric constants of the two layers is larger. (Figure 5c,d). In addition, when $R > 0$, the phase of the interface reflection wave is the same as that of the incident wave. When $R < 0$, the phase of the interface reflection wave is opposite to that of the incident wave [69].

The block-stone embankment is composed of multiple layers which are made up of different mediums, and the dielectric constant of each layer is also different. Therefore, different layers can be divided by the reflection of electromagnetic waves. A strong reflection interface, which will generate multiples, will be generated due to the large difference in the dielectric constant between the air and the block-stone. The generation of multiples is also an important method to recognize the block-stone layer.

3.4.4. Attenuation

GPR was used in the lossy medium which can cause signal attenuation. The attenuation of electromagnetic waves is related to the conductivity, dielectric, frequency scattering, and geometric spreading losses of the medium. The loss tangent, which measures the loss characteristics, can be calculated from [68]

$$tan\delta = \frac{\sigma}{\omega\varepsilon} \tag{5}$$

where δ is the loss angle, σ is the conductivity of a medium (S/m), ω is the angular frequency (rad/s), and ε is the permittivity (F/m).

In general, the air is a lossless medium, and another medium, whether a base course or a block-stone layer, will cause the attenuation of electromagnetic waves. Therefore, when the sand-gravel layer or the block-stone layer becomes loose, its attenuation will decrease. On the contrary, when the block-stone layer becomes dense, its attenuation will increase. This is also one of the methods to judge the abnormality of the sand-gravel layers and block-stone layers.

3.4.5. GPR Data Analogy

The comparison of GPR profile characteristics in similar studies is also an important method of GPR data interpretation, especially when there is a lack of certain data or controversy over the interpretation of the targets, and a comparison with previous research results can help us make judgments.

4. Results

4.1. Interpreted Results of GPR Profiles

Based on the structural characteristics of the block-stone embankment and the GPR data interpretation methods in 3.4, the damages were divided into: loosening of the upper sand-gravel layer; loosening of the block-stone layer; settlement of the block-stone layer; and dense filling of block-stone layer.

4.1.1. Loosening of the Upper Sand-Gravel Layer

Figure 8 shows the GPR images of the loosening of the upper sand-gravel layer at several different locations, and the sand-gravel layers were located at a depth of about 0.4–0.8 m. Figure 8a,b, and c all show the local weakening of the reflections at depths around 0.8-m (yellow dashed box), which was due to the damage of the geotextile. The damaged geotextile caused the upper sand-gravel to fall into the block-stone layer, which has been well-documented, and the low reflection area appeared in the block-stone layer (depth of about 0.8–2.2 m) in Figure 8c, which was the reason. From the analysis of the parameters of the GPR profile, the waveform showed low-frequency oscillation without a local strong amplitude, which was different from the waveform in the surrounding area. In general, the main reason for the loosening of the upper sand-gravel was the damage of the geotextile, which made the sand-gravel layer lack constraints.

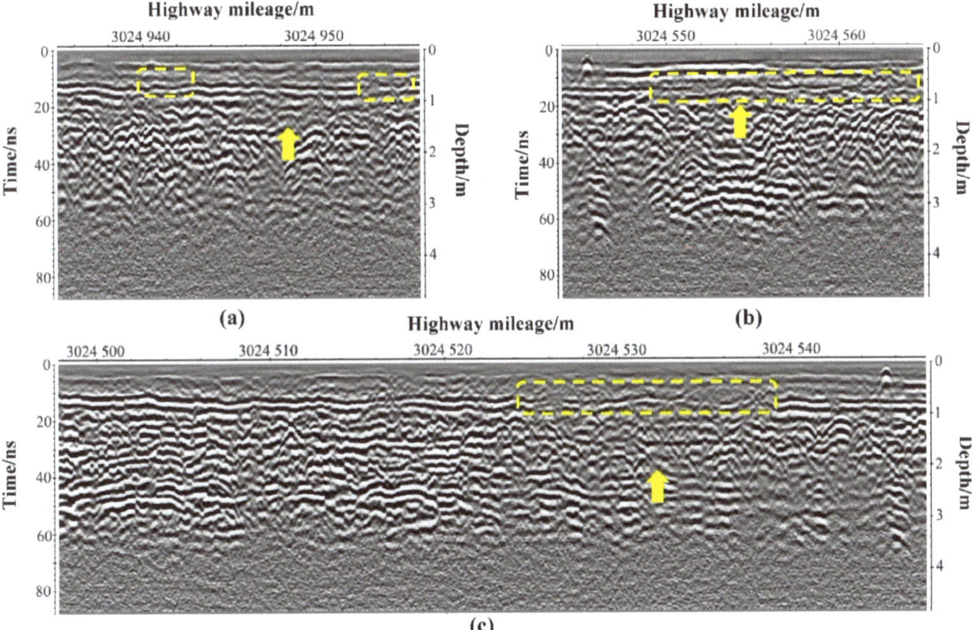

Figure 8. GPR images of the loosening of the upper sand-gravel layers. (**a**) K3024 + 935 to K3024 + 956, (**b**) K3024 + 544 to K3024 + 565; and (**c**) K3024 + 498 to K3024 + 548. The yellow dashed box shows the abnormal area of the sand-gravel layer, and the yellow arrow is used to facilitate reading.

4.1.2. Loosening of the Block-Stone Layer

Figure 9 shows two GPR images with the loosening of the block-stone layers, with a depth of about 0.8–2.4 m, which were mainly manifested as disordered electromagnetic waveforms in the block-stone layers, with a large number of short, discontinuous, and irregularly arranged reflections. This was significantly different from the GPR profile of the regularly arranged block-stone layers [74]. An abnormal signal was seen in the lower part of the block-stone layer (Figure 9a,b, red dashed box) of a strong reflection, and the amplitude increased significantly and was located in the abnormal area under the embankment, which may have been the reason leading to the loosening of the block-stone layer.

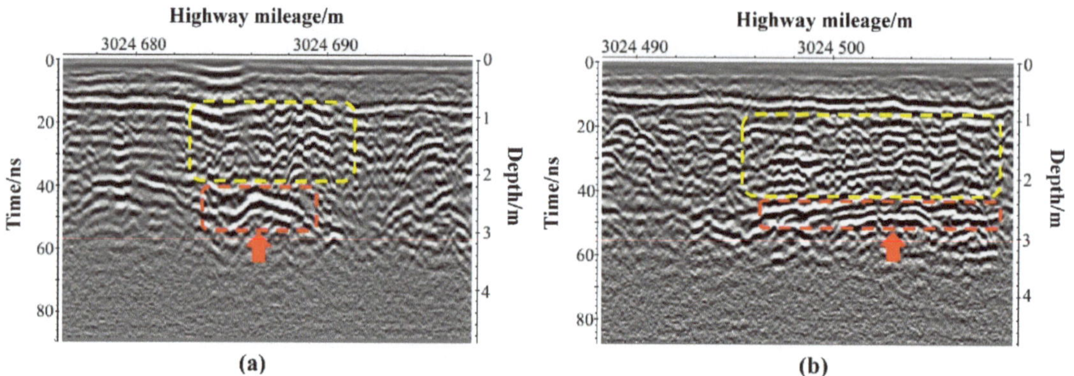

Figure 9. GPR images indicating looseness of the block-stone layers: (**a**) K3024 + 676 to K3024 + 697; (**b**) K3024 + 488 to K3024 + 509. The yellow dashed box shows the abnormal area of the block-stone layer, the red dashed box shows the abnormal area of the ground, and the yellow and red arrow is used to facilitate reading.

4.1.3. Settlement of the Block-Stone Layer

Figure 10 shows the GPR images of the block-stone layer at two locations, indicating the settlement of the block-stone layer. Figure 10a shows that the block-stone layer settled from a depth of 0.8-m on the left-hand side to a depth of 1.2-m on the right-hand side, and the sand-gravel layer also became loose with the settlement, and the geotextile was damaged on the right-hand side (yellow dashed box). Figure 10b shows that the whole block-stone layer had a settlement of about 0.2–0.4 m compared with other locations, and the settlement in the middle part was the largest. This settlement will lead to the loosening or settlement of the upper layers of the block-stone layer, which was also represented in the waveform diagram. Two irregular stone layers will be formed. The "wave" pavement will form due to these two irregular block-stone layer settlements (Figure 2c).

4.1.4. Dense Filling of Block-Stone Layer

Figure 11 shows the GPR images of the dense filling of the block-stone layer. The most notable feature is the large area of low reflection (yellow dashed box). The sand-gravel layer in both figures was abnormal, especially in Figure 11b. It is difficult to directly observe this layer in the GPR image. Only the embankment structure and GPR parameters (wave velocity, waveform, etc.) can determine the existence of this layer (shown by the red dotted line, with the bottom at a depth of about 1-m). In addition, the electromagnetic wave Rao arc generated by the block-stone was visible in the radar image.

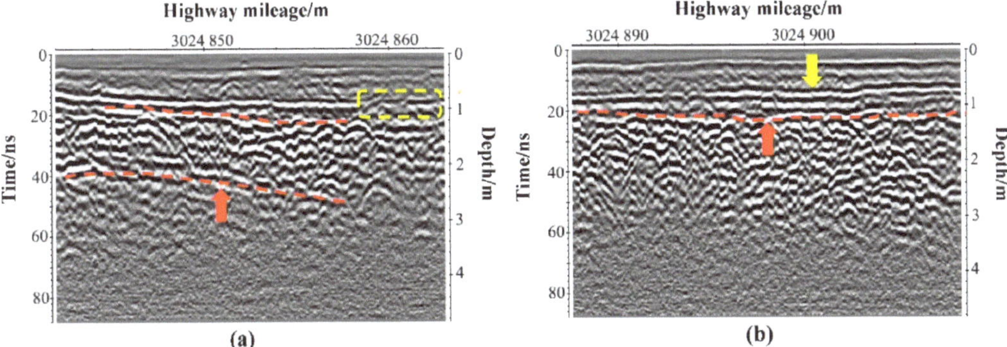

Figure 10. GPR images indicating the settlement of the block-stone layer: (**a**) K3024 + 842 to K3024 + 863; (**b**) K3024 + 887 to K3024 + 909. The yellow dashed box shows the abnormal area of the sand-gravel layer, the red dotted line indicates the settlement trend of the block-stone layer, and the yellow and red arrow is used to facilitate reading.

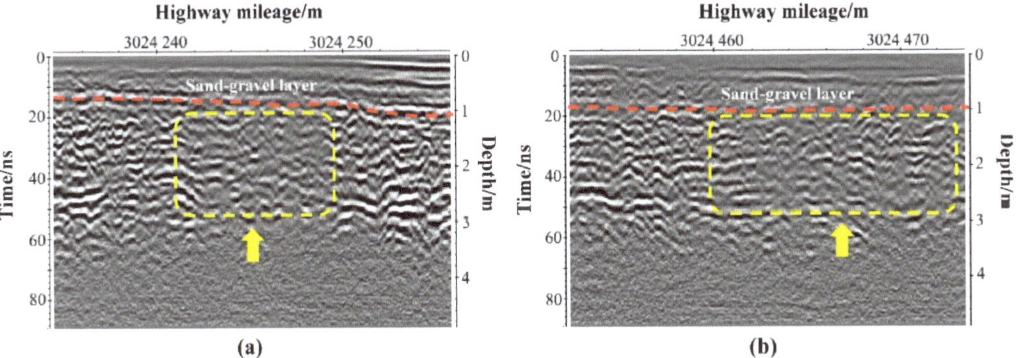

Figure 11. GPR images of an abnormally dense block-stone layer: (**a**) K 3024 + 235 to K3024 + 256; (**b**) K 3024 + 452 to K3024 + 474. The yellow dashed box shows the abnormal area of the block-stone layer, the red dotted line indicates the boundary between the sand-gravel layer and the block-stone layer, and the yellow arrow is used to facilitate reading.

GPR images cannot exhibit low reflections due to the different data processing steps, which are verified as follows (Figure 12). Figure 12a is an image processed according to the normal steps. In Figure 12b, only gain processing was performed on the image after direct wave removal. In Figure 12c, only horizontal signals were removed based on the last step. In Figure 12d, only moving-average processing was performed based on the last step. These results show that the different processing steps did not affect the results.

4.2. Distribution of Damages to the Block-Stone Embankment

The distribution of different types of damage, the overall condition of the embankment, and the relationship between various damages were studied in the K3024 to K3025 section of the block-stone embankment to further study the damage properties.

Figure 13 shows the basic properties of the embankment damage in the study section. Figure 13a shows the distribution of different types of damage. Loosening of the upper sand-gravel layer occurred most frequently. The densely filled block-stone layer occurred the least. The overall properties of the damage in the study section (Figure 13b) indicated that the damage ratio in this section was high, and the highway condition was poor.

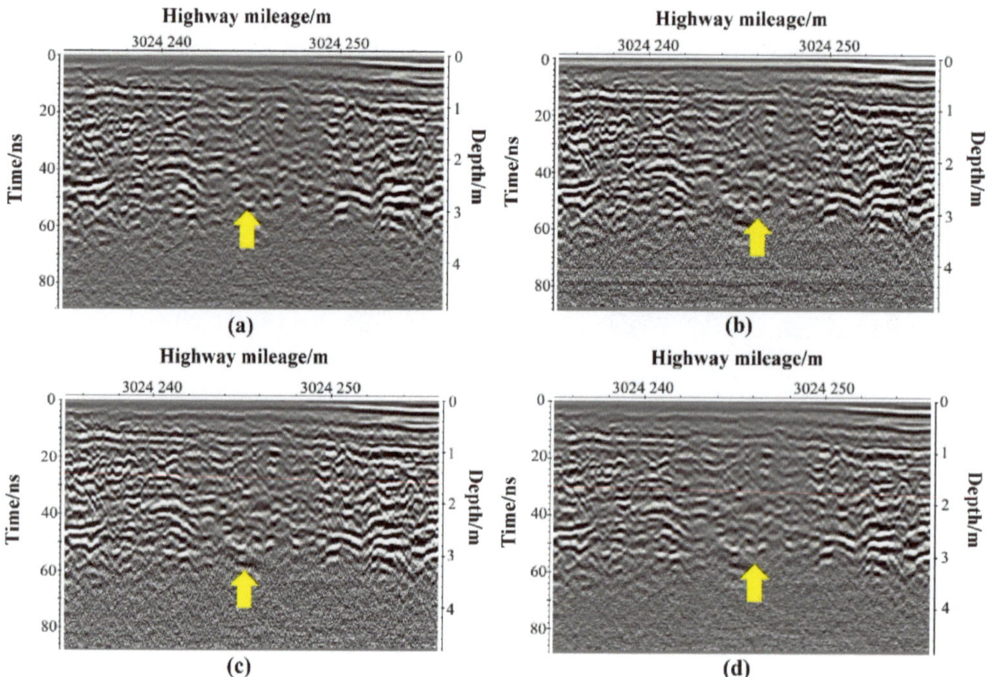

Figure 12. Comparison of the results of different processing methods for abnormally dense areas: (**a**) processed according to the normal flow; (**b**) after direct wave removal only gain processing was performed; (**c**) only horizontal signals were removed based on the previous step; (**d**) only moving-average processing was performed based on the previous step. The yellow arrow is used to increase visibility.

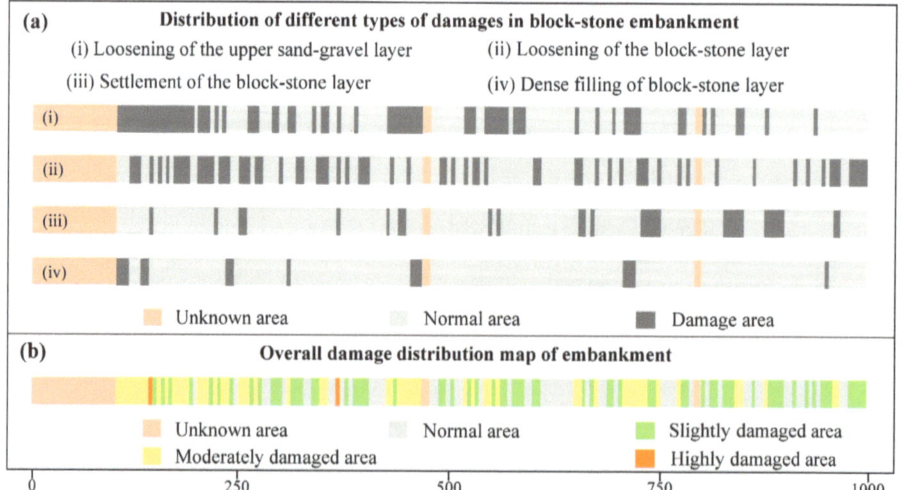

Figure 13. For the K3024 to K3025 section: (**a**) distribution of different types of damage in the block-stone embankment. Note: the unit (five meters as a unit) was considered damaged if the unit was damaged more than half, otherwise, it was a normal unit. (**b**) overall damage situation of the embankment. Note: The degree of embankment damage was based on the number of damaged species on each unit.

Figure 14 shows the distribution length of the different types of damage and the relationships between them. Figure 14a shows the damage length and damage ratio of the different types of damages. The damage of the loosening of the upper sand-gravel layer was the longest one, accounting for 40.9%. The damage of the dense filling of the block-stone layer was the shortest one, accounting for 7.39%. Figure 14b shows the length and proportion of highway sections with different damage ratios. The proportions of the normal area, slightly damaged area, and moderately damaged area were similar, accounting for 32.38%, 35.2%, and 31.25%, respectively. The proportions of the highly damaged area were equal, accounting for 1.14% of the study section.

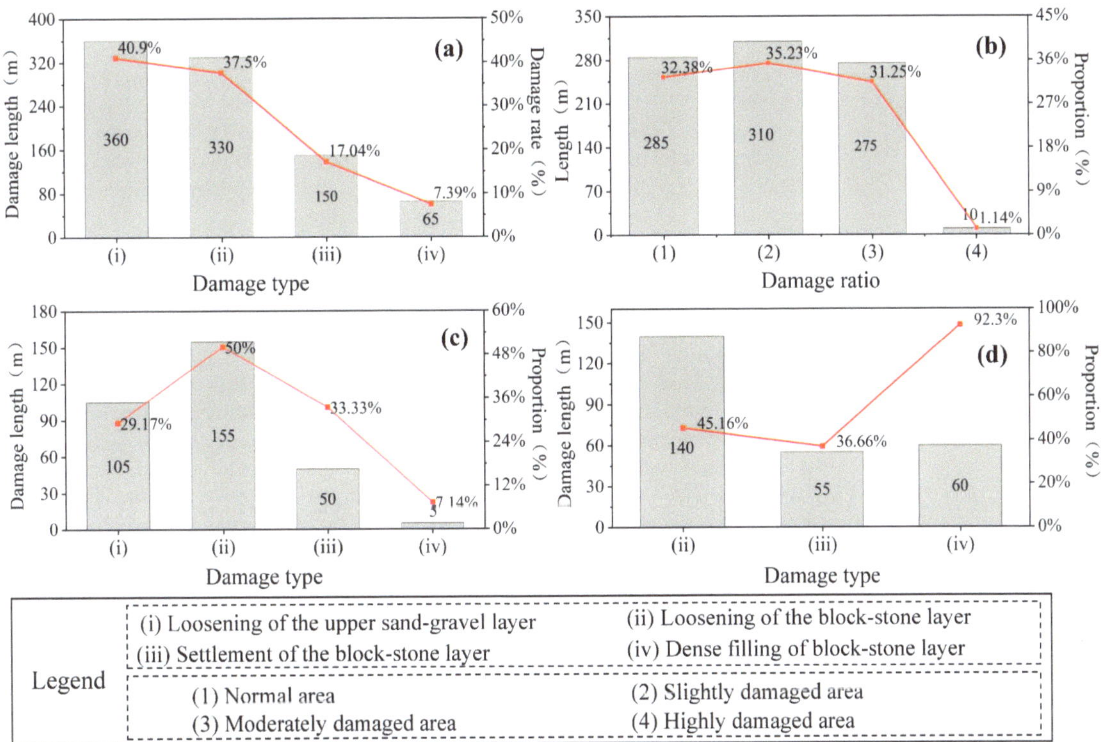

Figure 14. Relationship of different types of damage in the block-stone embankment: (**a**) shows the damaged length and rate of the different types of damage; (**b**) shows the length and proportion of highway sections with different degrees of damage; (**c**) shows the relationship between the different types of damage and the abnormality of the lower part of the embankment; (**d**) shows the relationship between the different types of damage and the loosening of the upper sand-gravel layer.

Figure 14c shows the relationship between the different types of damage and the anomaly of GPR images in the lower part (indicated by the red arrow in Figure 9) of the embankment. The damage of the loosening of the block-stone layers was the greatest one, accounting for about half. Figure 14d shows a very high correlation (92.3%) between the loosening of the upper sand-gravel layer and the dense filling of the block-stone layer, indicating that the primary reason for the dense filling was the falling and filling of the sand-gravel.

5. Discussion

The block-stone embankment is widely used in permafrost regions. However, studies to date have mainly focused on the design and optimization of the structure of block-stone embankments [28–32], with little research on their damages. The study on its damage reasons and properties is of great significance. In this study, GPR was used to conduct a relatively complete study on these damages.

GPR is a geophysical method for revealing underground medium details utilizing the reflection and refraction of electromagnetic waves within the medium [34–36]. Different parts and types of damages on the block-stone embankment have different characteristics of GPR signals, which can be observed in GPR images. For example, Figure 8 shows that the GPR images of the loosening of the upper sand-gravel layer were mainly manifested as the local weakening of the reflections at depths around 0.8-m, and the waveform showed low-frequency oscillation without local strong amplitude. Figure 9 shows GPR images of the loosening of the block-stone layers at depths of about 0.8–2.4 m, which were mainly manifested as disordered electromagnetic waveforms in the block-stone layers, with a large number of short, discontinuous, and irregularly arranged reflections. In this way, GPR imagery, together with the other data and methods (structural characteristics, field survey data, GPR parameters, etc.), indicated categories of damages. According to the above results, the distribution of different types of damages and the overall damages of the study section can be drawn (Figure 13).

The relationship between the different damages in a block-stone embankment was highly significant. The development process and causes of damages were roughly inferred according to their relationships. For example, Figure 14b shows that 50% of the loosening of the block-stone layer was related to the abnormality of the lower part of the embankment. Figure 14d shows that 92.3% of the dense filling of the block-stone layer was related to the loosening of the upper sand-gravel layer, indicating that most of the dense filling of the block-stone layer was caused by the falling and filling of the overlying sand-gravel soil. In addition, there were complex relationships between the damages and causes. For example, the thawing settlement of the permafrost under the embankment is the first prerequisite for the damage of block-stone embankment damages. These in turn cause the deterioration of the structure and function of the block-stone embankment, which leads to a breakdown in the engineering conditions of the permafrost beneath the embankment, thus completing a vicious cycle.

Based on the above research, this study has revealed a preliminary account of the formation processes of the damages (Figure 15), which are more complex and diverse in practice than the following simplified description. Figure 15a shows that the action of various influencing factors caused the thawing of the permafrost beneath the block-stone embankment, creating the hidden danger of embankment instability. Figure 15b shows that active layer changes affected the embankment, which was mainly manifested as the local loosening of the block-stone layer and local damage to the top sand-gravel layer, and partial longitudinal cracks then appeared on the pavement. These changes were exacerbated by the rolling action of vehicles on the damaged highway embankment. Figure 15c shows the worsening damage to the embankment as the block-stone layer was loosened in many places, and some of the block stones settled with the settlement of the embankment, further accelerating the local fragmentation of the top sand-gravel layer, with some of the material falling into the block-stone layer. The number and width of longitudinal cracks on the pavement increased, and transverse cracks began to appear [75]. Figure 15d shows that the interlayer space between each structural layer was further enlarged as the damage progressed, the cracks in the asphalt pavement further increased, and the embankment structure was greatly damaged. In addition, the uneven settlement of the embankment formed a "wavy" pavement, which severely affected the safety and comfortability of driving.

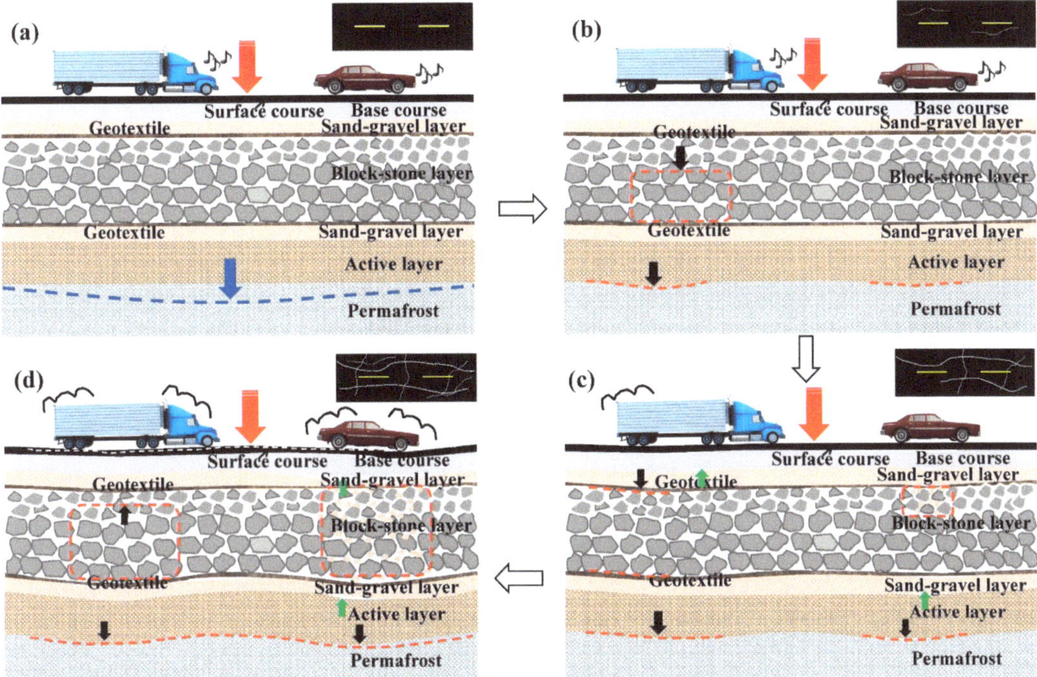

Figure 15. Progressive formation of block-stone embankment damage: (**a**) shows that the action of various influencing factors causes the thawing of the permafrost beneath the embankment; (**b**) shows the active layer changes that affected the embankment; (**c**) shows the worsening damage to the embankment; (**d**) shows that as the damage progressed, the embankment structure was greatly damaged. The red arrows indicate the influence of the external environment on the embankment (such as sun exposure and vehicle rolling), and the other dotted boxes and arrows are used to facilitate reading.

6. Conclusions

This study investigated the types, distribution, formation processes, and influencing factors of damage to block-stone embankments and their relationships using GPR imagery. The findings of the study led to the following conclusions:

(1) GPR efficiently and quickly detected and identified the damages in the study area of the block-stone embankment. Four categories of damage were determined: (i) loosening of the upper sand-gravel layer; (ii) loosening of the block-stone layer; (iii) settlement of the block-stone layer; and (iv) dense filling of the block-stone layer. Of these, (i) and (ii) were widely distributed, but (iii) and (iv) less so;

(2) Due to the complex structure of the block-stone embankment, in particular the block-stone layer itself, the attenuation of the electromagnetic wave signals was more noticeable than in the other embankment materials. Therefore, it was difficult to detect and study the deeper parts of the embankment. The quality of GPR data processing also played a very important role in image interpretation;

(3) Loosening of the sand-gravel layer was found to be the most widely distributed, and the least was the dense filling of the block-stone layer. However, the primary reason for the dense filling of the block-stone layer was the falling of the overlying sand-gravel soil. The studied section of the embankment was in the primary stage of deterioration. With further development of the damage, more of the block-stone layers will be filled with sand gravel soil;

(4) The formation of block-stone embankment damages is a complex process resulting from various factors that have close relationships with each other. The block-stone layer was put in the highway embankment to protect the underlying permafrost from the thaw. Once the damage occurs in the block stone embankment, it will weaken its cooling effect leading to worse damages.

Author Contributions: Conceptualization, S.Q., G.L., M.C. and D.C.; methodology, S.Q., M.C. and G.L.; validation, S.Q., G.L., D.C., Y.Z. and Q.D.; formal analysis, S.Q., G.L., D.C., Y.Z. and Q.D.; investigation, S.Q., G.L., D.C. and Q.D.; data curation, S.Q., G.L. and D.C.; writing—original draft preparation, S.Q. and G.L.; writing—review and editing, S.Q. and G.L.; visualization, S.Q. and G.L.; supervision, D.C., Y.Z., Q.D., G.L., M.C., Y.C., L.T. and H.J.; project administration, G.L., Y.C., L.T. and H.J.; funding acquisition, G.L. and D.C. All authors have read and agreed to the published version of the manuscript.

Funding: This research was funded by the Second Tibetan Plateau Scientific Expedition and Research Program (No.2019QZKK0905), the National Natural Science Foundation of China (No. U1703244), the Foundation of the State Key Laboratory of Frozen Soil Engineering (No. SKLFSE-ZY-20), the Foundation of the State Key Laboratory for Geomechanics and Deep Underground Engineering (No. SKLGDUEK 1904).

Data Availability Statement: Not applicable.

Acknowledgments: The data set in Figure 2a is provided by National Tibetan Plateau Data Center (http://data.tpdc.ac.cn). The authors especially thank Peifeng He and Jing Luo for their field assistance and Yan Zhang for his help with Figure 13.

Conflicts of Interest: The authors declare no conflict of interest.

References

1. Yu, S.; Zhang, J.; Liu, Y.; Wu, J. Thermal regime in the embankment of Qinghai–Tibetan Highway in permafrost regions. *Cold Reg. Sci. Technol.* **2002**, *35*, 35–44. [CrossRef]
2. Nelson, F.E.; Anisimov, O.A.; Shiklomanov, N.I. Subsidence risk from thawing permafrost. *Nature* **2001**, *410*, 889–890. [CrossRef]
3. Biskaborn, B.K.; Smith, S.L.; Noetzli, J.; Matthes, H.; Vieira, G.; Streletskiy, D.A.; Schoeneich, P.; Romanovsky, V.E.; Lewkowicz, A.G.; Abramov, A.; et al. Permafrost is warming at a global scale. *Nat. Commun.* **2019**, *10*, 264. [CrossRef] [PubMed]
4. Smith, S.L.; O'Neill, H.B.; Isaksen, K.; Noetzli, J.; Romanovsky, V.E. The changing thermal state of permafrost. *Nat. Rev. Earth Environ.* **2022**, *3*, 10–23. [CrossRef]
5. Zhongqiong, Z.; Qingbai, W.; Guanli, J.; Siru, G.; Ji, C.; Yongzhi, L. Changes in the permafrost temperatures from 2003 to 2015 in the Qinghai-Tibet Plateau. *Cold Reg. Sci. Technol.* **2019**, *169*, 102904. [CrossRef]
6. Yu, Q.; Fan, K.; You, Y.; Guo, L.; Yuan, C. Comparative analysis of temperature variation characteristics of permafrost roadbeds with different widths. *Cold Reg. Sci. Technol.* **2015**, *117*, 12–18. [CrossRef]
7. Yu, F.; Qi, J.; Lai, Y.; Sivasithamparam, N.; Yao, X.; Zhang, M.; Liu, Y.; Wu, G. Typical embankment settlement/heave patterns of the Qinghai–Tibet highway in permafrost regions: Formation and evolution. *Eng. Geol.* **2016**, *214*, 147–156. [CrossRef]
8. Peng, H.; Ma, W.; Mu, Y.-H.; Jin, L.; Yuan, K. Degradation characteristics of permafrost under the effect of climate warming and engineering disturbance along the Qinghai–Tibet Highway. *Nat. Hazards* **2014**, *75*, 2589–2605. [CrossRef]
9. Zhang, Z.; Wu, Q.; Xun, X. Radiation and energy balance characteristics of asphalt pavement in permafrost regions. *Environ. Earth Sci.* **2016**, *75*, 221. [CrossRef]
10. Zhang, Z.; Wu, Q.; Liu, Y.; Gao, S. Characteristics of water and heat changes in near-surface layers under influence of engineering interface. *Appl. Therm. Eng.* **2017**, *125*, 986–994. [CrossRef]
11. Zhang, Z.; Wu, Q.; Liu, Y.; Zhang, Z.; Wu, G. Thermal accumulation mechanism of asphalt pavement in permafrost regions of the Qinghai–Tibet Plateau. *Appl. Therm. Eng.* **2018**, *129*, 345–353. [CrossRef]
12. Alfaro, M.; Ciro, G.; Thiessen, K.; Ng, T. Case Study of Degrading Permafrost beneath a Road Embankment. *J. Cold Reg. Eng.* **2009**, *23*, 93–111. [CrossRef]
13. Ya-Ling, C.; Sheng, Y.; Ma, W. Study on the effect of the thermal regime differences in roadbed slopes on their thawing features in permafrost regions of Qinghai–Tibetan plateau. *Cold Reg. Sci. Technol.* **2008**, *53*, 334–345. [CrossRef]
14. O'Neill, B.; Burn, C. Impacts of variations in snow cover on permafrost stability, including simulated snow management, Dempster Highway, Peel Plateau, Northwest Territories. *Arct. Sci.* **2017**, *3*, 150–178. [CrossRef]
15. Wu, Q.; Zhang, Z.; Liu, Y. Long-term thermal effect of asphalt pavement on permafrost under an embankment. *Cold Reg. Sci. Technol.* **2010**, *60*, 221–229. [CrossRef]
16. Ma, W.; Mu, Y.; Wu, Q.; Sun, Z.; Liu, Y. Characteristics and mechanisms of embankment deformation along the Qinghai–Tibet Railway in permafrost regions. *Cold Reg. Sci. Technol.* **2011**, *67*, 178–186. [CrossRef]

17. Wang, S.; Niu, F.; Chen, J.; Dong, Y. Permafrost research in China related to express highway construction. *Permafr. Periglac.* **2020**, *31*, 406–416. [CrossRef]
18. Fan, K. Study on Design and Construction techniques of Special Subgrade for Permafrost Areas. Master's Thesis, Chang'an University, Xi'an, China, 2009.
19. Doré, G.; Niu, F.; Brooks, H. Adaptation Methods for Transportation Infrastructure Built on Degrading Permafrost. *Permafr. Periglac.* **2016**, *27*, 352–364. [CrossRef]
20. Zhi, W.; Yu, S.; Wei, M.; Jilin, Q. Evaluation of EPS application to embankment of Qinghai–Tibetan railway. *Cold Reg. Sci. Technol.* **2005**, *41*, 235–247. [CrossRef]
21. Qin, Y.; Zhang, J. A review on the cooling effect of duct-ventilated embankments in China. *Cold Reg. Sci. Technol.* **2013**, *95*, 1–10. [CrossRef]
22. Wang, S.; Li, Z.; Zhang, J.; Chen, J. *Highway Construction Technology on Permafrost Regions*; China Communications Press: Beijing, China, 2008; pp. 25–32.
23. Cheng, G.; Wu, Q.; Ma, W. Innovative designs of permafrost roadbed for the Qinghai-Tibet Railway. *Sci. China Ser. E Technol. Sci.* **2009**, *52*, 530–538. [CrossRef]
24. Fang, J.; Li, D.; Xu, A.; Tong, C. *Application Technology of Special Subgrade Engineering Measures in Permafrost Regions*; Lanzhou University Press: Lanzhou, China, 2016; pp. 25–32.
25. Mu, Y.; Ma, W.; Liu, Y.; Sun, Z. Monitoring investigation on thermal stability of air-convection crushed-rock embankment. *Cold Reg. Sci. Technol.* **2010**, *62*, 160–172. [CrossRef]
26. Xi, J.; Zhang, S.; Chen, J.; Jin, L.; Dong, Y. Analysis of the Cooling Effect of Block Stone Embankment at Wudaoliang Section of the Qinghai-Tibet Highway. *China J. Highw. Transport.* **2014**, *27*, 17–23. [CrossRef]
27. Liu, M.; Li, G.; Niu, F.; Lin, Z.; Shang, Y. Porosity of crushed rock layer and its impact on thermal regime of Qinghai−Tibet Railway embankment. *J. Cent. South Univ.* **2017**, *24*, 977–987. [CrossRef]
28. Liu, M.; Niu, F.; Ma, W.; Fang, J.; Lin, Z.; Luo, J. Experimental investigation on the enhanced cooling performance of a new crushed-rock revetment embankment in warm permafrost regions. *Appl. Therm. Eng.* **2017**, *120*, 121–129. [CrossRef]
29. Liu, M.; Niu, F.; Luo, J.; Yin, G.; Zhang, L. Performance, applicability, and optimization of a new slope cooling and protection structure for road embankment over warm permafrost. *Int. J. Heat Mass Transf.* **2020**, *162*, 120388. [CrossRef]
30. Zhang, M.; Lai, Y.; Li, S.; Zhang, S. Laboratory investigation on cooling effect of sloped crushed-rock revetment in permafrost regions. *Cold Reg. Sci. Technol.* **2006**, *46*, 27–35. [CrossRef]
31. Qian, J.; Yu, Q.; Guo, L.; Hu, J. Experimental study on convection characteristics of crushed-rock layer. *Can. Geotech. J.* **2013**, *50*, 834–840. [CrossRef]
32. Lai, Y.; Ma, W.; Zhang, M.; Yu, W.; Gao, Z. Experimental investigation on influence of boundary conditions on cooling effect and mechanism of crushed-rock layers. *Cold Reg. Sci. Technol.* **2006**, *45*, 114–121. [CrossRef]
33. Wang, S.; Xiong, L.; Zhang, C.; Mu, K.; Jin, L. Fuzzy expert prediction method for highway diseases in permafrost region. *J. Traffic Transp. Eng.* **2016**, *16*, 112–121. [CrossRef]
34. Davis, J.L.; Annan, A.P. Ground-Penetrating Radar for High-Resolution Mapping of Soil and Rock STRATIGRAPHY1. *Geophys. Prospect.* **1989**, *37*, 531–551. [CrossRef]
35. Blindow, N. Ground Penetrating Radar. In *Groundwater Geophysics: A Tool for Hydrogeology*; Kirsch, R., Ed.; Springer: Berlin/Heidelberg, Germany, 2006; pp. 227–252.
36. Liu, L.; Qian, R. Ground penetrating radar: A critical tool in near-surface geophysics. *Chin. J. Geophys.* **2015**, *58*, 2606–2617. [CrossRef]
37. Núñez-Nieto, X.; Solla, M.; Gómez-Pérez, P.; Lorenzo, H. GPR Signal Characterization for Automated Landmine and UXO Detection Based on Machine Learning Techniques. *Remote Sens.* **2014**, *6*, 9729–9748. [CrossRef]
38. Abouhamad, M.; Dawood, T.; Jabri, A.; Alsharqawi, M.; Zayed, T. Corrosiveness Mapping of Bridge Decks Using Image-Based Analysis of GPR Data. *Autom. Constr.* **2017**, *80*, 104–117. [CrossRef]
39. Zhang, F.; Xie, X.; Huang, H. Application of ground penetrating radar in grouting evaluation for shield tunnel construction. *Tunn. Undergr. Space Technol.* **2010**, *25*, 99–107. [CrossRef]
40. Diallo, M.; Cheng, L.; Rosa, E.; Gunther, C.; Chouteau, M. Integrated GPR and ERT data interpretation for bedrock identification at Cléricy, Québec, Canada. *Eng. Geol.* **2018**, *248*, 230–241. [CrossRef]
41. Pilecki, Z.; Krzysztof, K.; Elżbieta, P.; Andrzej, K.; Sylwia, T.-S.; Tomasz, Ł. Identification of buried historical mineshaft using ground-penetrating radar. *Eng. Geol.* **2021**, *294*, 106400. [CrossRef]
42. Iftimie, N.; Savin, A.; Steigmann, R.; Dobrescu, G.S. Underground Pipeline Identification into a Non-Destructive Case Study Based on Ground-Penetrating Radar Imaging. *Remote Sens.* **2021**, *13*, 3494. [CrossRef]
43. Forte, E.; Pipan, M.; Casabianca, D.; Di Cuia, R.; Riva, A. Imaging and characterization of a carbonate hydrocarbon reservoir analogue using GPR attributes. *J. Appl. Geophys.* **2012**, *81*, 76–87. [CrossRef]
44. Alani, A.M.; Aboutalebi, M.; Kilic, G. Applications of ground penetrating radar (GPR) in bridge deck monitoring and assessment. *J. Appl. Geophys.* **2013**, *97*, 45–54. [CrossRef]
45. Zhao, W.; Forte, E.; Pipan, M.; Tian, G. Ground Penetrating Radar (GPR) attribute analysis for archaeological prospection. *J. Appl. Geophys.* **2013**, *97*, 107–117. [CrossRef]

46. Hughes, L.J. Mapping contaminant-transport structures in karst bedrock with ground-penetrating radar. *Geophysics* **2009**, *74*, B197–B208. [CrossRef]
47. Solla, M.; Pérez-Gracia, V.; Fontul, S. A Review of GPR Application on Transport Infrastructures: Troubleshooting and Best Practices. *Remote Sens.* **2021**, *13*, 672. [CrossRef]
48. Peng, S.; Yang, F.; Xu, X. Application and Prospects of the GPR Technology on Road Damage Detection in China. In Proceedings of the Near-Surface Asia Pacific Conference, Waikoloa, Hawaii, 7–10 July 2015; pp. 476–479.
49. Krysiński, L.; Sudyka, J. GPR abilities in investigation of the pavement transversal cracks. *J. Appl. Geophys.* **2013**, *97*, 27–36. [CrossRef]
50. Liu, G.; Zhao, L.; Xie, C.; Zou, D.; Wu, T.; Du, E.; Wang, L.; Sheng, Y.; Zhao, Y.; Xiao, Y.; et al. The Zonation of Mountain Frozen Ground under Aspect Adjustment Revealed by Ground-Penetrating Radar Survey—A Case Study of a Small Catchment in the Upper Reaches of the Yellow River, Northeastern Qinghai–Tibet Plateau. *Remote Sens.* **2022**, *14*, 2450. [CrossRef]
51. Munroe, J.S.; Doolittle, J.A.; Kanevskiy, M.Z.; Hinkel, K.M.; Nelson, F.E.; Jones, B.M.; Shur, Y.; Kimble, J.M. Application of ground-penetrating radar imagery for three-dimensional visualisation of near-surface structures in ice-rich permafrost, Barrow, Alaska. *Permafr. Periglac.* **2007**, *18*, 309–321. [CrossRef]
52. Luo, J.; Niu, F.-J.; Lin, Z.-J.; Liu, M.-H.; Yin, G.-A. Variations in the northern permafrost boundary over the last four decades in the Xidatan region, Qinghai–Tibet Plateau. *J. Mt. Sci. Eng.* **2018**, *15*, 765–778. [CrossRef]
53. Sjöberg, Y.; Marklund, P.; Pettersson, R.; Lyon, S.W. Geophysical mapping of palsa peatland permafrost. *Cryosphere* **2015**, *9*, 465–478. [CrossRef]
54. Wang, Q.; Shen, Y. Calculation and Interpretation of Ground Penetrating Radar for Temperature and Relative Water Content of Seasonal Permafrost in Qinghai-Tibet Platea. *Electronics* **2019**, *8*, 731. [CrossRef]
55. Du, E.; Zhao, L.; Wu, T.; Li, R.; Yue, G.; Wu, X.; Li, W.; Jiao, Y.; Hu, G.; Qiao, Y.; et al. The relationship between the ground surface layer permittivity and active-layer thawing depth in a Qinghai–Tibetan Plateau permafrost area. *Cold Reg. Sci. Technol.* **2016**, *126*, 55–60. [CrossRef]
56. Stephani, E.; Fortier, D.; Shur, Y.; Fortier, R.; Doré, G. A geosystems approach to permafrost investigations for engineering applications, an example from a road stabilization experiment, Beaver Creek, Yukon, Canada. *Cold Reg. Sci. Technol.* **2014**, *100*, 20–35. [CrossRef]
57. Shen, Y.; Zuo, R.; Liu, J.; Tian, Y.; Wang, Q. Characterization and evaluation of permafrost thawing using GPR attributes in the Qinghai-Tibet Plateau. *Cold Reg. Sci. Technol.* **2018**, *151*, 302–313. [CrossRef]
58. Xiao, J.; Liu, L. Permafrost Subgrade Condition Assessment Using Extrapolation by Deterministic Deconvolution on Multifrequency GPR Data Acquired Along the Qinghai-Tibet Railway. *IEEE J. Sel. Top. Appl. Earth Obs. Remote Sens.* **2016**, *9*, 83–90. [CrossRef]
59. Wang, Y.; Jin, H.; Li, G. Investigation of the freeze–thaw states of foundation soils in permafrost areas along the China–Russia Crude Oil Pipeline (CRCOP) route using ground-penetrating radar (GPR). *Cold Reg. Sci. Technol.* **2016**, *126*, 10–21. [CrossRef]
60. Jørgensen, A.S.; Andreasen, F. Mapping of permafrost surface using ground-penetrating radar at Kangerlussuaq Airport, western Greenland. *Cold Reg. Sci. Technol.* **2007**, *48*, 64–72. [CrossRef]
61. Zou, D.; Zhao, L.; Sheng, Y.; Chen, J.; Hu, G.; Wu, T.; Wu, J.; Xie, C.; Wu, X.; Pang, Q.; et al. A new map of permafrost distribution on the Tibetan Plateau. *Cryosphere* **2017**, *11*, 2527–2542. [CrossRef]
62. Lin, Z. A new map of permafrost distribution on the Tibetan Plateau (2017). *Cryosphere* **2019**, *11*, 2527–2542. [CrossRef]
63. Cheng, G.; Zhao, L.; Li, R.; Wu, X.; Sheng, Y.; Hu, G.; Zhou, D.; Jin, H.; Li, X.; Wu, Q. Characteristic, changes and impacts of permafrost on Qinghai-Tibet Plateau. *Chin. Sci. Bull.* **2019**, *64*, 2783–2795. [CrossRef]
64. Wu, Q.; Zhang, Z.; Liu, G. Relationships between climate warming and engineering stability of permafrost on Qinghai-Tibet plateau. *J. Eng. Geol.* **2021**, *29*, 342–352. [CrossRef]
65. Wang, S.; Wang, Z.; Yuan, K.; Zhao, Y. Qinghai-Tibet Highway Engineering Geology in Permafrost Regions: Review and Prospect. *China J. Highw. Transp.* **2015**, *28*, 9. [CrossRef]
66. Song, Z. Study on the Adaptability of Block-Rock-Embankment in Permafrost Regions. Master's Thesis, Beijing Jiaotong University, Beijing, China, 2012.
67. Lai, Y.M.; Zhang, M.Y.; Li, S.Y. *Theory and Application of Cold Regions Engineering*; Science Press: Beijing, China, 2009; pp. 43–92.
68. Zeng, Z.F.; Liu, S.X.; Wang, Z.J.; Xue, J. *Principle and Application of Ground-Penetrating Radar*; Science Press: Beijing, China, 2006; 119p.
69. Benedetto, A.; Tosti, F.; Ciampoli, L.B.; D'Amico, F. An overview of ground-penetrating radar signal processing techniques for road inspections. *Signal Process.* **2017**, *132*, 201–209. [CrossRef]
70. Shu, Z.L.; Liu, B.X.; Liu, X.R.; Zhu, C.H. *Forward and Inverse Theory and Signal Processing of Ground-Penetrating Radar*; Science Press: Beijing, China, 2017; pp. 1–10.
71. Reflexw. *User Guide of Reflexw Computer Program*; Sandmeier Geophysical Research: Karlsruhe, Germany, 2018.
72. Sihvola, A.H.; Alanen, E. Studies of mixing formulae in the complex plane. *IEEE Trans. Geosci. Remote* **1991**, *29*, 679–687. [CrossRef]
73. Shivola, A.H. Self-consistency aspects of dielectric mixing theories. *IEEE Trans. Geosci. Remote* **2002**, *27*, 403–415. [CrossRef]

74. Wei, K.; Song, X.; Zhou, S.; Zhang, H. Characteristics of ground-penetrating radar response to abnormal defects of road gravel cushion. *J. Hohai. Univ.* **2015**, *43*, 133–138. [CrossRef]
75. Chai, M.; Li, G.; Ma, W.; Chen, D.; Du, Q.; Zhou, Y.; Qi, S.; Tang, L.; Jia, H. Damage characteristics of the Qinghai-Tibet Highway in permafrost regions based on UAV imagery. *Int. J. Pavement Eng.* **2022**, 247074867. [CrossRef]

Article

A Fast Inference Vision Transformer for Automatic Pavement Image Classification and Its Visual Interpretation Method

Yihan Chen [1,2], Xingyu Gu [1,2,*], Zhen Liu [1,2] and Jia Liang [1,2]

1   Department of Roadway Engineering, School of Transportation, Southeast University, Nanjing 211189, China; 220203239@seu.edu.cn (Y.C.); 230208344@seu.edu.cn (Z.L.); liangjiahs@seu.edu.cn (J.L.)

2   National Demonstration Center for Experimental Road and Traffic Engineering Education, Southeast University, Nanjing 211189, China

*   Correspondence: guxingyu1976@seu.edu.cn; Tel.: +86-025-86342563

Abstract: Traditional automatic pavement distress detection methods using convolutional neural networks (CNNs) require a great deal of time and resources for computing and are poor in terms of interpretability. Therefore, inspired by the successful application of Transformer architecture in natural language processing (NLP) tasks, a novel Transformer method called LeViT was introduced for automatic asphalt pavement image classification. LeViT consists of convolutional layers, transformer stages where Multi-layer Perception (MLP) and multi-head self-attention blocks alternate using the residual connection, and two classifier heads. To conduct the proposed methods, three different sources of pavement image datasets and pre-trained weights based on ImageNet were attained. The performance of the proposed model was compared with six state-of-the-art (SOTA) deep learning models. All of them were trained based on transfer learning strategy. Compared to the tested SOTA methods, LeViT has less than 1/8 of the parameters of the original Vision Transformer (ViT) and 1/2 of ResNet and InceptionNet. Experimental results show that after training for 100 epochs with a 16 batch-size, the proposed method acquired 91.56% accuracy, 91.72% precision, 91.56% recall, and 91.45% F1-score in the Chinese asphalt pavement dataset and 99.17% accuracy, 99.19% precision, 99.17% recall, and 99.17% F1-score in the German asphalt pavement dataset, which is the best performance among all the tested SOTA models. Moreover, it shows superiority in inference speed (86 ms/step), which is approximately 25% of the original ViT method and 80% of some prevailing CNN-based models, including DenseNet, VGG, and ResNet. Overall, the proposed method can achieve competitive performance with fewer computation costs. In addition, a visualization method combining Grad-CAM and Attention Rollout was proposed to analyze the classification results and explore what has been learned in every MLP and attention block of LeViT, which improved the interpretability of the proposed pavement image classification model.

Keywords: pavement distress; image classification; deep learning; vision transformer; LeViT; visual interpretation

Citation: Chen, Y.; Gu, X.; Liu, Z.; Liang, J. A Fast Inference Vision Transformer for Automatic Pavement Image Classification and Its Visual Interpretation Method. *Remote Sens.* **2022**, *14*, 1877. https://doi.org/10.3390/rs14081877

Academic Editors: Valerio Baiocchi, Alessandro Mei and Xianfeng Zhang

Received: 24 March 2022
Accepted: 11 April 2022
Published: 13 April 2022

Publisher's Note: MDPI stays neutral with regard to jurisdictional claims in published maps and institutional affiliations.

1. Introduction

Cracks are common pavement distress caused by vehicle loading and environmental conditions. Pavement cracking can cause damage to pavement structures, and the area of pavement structure damage can increase over time, affecting the performance of the pavement. Therefore, the timely detection and monitoring of pavement cracks and other distress is essential for road maintenance [1,2].

The traditional manual pavement distress detection method is inefficient and significantly affected by subjective factors, restricting the development of maintenance intelligence. In response to such problems, benefiting from the development of artificial intelligence technology, researchers have continuously applied the latest image-processing methods to pavement disease recognition [3]. Deep learning is the state-of-the-art (SOTA) artificial intelligence method. Compared with traditional image-processing and machine-learning methods, complex image preprocessing is not required in deep learning. Deep learning has the characteristics of automatic feature learning, which can significantly improve the efficiency of pavement distress recognition. It is a general trend to collect pavement images using automated detection equipment and adopt advanced algorithms to identify pavement distress [4,5].

Convolutional Neural Network (CNN) has been well established in image recognition among all the deep learning methods in pavement distress identification. Since AlexNet was proposed in 2012 [6], new CNN structures have been introduced every year. Breakthroughs in early CNN structures such as VGG [7] and MSRANet [8] were mainly in larger width and depth, resulting in more neurons and parameters. This may cause overfitting and requires a large number of computation resources. The application of residual blocks significantly improved the efficiency of training a CNN and was widely used in the architectures of some SOTA CNN, such as ResNet [9] and DenseNet [10]. In the last 5 years, extensive research has been conducted and has shown that the application of CNN in pavement distress identification is feasible and scientific. Wang's research team proposed several CNN-based crack detection models named CrackNet and improved versions [11–13]. Because training a CNN is time-consuming, the improvement direction of their proposed models is mainly higher detection accuracy and faster performance. Hou et al. [14] proposed an adaptive, lightweight CNN pavement object classification model named MobileCrack to achieve more efficient training. Ali et al. [15] introduced a customized CNN for concrete crack detection. By comparing existing well-known CNN models, they found that the number of learnable parameters of models can significantly affect the computational time. Kim et al. [16] employed a shallow CNN to acquire higher identification accuracy with minimum computation. All of this research showed that computation time and accuracy are critical for applying deep learning in pavement distress recognition. However, CNNs still have the problem that the lower-level features outside the effective receptive fields cannot be described [17,18]. It is not conducive to making full use of the context information to capture features of images. Continuously stacking deeper convolutional layers can help get more levels of image features, but it will cause a sharp increase in computation.

Recurrent neural networks (RNNs), including LSTM [19] and GRU [20], are also typical deep learning methods. RNNs have the advantage that they can combine contextual information and are widely used in speech recognition. Sequential data are required to conduct the RNN model, which differs from the pavement image dataset. Limited studies about RNN-based pavement image recognition have been completed. Zhang et al. [21] defined the sequence of pixels in an image as the input sequence of RNN. They proposed CrackNet-R for pavement crack detection, which revealed the potential of sequence-based deep learning in pavement image recognition. In the application of computer vision in pavement image recognition, CNNs remain dominant.

Self-attention is a structure to help deal with the problem of parallelization within training examples in recurrent models [22]. A network called Transformer entirely based on attention mechanism and eliminates recurrence and convolutions was proposed to solve the parallel processing of words in a sentence in RNN-based models and achieved considerable success in natural language processing [22,23]. Inspired by the successful application of Transformer in natural language processing, an image classification model that displaced the traditional convolutional networks and was based entirely on Transformer architecture called Vision Transformer (ViT) was initially introduced in computer vision and turned out to have competitive performance in a large-scale image database [24]. Nevertheless, the original ViT model still has some problems, such as its poor generalization ability on inadequate data. Thus, more and more improved versions of ViT-based image recognition methods were proposed recently [25–28].

A typical Transformer block contains multi-head self-attention, skip connection, and normalization. In terms of pavement image recognition, researchers have also made preliminary attempts to apply Transformer to pavement distress detection. Liu et al. [29] employed a network called CrackFormer using the self-attention modules embedded in the CNN structure. Guo et al. [30] adopted a vision transformer structure to detect the crack boundary and found it precedes several SOTA schemes. These works focus on the edge detection of crack images. Image classification is a fundamental task in deep learning and is the backbone of other complex tasks such as object detection and semantic segmentation. However, deep learning-based object classification in pavement images remains challenging because of the inconspicuous features, low contrast and noisy background, and high economic cost of computation. To solve the above problems, a faster inference vision transformer model called LeViT [31] was applied in pavement object classification in this study. It combined the convolutional operation in the original ViT structure and achieved competitive classification accuracy with a relatively higher processing speed.

On the other hand, as deep learning-based algorithms were once addressed as 'black box', the interpretability matters of deep learning have been a problem for many years [32,33]. Moreover, as deep learning models become increasingly complex, it is also becoming increasingly difficult for us to understand our model [34]. To build up trust in the AI-based pavement distress recognition models for engineers, ensuring the ability to understand why a particular prediction is made is critical. Therefore, in our study, a visual interpretation method for both multilayer perceptron (MLP) and self-attention blocks was proposed to better understand the mechanism of the employed deep learning model. Our work is a potential solution for the interpretability matters in deep learning-based pavement distress detection models.

The rest of this paper is structured as follows: Section 2 presents the methodology, including the introduction of the original ViT methods, the overall structure of the LeViT model, visual explanation methods, and the evaluation matrix; Section 3 presents the experimental results of the proposed classification model and the analysis of the visual interpretation results; Section 4 presents the discussion and comparison of other SOTA models; and, finally, Section 5 offers the research conclusion.

2. Methodology and Materials

2.1. Overall Procedure

Figure 1 shows the framework of the proposed work. We collected the pavement images from three distinct sources, forming three different datasets for comparison experiments. The original images were then preprocessed to a smaller size that meets the computer training requirements. Next, the three datasets were divided in the ratio of 6:2:2 to form the training set, validation set, and test set, respectively. A transformer-based deep learning approach called LeViT was used to automatically classify the pavement images. During training, pre-trained weights based on the ImageNet were obtained, and then the model was trained based on our datasets using the transfer learning strategy. The proposed method was also compared with other prevailing deep learning networks. Subsequently, a visualization strategy was proposed to visualize the classification results and what has been learned at each layer of our proposed network.

2.2. Data Acquisition

Three different pavement datasets are used in our experiments.

(1) JCAPs: The first dataset was collected from the multi-functional intelligent road detection vehicle on asphalt pavement on Lanhua Road, Nanjing, Jiangsu Province, China, in April 2018. This multi-functional pavement detection vehicle was equipped with onboard computers and embedded integrated multi-sensor synchronous control units to automatically capture pavement pictures. It took about half an hour for the vehicle to collect the original images. The original data collected were RGB images of 4096×2000 pixels. To satisfy the training requirements, the original images were processed to a proper size. First, the original images were horizontally

resized to 4000 × 2000 pixels via bilinear interpolation. Then, the resized images were continuously clipped to 400 × 400 pixels. Subsequently, the sub-images were resized to 224 × 224 pixels. To balance the number of images of various samples in the dataset, 1600 images were selected, including 400 pavement background images, pavement marking images, crack images, and sealed crack images. In terms of the crack images, transverse and longitudinal crack images are included in the dataset. This dataset is named as JCAPs. Example images of different classifications in JCAPs are shown in Figure 2. As the images were affected by the equipment and lighting conditions, the original dataset contained road images with good lighting exposure and poor lighting exposure. In addition, the cracks in this image dataset have a relatively smaller width, and the image feature is inconspicuous, which may increase the difficulty of classification for deep learning models.

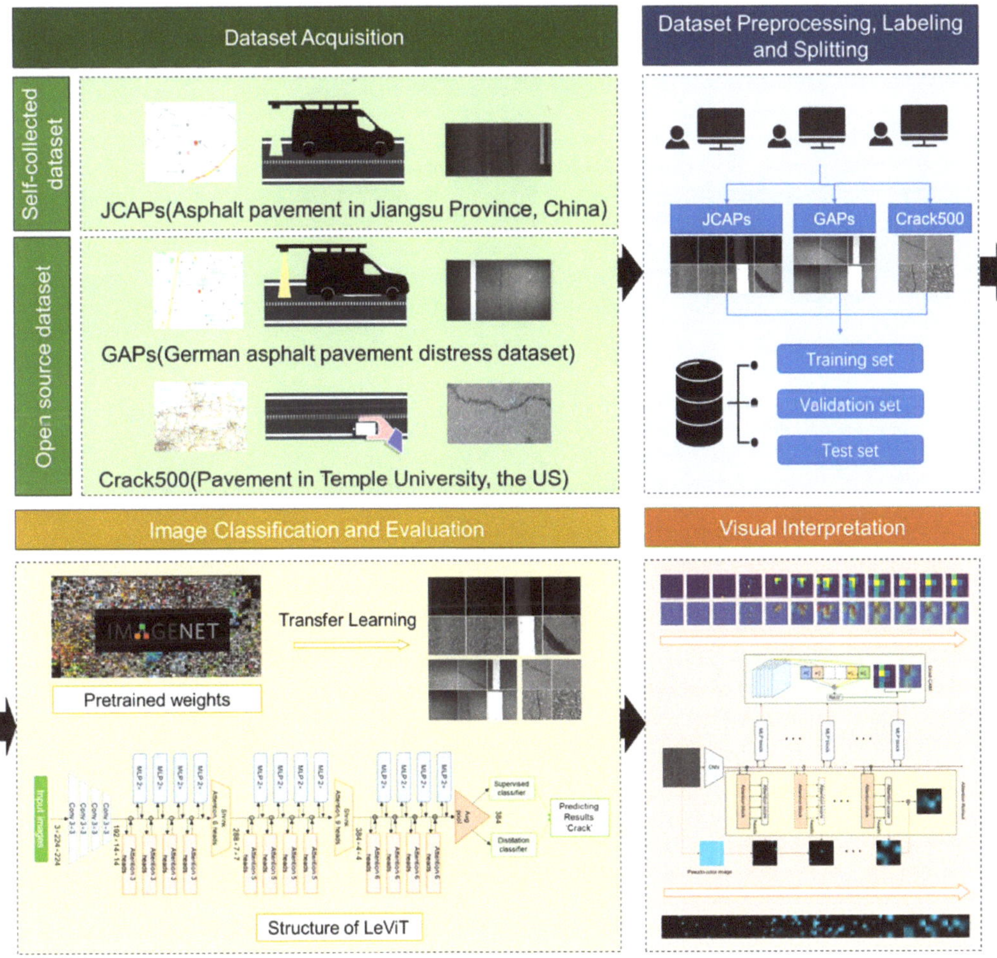

Figure 1. The framework of the proposed work.

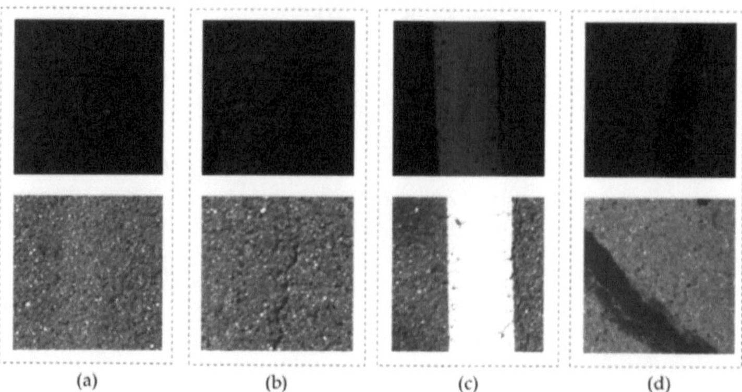

Figure 2. Example images from the JAPs dataset used in this study: (**a**) background, (**b**) crack, (**c**) pavement marking, and (**d**) sealed crack.

(2) GAPs: The German Asphalt Pavement Distress (GAPs) dataset is an open-source dataset with various classes of distress using a mobile mapping system. Four-year cycle images are contained in the original GAPs dataset. The resolution of the original images is 1920 × 1080 pixels. More details of the dataset are presented in [35]. We randomly selected several of the original images and cropped the original GAPs images into tiny images with 400 × 400 pixels using the sliding-window method and resized them into 224 × 224 pixels to prevent the problem of running out of memory while computing. A total of 1200 images with the size of 224 × 224 were manually labeled, including 400 pavement background images, pavement marking images, and crack images. Longitudinal and transverse cracking are included in the crack images. Figure 3 shows the example images from the GAPs dataset.

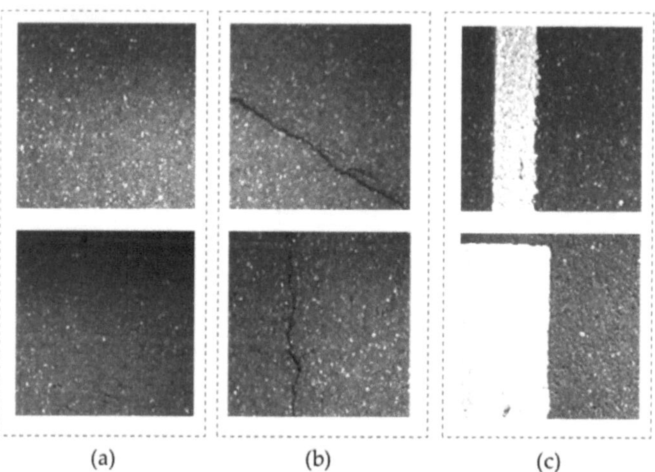

Figure 3. Example images from the GAPs dataset used in this study: (**a**) background, (**b**) crack, and (**c**) pavement marking.

(3) Crack500: This dataset is also an open-source dataset collected from the main campus of Temple University, Philadelphia, Pennsylvania, U.S. The resolution of the original images in the Crack500 dataset is 2000 × 1500 pixels. More details of the dataset are presented in [36,37]. We also randomly selected several of the original images and cropped the original images into 500 × 500 pixels using the sliding-window method. The tiny images were also resized to 224 × 224 pixels before inputting into

the training networks. A total of 2000 images with the size of 224 × 224 were manually categorized, including 1000 pavement background images and 1000 crack images. The crack images include longitudinal and transverse cracks. Figure 4 shows the example preprocessed images in the Crack500 dataset.

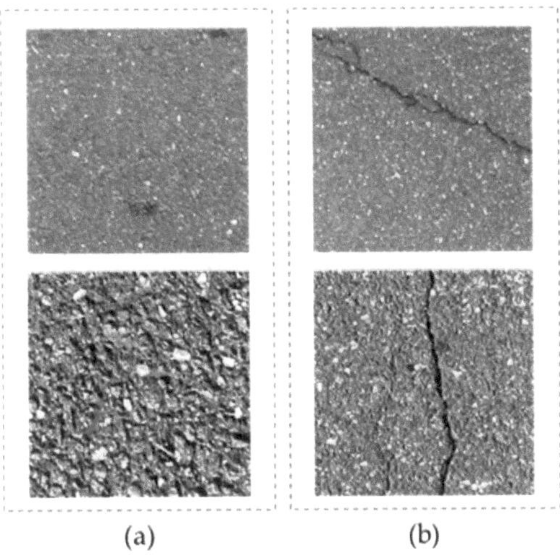

(a)   (b)

Figure 4. Example images from the Crack500 dataset used in this study: (**a**) background and (**b**) crack.

In this study, 60% of every dataset was used as the training set for network training, 20% of every dataset was used as the validation set for network validation and monitoring, and 20% was allocated as the test set to evaluate different models. The different classes of samples in the dataset had a balanced distribution. Table 1 describes the distribution of pavement images in each classification in every dataset.

Table 1. Distribution of different datasets used in this study.

	Datasets	Number of Images	Background	Crack	Pavement Marking	Sealed Crack
JCAPs	Training set	960	240	240	240	240
	Validation set	320	80	80	80	80
	Test set	320	80	80	80	80
	All	1600	400	400	400	400
GAPs	Training set	720	240	240	240	-
	Validation set	240	80	80	80	
	Test set	240	80	80	80	-
	All	1200	400	400	400	-
Crack500	Training set	1200	600	600	-	-
	Validation set	400	200	200	-	
	Test set	400	200	200	-	-
	All	2000	1000	1000	-	-

2.3. Vision Transformer

The structure of ViT entirely replaces the convolution operation in image recognition. The original architecture of ViT splits images into patches. Then, the patches are flattened and linear projected into learnable features. To acquire the position information of image patches, position encoding is added to the Patch embedding. Moreover, an extra learnable class embedding, named class token, is also input into the transformer encoder. Layer normalization and two fully connected layers are included in the classifier. The output of the class token is taken as the classification result of the whole network. Figure 5 shows the framework of the original ViT.

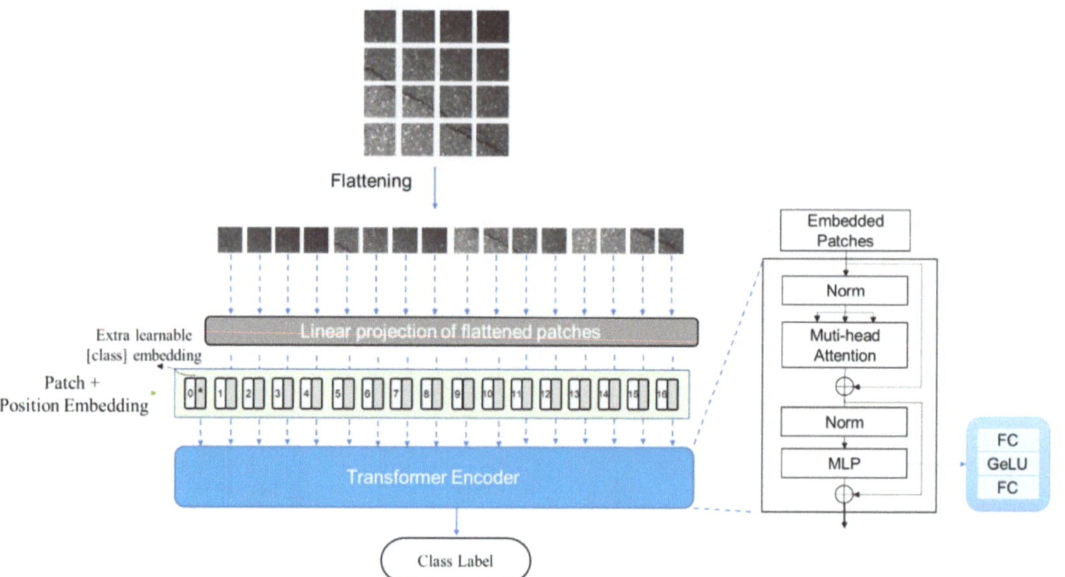

Figure 5. Structure of Vision Transformer (ViT).

Residual connection is applied in the transformer encoder. This structure was first proposed in ResNet [9]. It allows lower layers to directly connect to higher layers and skip one or more middle layers. It does not introduce additional parameters or add computation complexity but can help improve the accuracy with the same computation resources.

Multi-head attention is the core block in transformer models. The architecture of attention was firstly introduced in 2014 [38]. The neural network can pay more attention to relevant information in input vectors using an attention mechanism. An illustration of the multi-head self-attention is presented in Figure 6. The input vector of embedded patches after normalization is I. By multiplying with three different trainable weight matrices W^q, W^k, and W^v, the input of multi-head attention is query Q, key K, and value V. All of them are linear projected, and then scaled dot-product is applied to calculate attention score, seen in Equation (1) [22], and SoftMax function is used as alignment function, and Equation (2).

$$Attention(Q, K, V) = softmax\left(\frac{QK^T}{\sqrt{d_k}}\right) V \tag{1}$$

$$softmax(x_i) = e^{x_i} / \sum_j^n x_j \tag{2}$$

where d_k is the dimension of K. x_i is the input activation of attention blocks. n is the dimension of the input vector.

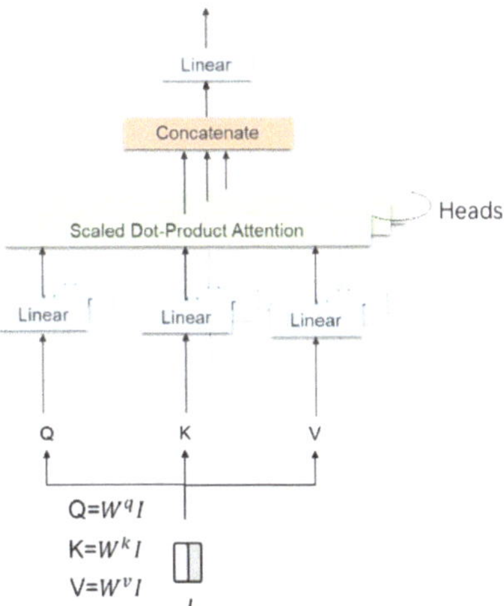

Figure 6. Illustration of multi-head self-attention in Transformers.

Multi-head attention is to conduct different linear transformations by h times to project Q, K, and V. After that, attention results of different heads are concatenated, shown in Equation (3) [22] and Equation (4) [22]:

$$head_i = Attention(QW_i^Q, KW_i^k, VW_i^V) \tag{3}$$

$$MultiHead(Q, K, V) = Concat(head_1, \ldots, head_h)W^o \tag{4}$$

where W_i^Q, W_i^k and W_i^V present learnable weight matrix in the ith head, and h is the number of head.

2.4. Levit Structure

LeViT is a newly developed approach proposed in 2021 [31]. It reintroduced convolutional operations, which replaced some Transformer components to achieve a faster inference speed. The architecture of LeViT is similar to LeNet [39], where the whole network presents a pyramid shape with pooling. Specifically, the difference between LeViT and original ViT can be summarized in the following aspects.

(1)   Convolutional layers

The convolution operation is adopted in this network to extract features. Four convolutional layers with the kernel size of 3×3 are added to process the input images. After the convolutional layers, a tensor with more channels and smaller width and height is acquired as the input of Transformer Blocks. Batch normalization [40] is used to stabilize the training process. All of the activation function in LeViT is *Hardswish*, shown in Equation (5) [41]:

$$Hardswish(x) = \begin{cases} 0 & x \leq -3 \\ x & x \geq +3 \\ x \cdot (x + 3)/6 & -3 < x < +3 \end{cases} \tag{5}$$

The size of the output feature map of the last convolutional layers is 13×13. The number of the channels of the output of the previous convolutional layer determines the input size of the transformer blocks. LeViT-192 represents that the number of channels on the input of transformer stage is 192. The architecture of the LeViT-192 is presented in Figure 7.

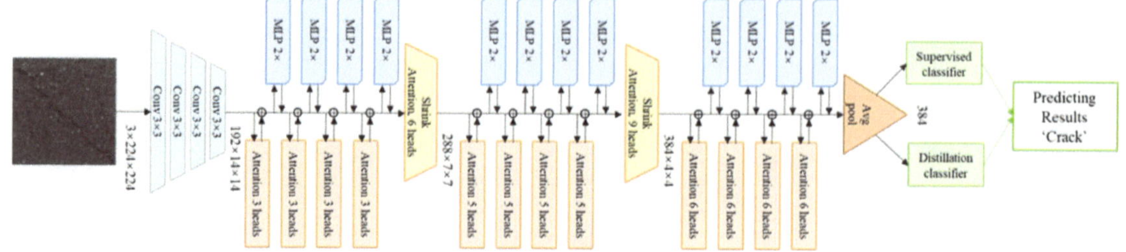

Figure 7. The architecture of the LeViT-192.

(2)   Transformer stages

There are three Transformer stages in LeViT. Attention and MLP blocks alternate with a residual structure in every transformer stage. The following is the description of the attention and MLP blocks.

Attention blocks—There are two kinds of attention blocks in LeViT, seen in Figure 8. One is in Transformer Blocks, seen in Figure 8a; another is between the stages, seen in Figure 8b, and has the function of subsampling. Position embeddings are replaced by attention bias to get position information not only in the input layer but also in the inner layers of transformer blocks. Assume that C × H × W is the input activation map. N is the number of heads. D is the output dimension of Q and K. Operator V has 2D channels. For two pixels (x, y) and (x', y') belonging to H × W for one head h, their attention value can be calculated as Equation (6) [31]:

$$A^h_{(x,y),(x',y')} = Q_{(x,y)} \bullet K_{(x',y')} + B^h_{|x-x'|,|y-y'|} \tag{6}$$

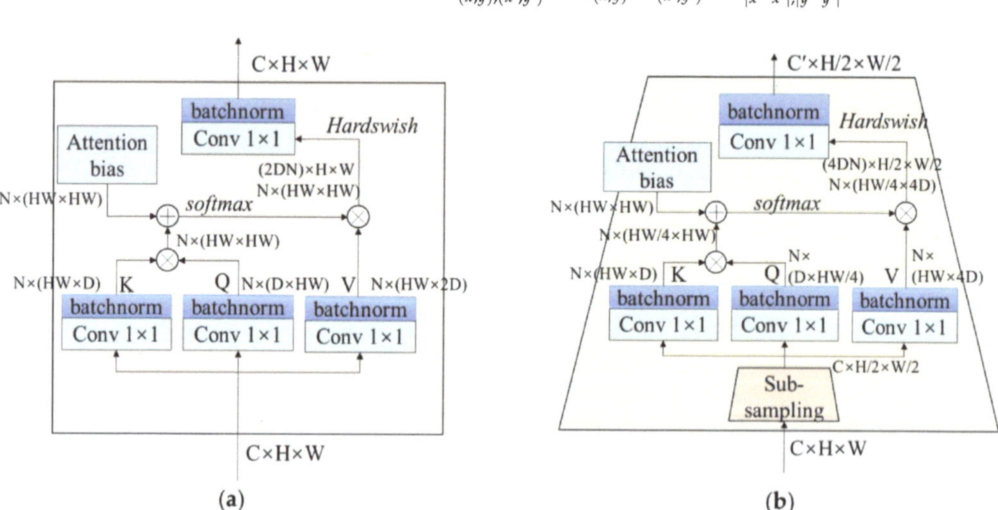

(a)   (b)

Figure 8. The structure of multi-head attention in LeViT: (**a**) Regular attention block; (**b**) Attention block with subsampling function (shrink attention blocks). C is the channel of feature maps, and H and W represent the height and width, respectively, of the feature map. N is the number of self-attention heads. D is the output dimension of key K and query Q. Operator V has 2D channels.

MLP blocks—The structure of MLP in traditional ViT is linear transformation. In LeViT, a convolutional kernel with the size of 1 × 1 is adopted. Every convolution operation is followed by batch normalization. The expansion factor of the convolution is 2, which means the number of neurons will double in the first layer of MLP and return the same number of neurons as the input vector in the second layer of MLP.

(3) Classification layers

In classification layers of LeViT, Globel Average Pooling (GAP) [42] is used. All pixel values of the feature map in one channel are added to get a vector with the same dimension as the channel of the feature maps. The output vector of the GAP is the basis of the supervised classifier and distillation classifier. The supervised classifier compares the loss between the predicted label and the ground truth label to calculate the cross-entropy loss (L_{CE}), shown in Equation (7). The distillation classifier will calculate the L_{globle}, shown in Equation (8) [26], the loss between the predicted label and the teacher-predicted label. The average of these two predicted labels is set as the ultimate label of this model.

$$L_{CE}(\hat{y}, y) = -(y \log(\hat{y}) + (1 - y) \log(1 - \hat{y})) \tag{7}$$

where $\hat{y}$ represents the prediction result of the model and y represents the ground truth labels.

$$L_{global} = (1 - \lambda) L_{CE}(\psi(Z_s), y) + \lambda \tau^2 KL(\psi(Z_s/\tau), \psi(Z_t/\tau)) \tag{8}$$

where Z_t is the outputs of the teacher model, Z_s is the outputs of the student model, τ is distillation temperature, and λ is the coefficient to balance KL Loss and the cross-entropy loss.

The number of parameters varies with the different sizes of the LeViT network. Table 2 shows the parameters of LeViT models.

Table 2. Parameters of LeViT-192. Each stage contains pairs of attention and MLP blocks. C is the number of channels in MLP blocks. N is the number of heads in attention blocks. n is the number of the classifications.

Layers		Operation	Output Size
Convolutional layers		$4 \times$ [Conv 3×3, stride = 2]	$14 \times 14 \times 192$
Transformer stages	Stage 1	$4 \times \begin{bmatrix} MLP, C = 192 \\ Attention, N = 3 \end{bmatrix}$	$14 \times 14 \times 192$
	Subsample	$[Shrink\ Attention, N = 6]$	$7 \times 7 \times 288$
	Stage 2	$4 \times \begin{bmatrix} MLP, C = 288 \\ Attention, N = 5 \end{bmatrix}$	$7 \times 7 \times 288$
	Subsample	$[Shrink\ Attention, N = 9.]$	$4 \times 4 \times 384$
	Stage 3	$4 \times \begin{bmatrix} MLP, C = 384 \\ Attention, N = 6 \end{bmatrix}$	$4 \times 4 \times 384$
Classification layers		Average Pooling Classifiers	384 $[n, n]$

2.5. Visual Interpretation Methods

As the deep learning algorithms have been criticized as black boxes, this section will try to interpret the proposed models visually and help us better understand the proposed algorithms' operation mode. Figure 9 shows the visual explanation methods used in this study. The visual interpretation of MLP blocks and the attention blocks are both explored.

For MLP layers, a technique called Gradient-weighted Class Activation Mapping (Grad-CAM) [33] was employed to explain what image features are learned at each network layer.

The gradient information in the specific convolutional layers was attained to assign an importance score for every neuron to acquire the class activation maps in Grad-CAM. Assume the output logits of a specific layer of class c is y^c and the activation of feature map is A^k, the gradient of class c is calculated. Global average pooling is conducted in the gradient of width and height. Then the importance weight of neurons is attained as α_k^c, shown in Equation (9) [33]:

$$\alpha_k^c = \frac{1}{Z} \sum_i \sum_j \frac{\partial y^c}{\partial A_{ij}^k} \tag{9}$$

where i and j represent the index of gradients in width and height. A represents the output feature map of this layer. k represents the channel of the feature map. $\frac{\partial y^c}{\partial A_{ij}^k}$ is the gradients via back-propagation. Z is the total number of pixels in this feature map. $\frac{1}{Z}\sum_i\sum_j \cdot$ is the function of global average pooling.

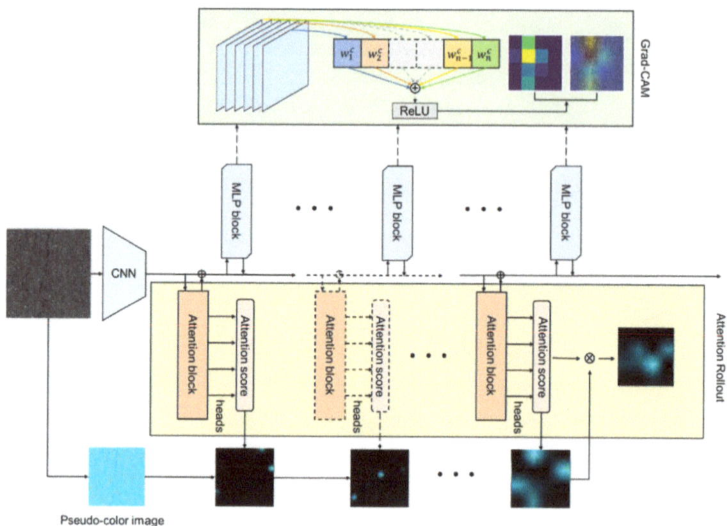

Figure 9. Visual interpretation methods used in this study.

All the channels of feature maps are linear weighted and sum up to get the heat map, as shown in Equation (10) [33]. Only regions that have a positive effect on class c are reserved using the activation function *ReLU*, seen in Equation (11).

$$L_{Grad-CAM}^c = ReLU\left(\sum_k \alpha_k^c A^k\right) \tag{10}$$

$$ReLU(x) = \begin{cases} x & x > 0 \\ 0 & x \leq 0 \end{cases} \tag{11}$$

For attention layers, the attention maps of each attention lay were computed using the Attention Rollout [43]. It can be described as the following steps. First, the attention weights from each attention layer of LeViT were attained. Then, the attention weights across all heads of this layer were averaged. Next, recursively multiply the weight matrices of all layers, as seen in Equation (12). The attention from all positions in layer l_i to all positions in layer l_j ($j < i$) is calculated:

$$\widetilde{A}(l_i) = \begin{cases} A(l_i)\widetilde{A}(l_{i-1}) & if\ i > j \\ A(l_i) & if\ i = j \end{cases} \tag{12}$$

where $\widetilde{A}$ is attention rollout. A is raw attention. Matrix multiplication is conducted.

The original input images were transformed into pseudo-color images to highlight the attention mask applied to the input images. Finally, apply masks from the output token to the input space.

2.6. Evaluation Indexes

To evaluate the performance of the proposed model, the indices of accuracy, precision, recall rate, F1-score, and processing speed were calculated to evaluate the experimental results.

Accuracy (ACC) is the proportion of correctly classified samples in the total number of samples, seen in Equation (13):

$$ACC = \frac{n_c}{n_t} \tag{13}$$

where n_c is the number of samples correctly predicted and n_t is the total number of samples.

Precision (P) is the proportion of the samples predicted to be positive and is actually a positive sample (Equation (14)):

$$P = \frac{TP}{TP + FP} \tag{14}$$

where TP is the positive samples that are correctly predicted and FP is the negative samples that are wrongly predicted to positive samples.

Recall (R) is the proportion of correctly predicted positive samples (Equation (15)):

$$R = \frac{TP}{TP + FN} \tag{15}$$

where TN is the negative samples that are correctly predicted and FN is the negative samples that are wrongly predicted to positive cracks.

As P and R were unable to achieve high performance simultaneously, F1-score was used to take the harmonic mean of the two to measure the effectiveness of the model (Equation (16)):

$$F_1 = 2 \cdot \frac{P \cdot R}{P + R} \tag{16}$$

As each class has equal weight, the Macro-Average principle is used when computing the evaluation metric. Metric within each category was calculated first, then the average resulting metrics across categories.

In addition, the computational costs were also evaluated for different methods. The number of the parameters, average training time per epoch, and inference time per step were set as the evaluation indexes.

3. Experimental Results and Analysis

3.1. Experimental Environment and Hyperparameters

The training software environment in this study was python 3.7, keras 2.3 with TensorFlow-2.5.0 backend. The operating system is Linux. All the experiments in this study were conducted in a supercomputing cluster equipped with a 7185 32C 2.0 GHz CPU with a 64-core Processor, 256 GB memory, and 8 DCU accelerator card.

A total of 100 training epochs were conducted with a batch size of 16 to improve the computation memory utilization. *Adam* [44] is the optimization algorithm for gradient descent in back-propagation in this study. The hyperparameters of the Adam algorithm were set as follows.

(1)   The learning rate was 1×10^{-4}. The learning rate can control the weight update ratio and a lower learning rate allows the model to achieve better convergence.
(2) $\beta 1$ was 0.9 and $\beta 2$ was 0.999. $\beta 1$ and $\beta 2$ can control the decay rates of the first and second moment means, respectively.
(3) ε was 1×10^{-8}, which prevented a relatively fixed value division by zero in the implementation.

Before training, the dataset was randomly shuffled before imputing into the networks. To reduce the training time, the transfer learning method was employed. The models were pretrained based on ImageNet [45], a large public dataset with approximately 15 million images.

3.2. Training Results of LeViT

The training loss curves of two classification heads in the different datasets are presented in Figures 10a, 11a and 12a. During training, the supervised classifier and the distillation classifier work together, reaching convergence by the end of the training process. The training accuracy curves of the training and validation sets of different datasets based on LeViT are displayed in Figures 10b, 11b and 12b. The curves show that the training and validation loss drops rapidly at the beginning epochs and, after several epochs, they stabilize around a value with slight fluctuation.

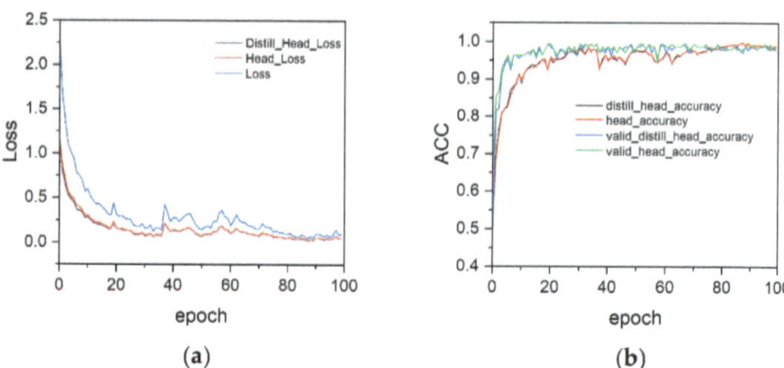

Figure 10. Training curves of JCAPs. (**a**) Loss curves of two classifier heads, (**b**) Training curves of training and validation set.

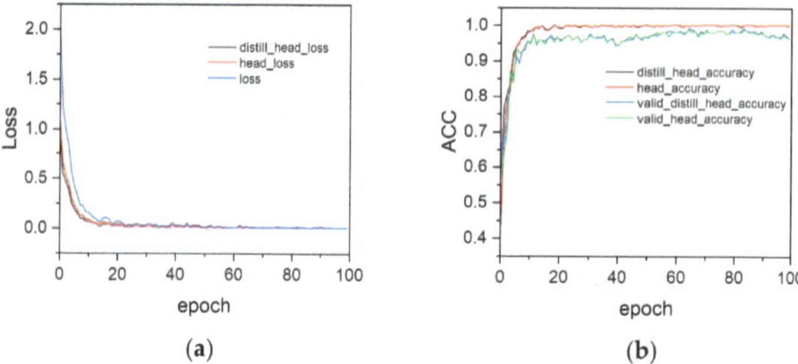

Figure 11. Training curves of GAPs. (**a**) Loss curves of two classifier heads, (**b**) Training curves of training and validation set.

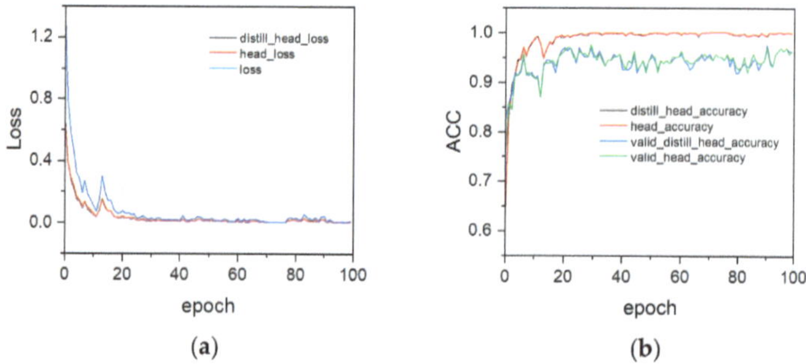

Figure 12. Training curves of Crack500. (**a**) Loss curves of two classifier heads, (**b**) Training curves of training and validation set.

After 100 epochs of training, the LeViT model was tested by different test sets. Figure 13 shows the classification confusion matrix of different datasets. The overall classification accuracy of different test sets was 91.56%, 99.17%, and 94.50%, respectively. LeViT appears to have an excellent performance in classifying the images in dataset GAPs and Crack500. However, there are certain fusions of crack images. Among 80 crack images in the test set of JCAPs, eight images are misclassified as background, and eight images are misclassified as sealed crack images. The reason for these misclassifications could be the quality of the images. To obtain a better classification result, image augmentation methods can be used for further study.

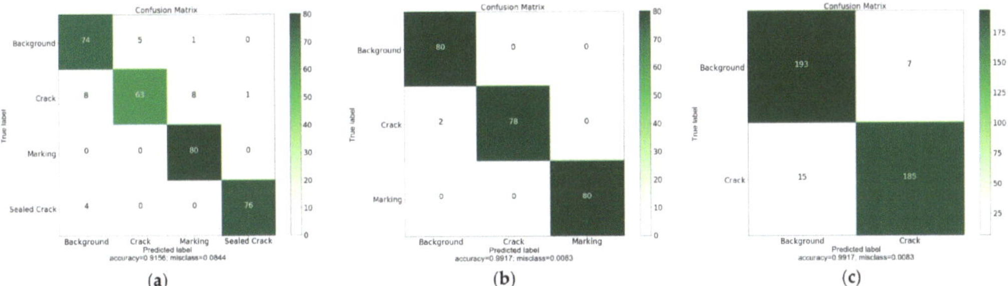

Figure 13. Confusion Matrix of LeViT in different test set: (**a**) JACPs, (**b**) GAPs, and (**c**) Crack500.

3.3. Visual Interpretation

Visual interpretation results of both MLP blocs and attention blocks were explored. Figure 14 shows the visual explanation results of MLP blocks based on Grad-CAM. Figure 15 displays the interpretation results of attention blocks based on Attention Rollout.

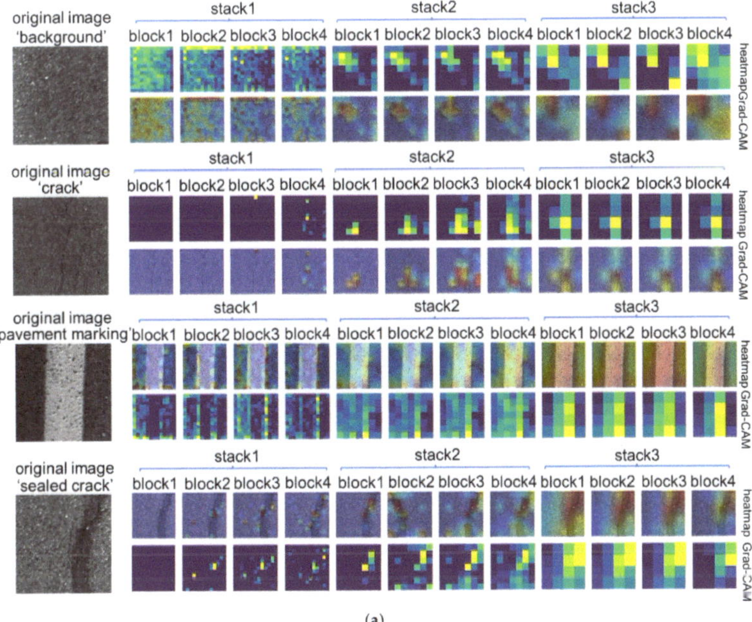

(**a**)

Figure 14. *Cont.*

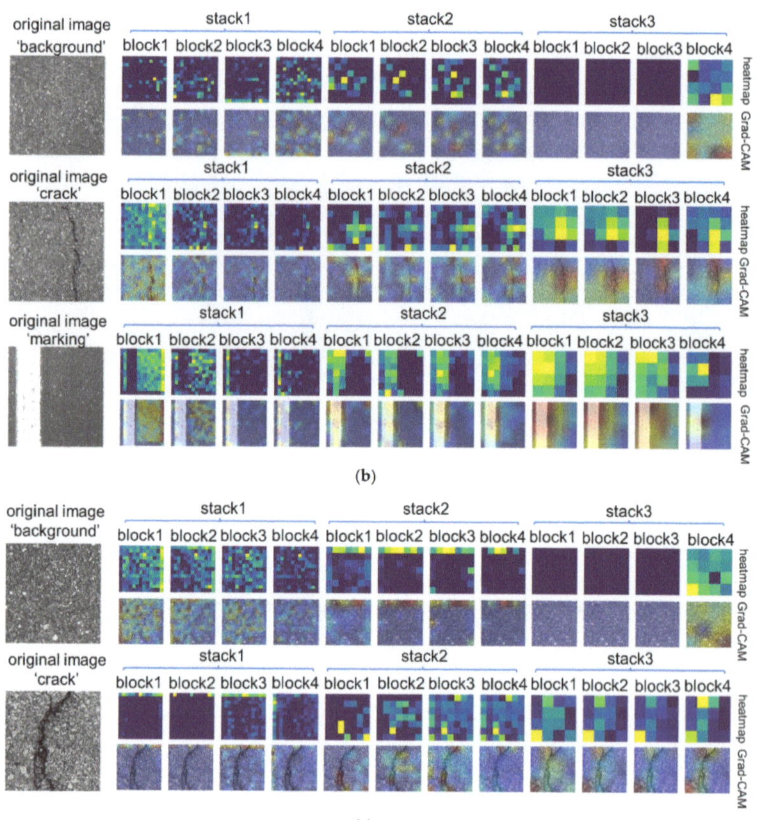

Figure 14. The heat maps and Grad-CAM images of different classifications in different datasets. (**a**) JACPs, (**b**) GAPs, (**c**) Crack500.

The heat maps and Grad-CAM outputs from different MLP blocks of LeViT networks in different datasets are illustrated in Figure 14. In heat maps, the brighter color indicates regions that contribute more to the classification results in heat maps. Correspondingly in Grad-CAM output images, the red areas present higher scores for the class. As can be seen, for the 'background' class of the model, MLP blocks at the first stack have a wide range of focuses. In the following stack, blocks start to focus on some small regions. At the last stack, the blocks return to a broader focus and tend not to concentrate on small spots in the 'background' images. For the 'crack' and 'sealed crack' classes of the networks, as the layer of the LeViT network goes deeper, it generally learns to focus on the regions related to the specific objects and focus more on some small spots. As for the 'pavement marking', the blocks at the first stack concentrate mainly on the pavement background regions. In the following stacks, MLP blocks start to learn a more extensive range of area at the second stack. Finally, at the last stack, blocks gradually shift their attention to areas relevant to pavement marking and ignore the irrelevant regions. Overall, from the interpretation results, we can find that as the layer of the network goes deeper, it will focus more on the related part of the images. In addition, as can be seen in the heat maps of different classifications, the 'crack' and the 'sealed crack' have similar features with a long and wide light spot. To some extent, it explained why these two classifications of images were misclassified in the same class.

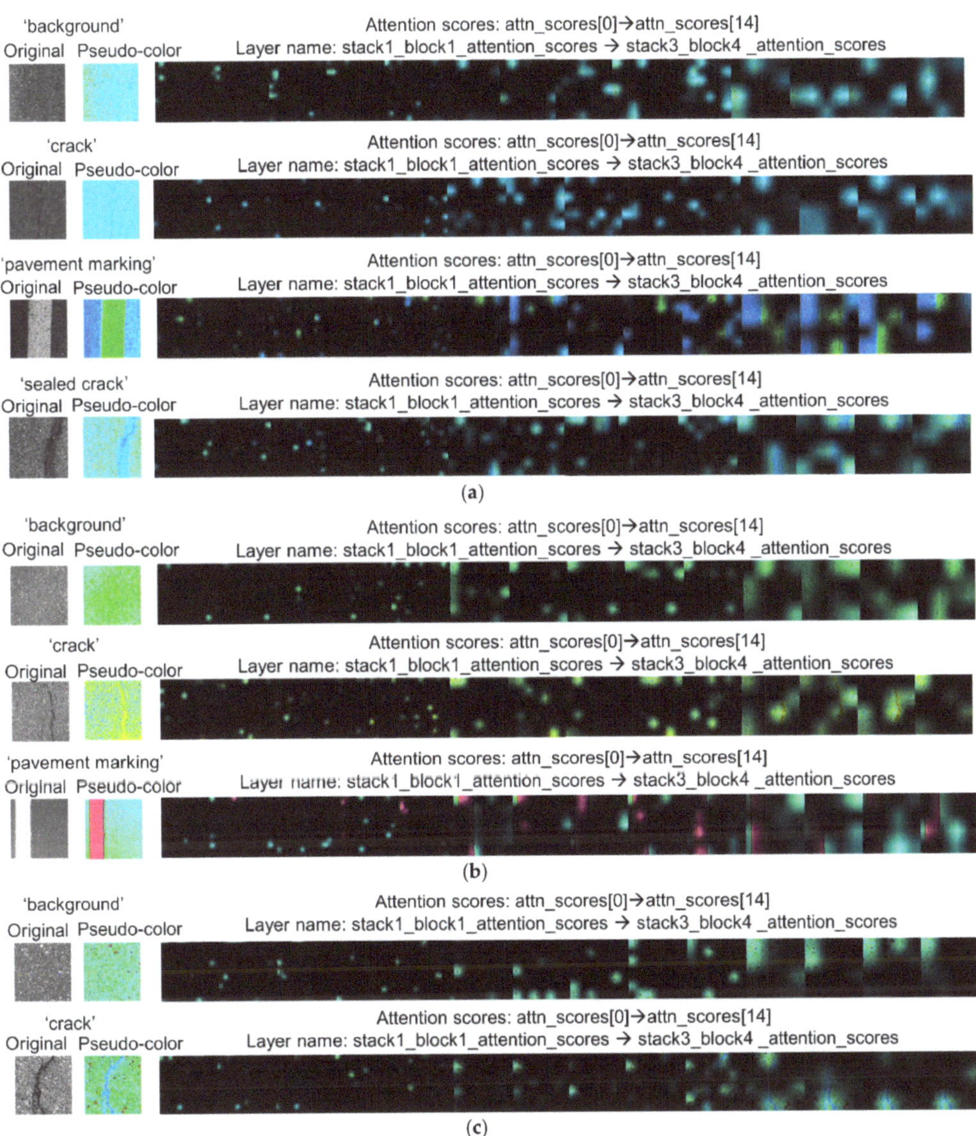

Figure 15. The attention score maps of different classifications. (**a**) JCAPs, (**b**) GAPs, (**c**) Crack500.

Figure 15 shows the attention score map from the output of different attention blocks. The regions with higher attention scores were highlighted by applying masks to areas with lower attention scores. From the visual interpretation results of the attention layers, in the first few attention blocks, the network pays almost equal attention to all regions in the images. Subsequently, in the following few regions, the network begins to focus on some separated and scattered regions in the images. As the layers get deeper, the regions of the posterior attention layers are connected to a larger area. Overall, there is a trend that the areas with higher attention scores will cover more effective feature areas in the deeper layers.

4. Discussion and Comparison

To validate the proposed methods, six popular models, including VGG-16 [7], ResNet-50 [9], InceptionV3 [46], DenseNet121 [10], MobileNetV1 [47], and ViT-b16 [24], were conducted for comparison. The basic principles of each method are as follows.

VGG-16 [7]: This adopts a convolution kernel with a smaller size (3×3) and uses deeper layers to achieve good performance. VGG-16 means the number of its weight layer is 16. ResNet-50 [9]: The core idea of this network is its shortcut or skip connections. In residual blocks, the input information can detour to the output, which is helpful to decrease learning difficulties. ResNet-50 represents that the number of its weight layer is 50. InceptionNet-V3 [46]: Its main idea is to find out how to approximate the optimal locally sparse junction with dense components. Densenet-121 [10]: The core of the network is a dense block. The input to each layer of the network in the structure comes from the outputs of all the previous layers. MobileNet-V1 [47]: It is a lightweight network. A structure of depth-wise separable convolutions was introduced in this network. ViT-b16 [24]: The framework of ViT has illustrated in Figure 5. ViT-b16 means its base version with 16×16 input image patches.

Before training on our experimental dataset, pre-trained weights based on ImageNet of all the models were obtained to speed up the training process. During training, the hyperparameters were kept the same with LeViT, seen in Section 3.1. Sparse categorical cross-entropy was adopted to monitor the training process. The pre-trained weights of all the comparison models were obtained. The test results are provided in Table 3.

Table 3. Classification results of different methods based on different datasets.

Dataset	Method	All Classifications				Background		Crack		Pavement Marking		Sealed Crack	
		ACC (%)	P (%)	R (%)	F1 (%)	ACC (%)	F1 (%)	ACC (%)	F1 (%)	ACC (%)	F1 (%)	ACC (%)	F1 (%)
JCAPs	ViT	88.13	89.33	88.13	88.21	83.75	88.16	77.50	82.67	93.75	96.77	97.50	85.25
	ResNet	88.75	89.12	88.75	88.80	88.75	87.65	85.00	85.54	87.50	92.71	93.75	89.29
	DenseNet	88.75	89.55	88.75	88.61	97.50	96.30	70.00	79.43	97.50	96.89	90.00	81.81
	VGG	89.38	90.16	89.38	89.45	92.50	82.50	87.50	87.50	91.25	94.81	96.25	87.50
	InceptionNet	89.69	90.12	89.69	89.64	91.25	87.95	78.75	85.71	93.75	94.94	95.50	89.94
	MobileNet	90.31	91.03	90.31	90.35	96.25	91.12	82.50	82.50	88.75	94.04	93.75	88.24
	Levit(ours)	91.56	91.72	91.56	91.45	92.50	89.16	78.75	85.14	100.00	99.38	95.00	92.12
GAPs	InceptionNet	97.08	97.15	97.08	97.10	96.25	96.86	97.50	95.71	97.50	98.73	-	-
	DenseNet	98.33	98.38	98.33	98.32	100.00	98.16	95.00	97.44	100.00	99.38	-	-
	ViT	98.33	98.38	98.33	98.32	100.00	98.16	95.00	97.44	100.00	99.38	-	-
	ResNet	98.75	98.80	98.75	98.75	100.00	98.16	97.50	98.73	98.75	99.37	-	-
	MobileNet	98.75	98.80	98.75	98.75	100.00	98.16	96.25	98.09	100.00	100.00	-	-
	VGG	98.75	99.17	99.17	99.17	100.00	99.38	98.75	98.75	98.75	99.37	-	-
	Levit(ours)	99.17	99.19	99.17	99.17	100.00	100.00	97.50	98.73	100.00	100.00	-	-
Crack500	InceptionNet	92.00	92.60	92.00	91.97	98.00	92.45	86.00	91.49	-	-	-	-
	MobileNet	93.75	93.88	93.75	93.75	96.50	93.92	91.00	93.57	-	-	-	-
	VGG	94.00	94.02	94.00	94.00	95.00	94.06	93.00	93.94	-	-	-	-
	ViT	94.00	94.07	94.00	94.00	96.00	94.12	92.00	93.88	-	-	-	-
	LeViT(ours)	94.50	94.57	94.50	94.50	96.50	94.61	92.50	94.39	-	-	-	-
	ResNet	94.75	95.08	94.75	94.74	99.00	94.96	90.50	94.52	-	-	-	-
	DenseNet	95.00	95.11	95.00	95.00	97.50	95.12	92.50	94.87	-	-	-	-

In the classification of dataset JCAPs, our model acquired the highest classification accuracy (91.56%), precision (91.72%), recall (91.56%), and F1-score (91.45%), while the original ViT network did not show competitive advantages compared with the equivalent CNN-based image classification model. Among all the classifications, the 'crack' images of our JCAPs dataset remain the biggest challenge for classification tasks.

In the test of dataset GAPs, our model also had the best performance with the highest classification accuracy (99.17%), precision (99.19%), recall (99.17%), and F1-score (99.17%). Among all the classifications, the 'background' images and the 'pavement marking' images reached the classification accuracy of 100.00%.

In the classification tasks of dataset Crack500, the accuracy, precision, recall, and F1-score of LeViT are 94.50%, 94.57%, 94.50%, and 94.50%, while the counterpart of the best-performed model is 95.00%, 95.11%, 95.00%, and 95.00%. The gap between them is acceptable.

Table 4 presents the comparison of training costs in dataset JCAPs. As the batch size is 16, the inference time per step is the processing time for every 16 images. From the table, it can be found that the proposed LeViT model has the fastest inference speed (86 ms/step) among all the comparison models. The average training time of LeViT for every epoch is 6.21 s, which is approximately 25% of the original ViT method and 80% of some prevailing CNN-based models, including DenseNet, VGG, and ResNet. Furthermore, the number of parameters in LeViT-192 is less than half of ResNet and InceptionNet, three-quarters of VGG, and one-eighth of original ViT. Although MobileNet has the fewest number of parameters among these methods, its inference speed is the lowest in our test. Overall, LeViT achieved competitive results with relatively fewer computation costs.

Table 4. Comparison of training costs in JCAPs.

Method	Number of Parameters	Inference Time/Step	Average Training Time/Epoch
MobileNetV1	3.2 M	3 s	210.83 s
ViT-b16	85.8 M	333 ms	24.08 s
DenseNet-121	7.0 M	103 ms	7.45 s
VGG-16	14.7 M	107 ms	7.74 s
ResNet-50	23.6 M	108 ms	7.94 s
InceptionNetV3	21.8 M	95 ms	6.90 s
LeViT(ours)	10.2 M	86 ms	6.21 s

5. Conclusions

To solve the problem of high computation resources cost and poor interpretation of deep learning-based pavement image recognition models, this paper introduced a fast inference Transformer-based deep learning method called LeViT for the automatic object classification of pavement images. A visual explanation method was also employed to improve the interpretability of the proposed deep learning model. The performance of the proposed methods was evaluated on three different sources of pavement datasets. Compared to several state-of-the-art methods, LeViT attained the best classification accuracy, precision, recall, and F1-score in two pavement image datasets tested in this study. Moreover, the proposed LeViT model has a relatively faster inference speed (86 ms/step) among all the comparison models, which is approximately 80% of some prevailing CNN-based models, including DenseNet, VGG, and ResNet. Overall, the proposed method can achieve competitive performance with fewer computation costs. In addition, the proposed visual interpretation methods combining Grad-CAM and Attention Rollout can well visually explain the feature extraction mechanism of both MLP and attention blocks. Several issues that were not addressed in this study need to be noted. First, the impact of the number of the training images on the performance of the model was not be considered. Second, only longitudinal and transverse cracks are included in this research. For further study, more data containing alligator cracking can be added, and data augmentation methods can also be applied to improve the prediction performance of pavement images with poor quality.

Author Contributions: The authors confirm contributions to the paper as follows: study conception and design: Y.C. and X.G.; data collection: X.G. and J.L.; analysis and interpretation of results: Y.C. and Z.L.; draft manuscript preparation: Y.C., X.G., Z.L. and J.L. All authors have read and agreed to the published version of the manuscript.

Funding: This research received no external funding.

Institutional Review Board Statement: Not applicable.

Informed Consent Statement: Not applicable.

Data Availability Statement: Not applicable.

Conflicts of Interest: The authors declare no conflict of interest.

References

1. Chen, C.; Chandra, S.; Han, Y.; Seo, H. Deep Learning-Based Thermal Image Analysis for Pavement Defect Detection and Classification Considering Complex Pavement Conditions. *Remote Sens.* **2021**, *14*, 106. [CrossRef]
2. Liu, Z.; Wu, W.; Gu, X.; Li, S.; Wang, L.; Zhang, T. Application of combining YOLO models and 3D GPR images in road detection and maintenance. *Remote Sens.* **2021**, *13*, 1081. [CrossRef]
3. Dorafshan, S.; Thomas, R.J.; Maguire, M. Comparison of deep convolutional neural networks and edge detectors for image-based crack detection in concrete. *Constr. Build. Mater.* **2018**, *186*, 1031–1045. [CrossRef]
4. Hou, Y.; Li, Q.; Zhang, C.; Lu, G.; Ye, Z.; Chen, Y.; Wang, L.; Cao, D. The state-of-the-art review on applications of intrusive sensing, image processing techniques, and machine learning methods in pavement monitoring and analysis. *Engineering* **2021**, *7*, 845–856. [CrossRef]
5. Liu, Z.; Gu, X.; Dong, Q.; Tu, S.; Li, S. 3D visualization of airport pavement quality based on BIM and WebGL integration. *J. Transp. Eng. Part B Pavements* **2021**, *147*, 04021024. [CrossRef]
6. Krizhevsky, A.; Sutskever, I.; Hinton, G.E. Imagenet classification with deep convolutional neural networks. *Adv. Neural Inf. Process. Syst.* **2012**, *25*, 1097–1105. [CrossRef]
7. Simonyan, K.; Zisserman, A. Very deep convolutional networks for large-scale image recognition. *arXiv* **2014**, arXiv:1409.1556.
8. He, K.; Zhang, X.; Ren, S.; Sun, J. Delving deep into rectifiers: Surpassing human-level performance on imagenet classification. In Proceedings of the IEEE International Conference on Computer Vision, Santiago, Chile, 7–13 December 2015; pp. 1026–1034.
9. He, K.; Zhang, X.; Ren, S.; Sun, J. Deep residual learning for image recognition. In Proceedings of the IEEE Conference on Computer Vision and Pattern Recognition, Vegas, NV, USA, 27–30 June 2016; pp. 770–778.
10. Huang, G.; Liu, Z.; Van Der Maaten, L.; Weinberger, K.Q. Densely connected convolutional networks. In Proceedings of the IEEE Conference on Computer Vision and Pattern Recognition, Honolulu, HI, USA, 21–26 July 2017; pp. 4700–4708.
11. Zhang, A.; Wang, K.C.; Li, B.; Yang, E.; Dai, X.; Peng, Y.; Fei, Y.; Liu, Y.; Li, J.Q.; Chen, C. Automated pixel-level pavement crack detection on 3D asphalt surfaces using a deep-learning network. *Comput.-Aided Civil Infrastruct. Eng.* **2017**, *32*, 805–819. [CrossRef]
12. Zhang, A.; Wang, K.C.; Fei, Y.; Liu, Y.; Tao, S.; Chen, C.; Li, J.Q.; Li, B. Deep learning–based fully automated pavement crack detection on 3D asphalt surfaces with an improved CrackNet. *J. Comput. Civil. Eng.* **2018**, *32*, 04018041. [CrossRef]
13. Fei, Y.; Wang, K.C.; Zhang, A.; Chen, C.; Li, J.Q.; Liu, Y.; Yang, G.; Li, B. Pixel-level cracking detection on 3D asphalt pavement images through deep-learning-based CrackNet-V. *IEEE Trans. Intell. Transp. Syst.* **2019**, *21*, 273–284. [CrossRef]
14. Hou, Y.; Li, Q.; Han, Q.; Peng, B.; Wang, L.; Gu, X.; Wang, D. MobileCrack: Object classification in asphalt pavements using an adaptive lightweight deep learning. *J. Transp. Eng. Part B Pavements* **2021**, *147*, 04020092. [CrossRef]
15. Ali, L.; Alnajjar, F.; Jassmi, H.A.; Gochoo, M.; Khan, W.; Serhani, M.A. Performance Evaluation of Deep CNN-Based Crack Detection and Localization Techniques for Concrete Structures. *Sensors* **2021**, *21*, 1688. [CrossRef] [PubMed]
16. Kim, B.; Yuvaraj, N.; Preethaa, K.S.; Pandian, R.A. Surface crack detection using deep learning with shallow CNN architecture for enhanced computation. *Neural Comput. Appl.* **2021**, *33*, 9289–9305. [CrossRef]
17. Wu, Y.; Qi, S.; Sun, Y.; Xia, S.; Yao, Y.; Qian, W. A vision transformer for emphysema classification using CT images. *Phys. Med. Biol.* **2021**, *66*, 245016. [CrossRef]
18. Liu, Z.; Chen, Y.; Gu, X.; Yeoh, J.K.; Zhang, Q. Visibility classification and influencing-factors analysis of airport: A deep learning approach. *Atmos. Environ.* **2022**, *278*, 119085. [CrossRef]
19. Xingjian, S.; Chen, Z.; Wang, H.; Yeung, D.-Y.; Wong, W.-K.; Woo, W.-C. Convolutional LSTM network: A machine learning approach for precipitation nowcasting. In Proceedings of the Advances in Neural Information Processing Systems, Montreal, QC, Canada, 7–12 December 2015; pp. 802–810.
20. Cho, K.; Van Merriënboer, B.; Gulcehre, C.; Bahdanau, D.; Bougares, F.; Schwenk, H.; Bengio, Y. Learning phrase representations using RNN encoder-decoder for statistical machine translation. *arXiv* **2014**, arXiv:1406.1078.
21. Zhang, A.; Wang, K.C.; Fei, Y.; Liu, Y.; Chen, C.; Yang, G.; Li, J.Q.; Yang, E.; Qiu, S. Automated pixel-level pavement crack detection on 3D asphalt surfaces with a recurrent neural network. *Comput.-Aided Civil Infrastruct. Eng.* **2019**, *34*, 213–229. [CrossRef]
22. Vaswani, A.; Shazeer, N.; Parmar, N.; Uszkoreit, J.; Jones, L.; Gomez, A.N.; Kaiser, Ł.; Polosukhin, I. Attention is all you need. In Proceedings of the Advances in Neural Information Processing Systems, Long Beach, CA, USA, 4–9 December 2017; pp. 5998–6008.
23. Bazi, Y.; Bashmal, L.; Rahhal, M.M.A.; Dayil, R.A.; Ajlan, N.A. Vision transformers for remote sensing image classification. *Remote Sens.* **2021**, *13*, 516. [CrossRef]
24. Dosovitskiy, A.; Beyer, L.; Kolesnikov, A.; Weissenborn, D.; Zhai, X.; Unterthiner, T.; Dehghani, M.; Minderer, M.; Heigold, G.; Gelly, S. An image is worth 16x16 words: Transformers for image recognition at scale. *arXiv* **2020**, arXiv:2010.11929.
25. Zhou, D.; Kang, B.; Jin, X.; Yang, L.; Lian, X.; Jiang, Z.; Hou, Q.; Feng, J. Deepvit: Towards deeper vision transformer. *arXiv* **2021**, arXiv:2103.11886.
26. Touvron, H.; Cord, M.; Douze, M.; Massa, F.; Sablayrolles, A.; Jégou, H. Training data-efficient image transformers & distillation through attention. In Proceedings of the International Conference on Machine Learning, online, 18–24 July 2021; pp. 10347–10357.
27. Chen, C.-F.; Fan, Q.; Panda, R. Crossvit: Cross-attention multi-scale vision transformer for image classification. *arXiv* **2021**, arXiv:2103.14899.
28. Mehta, S.; Rastegari, M. MobileViT: Light-weight, General-purpose, and Mobile-friendly Vision Transformer. *arXiv* **2021**, arXiv:2110.02178.

29. Liu, H.; Miao, X.; Mertz, C.; Xu, C.; Kong, H. CrackFormer: Transformer Network for Fine-Grained Crack Detection. In Proceedings of the IEEE/CVF International Conference on Computer Vision, Montreal, QC, Canada, 10–17 October 2021; pp. 3783–3792.
30. Guo, J.-M.; Markoni, H. Transformer based Refinement Network for Accurate Crack Detection. In Proceedings of the 2021 International Conference on System Science and Engineering (ICSSE), Ho Chi Minh City, Vietnam, 26–28 August 2021; pp. 442–446.
31. Graham, B.; El-Nouby, A.; Touvron, H.; Stock, P.; Joulin, A.; Jégou, H.; Douze, M. LeViT: A Vision Transformer in ConvNet's Clothing for Faster Inference. *arXiv* **2021**, arXiv:2104.01136.
32. Castelvecchi, D. Can we open the black box of AI? *Nat. News* **2016**, *538*, 20–23. [CrossRef]
33. Selvaraju, R.R.; Cogswell, M.; Das, A.; Vedantam, R.; Parikh, D.; Batra, D. Grad-cam: Visual explanations from deep networks via gradient-based localization. In Proceedings of the IEEE International Conference on Computer Vision, Venice, Italy, 22–29 October 2017; pp. 618–626.
34. Serrano, S.; Smith, N.A. Is attention interpretable? *arXiv* **2019**, arXiv:1906.03731.
35. Eisenbach, M.; Stricker, R.; Seichter, D.; Amende, K.; Debes, K.; Sesselmann, M.; Ebersbach, D.; Stoeckert, U.; Gross, H.-M. How to get pavement distress detection ready for deep learning? A systematic approach. In Proceedings of the 2017 International Joint Conference on Neural Networks (IJCNN), Anchorage, AK, USA, 14–19 May 2017; pp. 2039–2047.
36. Yang, F.; Zhang, L.; Yu, S.; Prokhorov, D.; Mei, X.; Ling, H. Feature pyramid and hierarchical boosting network for pavement crack detection. *IEEE Trans. Intell. Transp. Syst.* **2019**, *21*, 1525–1535. [CrossRef]
37. Zhang, L.; Yang, F.; Zhang, Y.D.; Zhu, Y.J. Road crack detection using deep convolutional neural network. In Proceedings of the 2016 IEEE International Conference on Image Processing (ICIP), Phoenix, AZ, USA, 25–28 September 2016; pp. 3708–3712.
38. Bahdanau, D.; Cho, K.; Bengio, Y. Neural machine translation by jointly learning to align and translate. *arXiv* **2014**, arXiv:1409.0473.
39. LeCun, Y.; Boser, B.; Denker, J.S.; Henderson, D.; Howard, R.E.; Hubbard, W.; Jackel, L.D. Backpropagation applied to handwritten zip code recognition. *Neural Comput.* **1989**, *1*, 541–551. [CrossRef]
40. Ioffe, S.; Szegedy, C. Batch normalization: Accelerating deep network training by reducing internal covariate shift. In Proceedings of the International Conference on Machine Learning, Lile, France, 6–11 July 2015; pp. 448–456.
41. Howard, A.; Sandler, M.; Chu, G.; Chen, L.-C.; Chen, B.; Tan, M.; Wang, W.; Zhu, Y.; Pang, R.; Vasudevan, V. Searching for mobilenetv3. In Proceedings of the IEEE/CVF International Conference on Computer Vision, Seol, Korea, 27–28 October 2019; pp. 1314–1324.
42. Lin, M.; Chen, Q.; Yan, S. Network in network. *arXiv* **2013**, arXiv:1312.4400.
43. Abnar, S.; Zuidema, W. Quantifying attention flow in transformers. *arXiv* **2020**, arXiv:2005.00928.
44. Kingma, D.P.; Ba, J. Adam: A method for stochastic optimization. *arXiv* **2014**, arXiv:1412.6980.
45. Deng, J.; Dong, W.; Socher, R.; Li, L.-J.; Li, K.; Fei-Fei, L. Imagenet: A large scale hierarchical image database. In Proceedings of the 2009 IEEE Conference on Computer Vision and Pattern Recognition, Miami, FL, USA, 20–25 June 2009; pp. 248–255.
46. Szegedy, C.; Vanhoucke, V.; Ioffe, S.; Shlens, J.; Wojna, Z. Rethinking the inception architecture for computer vision. In Proceedings of the IEEE Conference on Computer Vision and Pattern Recognition, Las Vegas, NV, USA, 27–30 June 2016; pp. 2818–2826.
47. Howard, A.G.; Zhu, M.; Chen, B.; Kalenichenko, D.; Wang, W.; Weyand, T.; Andreetto, M.; Adam, H. Mobilenets: Efficient convolutional neural networks for mobile vision applications. *arXiv* **2017**, arXiv:1704. 04861.

remote sensing

MDPI

Technical Note

Fast Segmentation and Dynamic Monitoring of Time-Lapse 3D GPR Data Based on U-Net

Ke Shang [1,2], Feizhou Zhang [2], Ao Song [3], Jianyu Ling [1], Jiwen Xiao [4], Zihan Zhang [2] and Rongyi Qian [1,*]

[1]   School of Geophysics and Information Technology, China University of Geosciences, Beijing 100083, China
[2]   School of Earth and Space Sciences, Peking University, Beijing 100871, China
[3]   National Institute of Natural Hazards, Ministry of Emergency Management of China (MEMC), Beijing 100085, China
[4]   Institute of Software Chinese Academy of Sciences, Beijing 100190, China
[*]   Correspondence: rongyiqian@cugb.edu.cn

Abstract: As the amount of ground-penetrating radar (GPR) data increases significantly with the high demands of nondestructive detection methods under urban roads, a method suitable for time-lapse data dynamic monitoring should be developed to quickly identify targets on GPR profiles and compare time-lapse datasets. This study conducted a field experiment aiming to monitor one backfill pit using three-dimensional GPR (3D GPR), and the time-lapse data collected over four months were used to train U-Net, a fast neural network based on convolutional neural networks (CNNs). Consequently, a trained network model that could effectively segment the backfill pit from inline profiles was obtained, whose Intersection over Union (IoU) was 0.83 on the test dataset. Moreover, segmentation masks were compared, demonstrating that a change in the southwest side of the backfill pit may exist. The results demonstrate the potential of machine learning algorithms in time-lapse 3D GPR data segmentation and dynamic monitoring.

Keywords: 3D ground-penetrating radar; time-lapse monitoring; U-Net; data segmentation

Citation: Shang, K.; Zhang, F.; Song, A.; Ling, J.; Xiao, J.; Zhang, Z.; Qian, R. Fast Segmentation and Dynamic Monitoring of Time-Lapse 3D GPR Data Based on U-Net. *Remote Sens.* **2022**, *14*, 4190. https://doi.org/10.3390/rs14174190

Academic Editors: Valerio Baiocchi, Alessandro Mei and Xianfeng Zhang

Received: 9 July 2022
Accepted: 24 August 2022
Published: 25 August 2022

Publisher's Note: MDPI stays neutral with regard to jurisdictional claims in published maps and institutional affiliations.

1. Introduction

Ground-penetrating radar (GPR) is a nondestructive detection technique that transmits high-frequency electromagnetic waves and receives the backscattering from the medium, which is suitable for defect inspection and dynamic monitoring of urban roads. To avoid road collapses, GPR is used to detect underground targets, including pipelines, voids, backfill pits, and high-water-cut areas. As it is known, it is not difficult to distinguish strong reflections from other signals, such as the reflections of metal pipes and clear boundaries of voids [1–3]. However, in many cases, the reason defects under roads are not discovered in time is that the radar data do not provide a strong reflection feature before the defect is generated, but cluttered weak scattering caused by loosening or deformation of the medium may be evident. Therefore, cases containing weak reflections with extended ranges should be approached seriously. Furthermore, to satisfy the goal of road safety maintenance, defects should be identified and monitored timely and properly to know if any changes occur or if actions are required to avoid road collapses. Identifying and locating regional features on GPR data are crucial and remain challenging tasks, because of the amplitude of the reflections as well as the regional characteristics that must be considered, which cannot be measured by mathematical formulas or specific values. In addition, the large amount of time-lapse data generated by GPR monitoring introduces difficulties in data processing and target identification. Therefore, the development of a method for the quick identification and monitoring of targets with regional features from time-lapse GPR data is necessary.

In recent years, convolutional neural networks (CNNs) and their neural network models have been widely applied in object recognition, image segmentation, and face and speech recognition owing to their powerful image processing abilities. Their advantage is that it is not necessary to know the quantitative characteristic parameters of the target; the network can learn the discriminant basis of the target from data from multiple trainings. In the user layer, this is represented as end-to-end mapping from the data to the target, hiding the complex judgment basis and quantitative criteria in the middle [4–8]. In recent years, CNNs have been used to identify hyperbolic features [9,10], manhole covers, reflective interfaces, soil [11,12], and landmines [13,14] in GPR data images. However, for defects in urban roads, research on object recognition and dynamic-change monitoring using machine learning is not comprehensive. Studies based on simulated data are insufficient, but the plausibility of real data must be discussed. In addition, both strongly and weakly reflected local signals should be considered, because any change in the data might indicate a change in the dielectric constant of the underground material. Therefore, with the necessity to identify specific obvious features and weak signals with regional characteristics, an appropriate neural network is required, which can achieve satisfactory recognition and segmentation results from limited GPR datasets.

U-Net is a deep learning network based on CNNs. Through powerful data augmentation, a reasonable weighted loss function, and full utilization of subunit neighborhood information, it can accurately identify image features with available annotated samples and realize end-to-end mappings between input data and segmentation units. Compared with other networks, it has the advantages of a simple structure, faster operation speed, and higher accuracy [15,16].

The main purpose of this study is to discuss a fast method for interpreting and comparing time-lapse GPR data in monitoring weak signal targets with regional characteristics. Firstly, the 3D GPR data acquisition and preprocessing methods are presented. Afterwards, the used neural network, U-Net, is presented along with how we used our data to train it. Finally, the detection results from the time-lapse dataset are compared and the results are discussed.

2. Experiment and Methods

2.1. Field Experiment and Data

As shown in Figure 1, the experiment was conducted at the gate of a construction site with a fresh backfill pit. Considering the material in the backfill pit may be deformed after being crushed, a time-lapse full-coverage (TLFC) 3D GPR acquisition containing four group lines was arranged on the sidewalk, covering the backfill pit from inline profiles P40 to P91, as shown in Figure 2. The MobyScan-V 3D GPR system with the 30-channel antenna arrays (Figure 3) produced by DECOD Science & Technology Pte. Ltd. (Singapore). was used in this experiment, and Table 1 lists the acquisition parameters. Fifteen time-lapse detection sessions were conducted from 20 May to 25 October 2019 (Table A1).

The 3D data preprocessing methods listed in Table 2 were applied to the raw 3D GPR data, in which the first arrive disagreed with the true time zero, and useful reflection signals were buried by strong noise (Figure A1). Firstly, an antenna lifting test was carried out to find the position of the ground as the true zero, as shown in Figure A2. Then, time zero correction, background removal (BGR), and frequency filtering were applied in turn (Figure A3). Finally, correlation was used to carry out the 3D time zero normalization and 3D data combination, which can be found in Figures A4 and A5 and [17]. The 3D GPR data cube after preprocessing is shown in Figure A6.

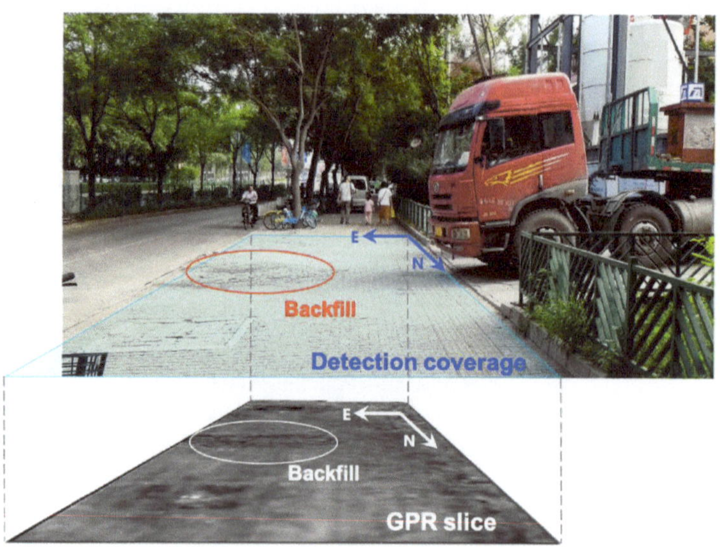

Figure 1. Detection coverage and 3D GPR slice. Data were collected on 8 August 2019. The double travel time of the slice is 1.464 ns.

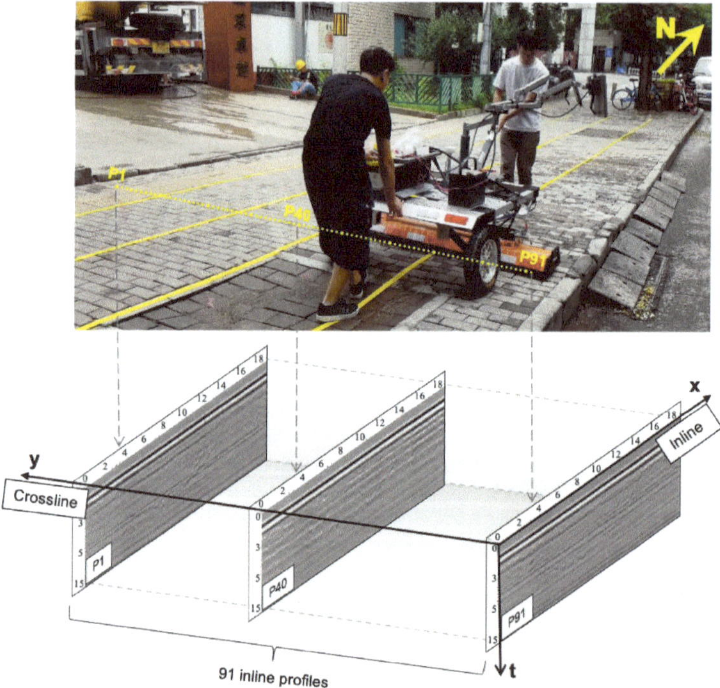

Figure 2. Survey lines of the TLFC 3D GPR acquisition and the raw data of inline profiles. Data were collected on 8 August 2019.

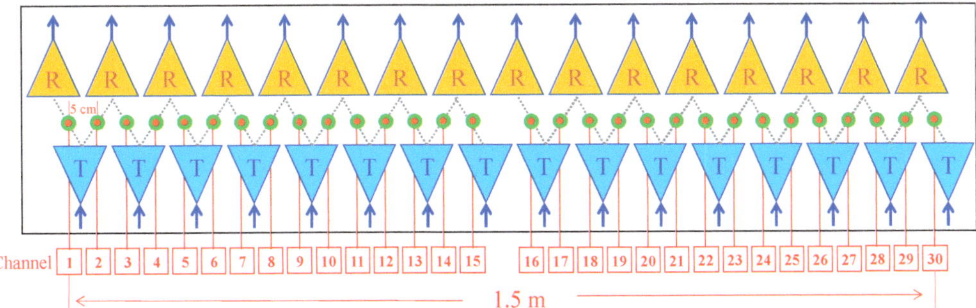

Figure 3. Antenna array of the MobyScan-V. There are 16 transmitters (blue triangles) and 16 receivers (yellow triangles). Red dots with green borders indicate the GPR channels between transmitters and receivers. The distance between channels 15 and 16 is 10 cm, and the data here were interpolated to achieve uniform spatial sampling.

Table 1. The technical specifications of the MobyScan-V and data acquisition parameters.

Indicators	Parameters
Antenna type	Ground-coupled antenna arrays
Number of channels	30
Center Frequency	2 GHz
Channels spacing (crossline)	0.05 m
Scans spacing (inline)	0.02 m
Sampling points	512
Time window	15 ns

Table 2. Preprocessing steps and parameters.

Step	Processing Method	Parameter/Object
Time zero correction	Antenna lifting test	2 ns
Background removal (BGR)	Sliding window	150 scans
Frequency filtering	Butterworth filter	500–1500 MHz
3D time zero normalization	Correlation	Direct-coupling
3D data combination	Correlation	Overlapping data

2.2. U-Net Structure and Model Training

Figure 4 illustrates the U-Net architecture. The entire network presents a symmetrical "U" shape. It consists of contracting and expansive paths, which correspond to the left and right sides of the figure, respectively.

- Contracting path: This path is used to obtain the context information. It has four layers, and each layer consists of two identical 3×3 convolutions and rectified linear unit (ReLU) activation functions. Downsampling is performed through a 2×2 max pooling operation. The number of feature channels is doubled after each downsampling. On this path, four max pooling operations are performed to extract the feature information from the sample.
- Expansive path: This is the upsampling part used to locate the target. A 2×2 convolution is applied to half of the feature channels, and then two 3×3 convolutions are used, each followed by a ReLU function. In the last layer, a 1×1 convolution is used to map the feature vectors to the corresponding prediction classification to complete the data segmentation and make the size of the output data consistent with that of the input data.

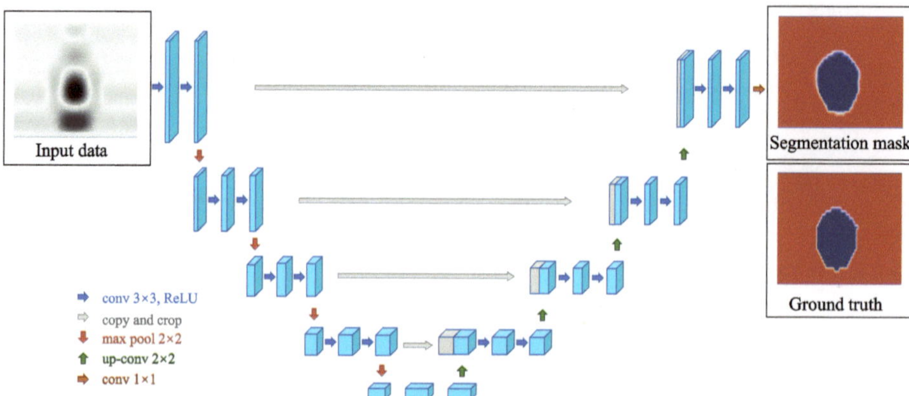

Figure 4. The classic U-Net architecture. The model comprises contracting and expansive pathways, with skip connections between the corresponding layers.

2.3. Data Segmentation

For 3D GPR, data can be interpreted and annotated from different profiles, such as inline, crossline, and horizontal slices. In this study, the spatial resolution of inline profiles along the acquisition direction was higher. Therefore, the target areas in the inline profiles of the 3D GPR data were segmented pixel-wise to generate ground truths and show the clear borders of the target. These inline profiles with ground truths were used as the training and testing datasets.

2.4. Accuracy Evaluation

The IoU between the segmentation and manual ground truth was applied to evaluate the network availability, as shown in Figure 5. The prediction results can be divided into the following two grades:

- IoU > 0.6, good result;
- IoU < 0.6, unsatisfactory result, and the segmentation is invalid.

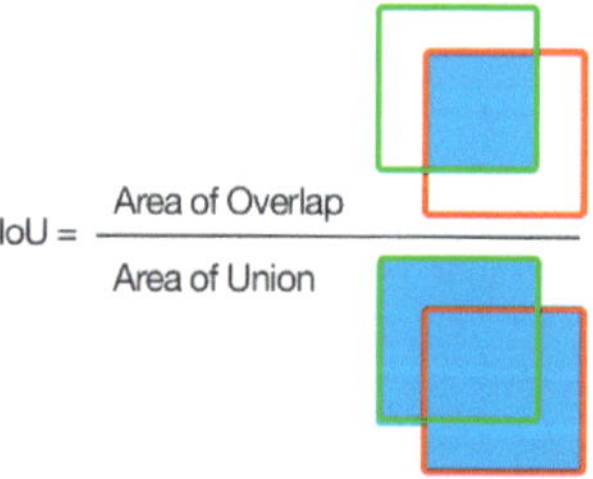

Figure 5. IoU is the ratio of overlap to union of the ground truth (green box) and the segmentation result (red box).

2.5. Mask Comparison

The changes in the targets can be determined by overlapping the masks of the time-lapse GPR data. P1 and P2 are assumed to be masks of the same target "X" at different times (t1 and t2, as shown in Figure 6a,b, respectively). By taking P1 as a reference overlaid with P2, the changes are determined as two types: changed (blue and yellow masks in Figure 6c) and unchanged areas (red masks in Figure 6c).

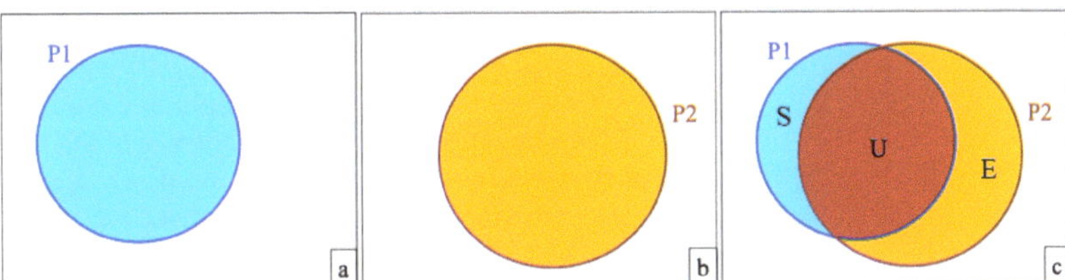

Figure 6. Changes in the target are shown by overlapping the masks. (**a**) Segmentation mask P1; (**b**) segmentation mask P2; (**c**) results of comparison by overlapping P1 and P2. The blue mask marked with "S" indicates the part in P1 which is not seen at the same position in P2; the yellow mask marked with "E" indicates the part that does not appear in P1 but is seen in P2.

3. Results

The aim of data segmentation is to track the backfill pit in the inline profiles. The masks were compared to determine the changes in the backfill pit from 20 May to 26 September 2019.

3.1. Training and Segmentation

Data annotation was performed on the inline GPR profiles after 3D preprocessing, as shown in Figure 7. The blue borders indicate the boundaries of the backfill pit. The mask value within the border was set to 1, and the background was set to 0. In this study, 249 annotated profiles from Datasets (a) to (n) in Table A1 were randomly selected to train the network, as shown in Figure 3. After training for 54.2 h on a PC with a CPU Intel(R) Core(TM) I5-4460 @ 3.20 GHz and 8 GB of RAM (Santa Clara, CA, USA), the average IoU of 0.96 was obtained on the 432nd training model, indicating a satisfactory segmentation of the training dataset.

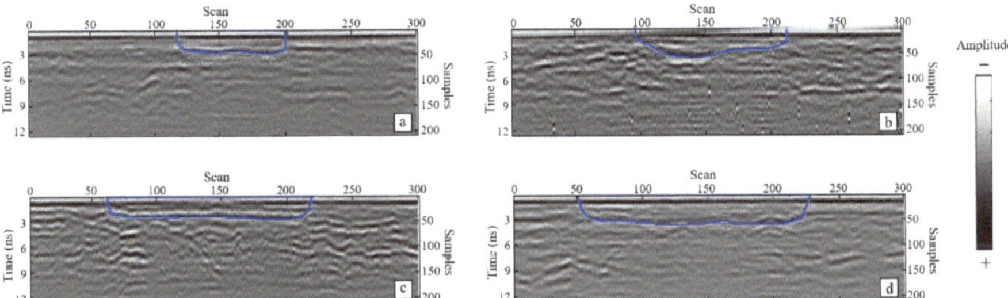

Figure 7. Data annotation of the target backfill pit: (**a**) profile P44; (**b**) profile P54; (**c**) profile P69; (**d**) profile P87.

All data collected and obtained on 25 October 2019 (Dataset (o) in Table A1) and their manual annotations were used to test the 432nd network, as shown in Table 3.

Table 3. Information of the training and testing datasets.

	Training Dataset	**Testing Dataset**
Number	249	52
Sampling points of GPR data	16,434,000	3,432,000
Proportion of total data	19.5%	7.1%
Data acquisition date	20 May 2019–26 September 2019	25 October 2019

Furthermore, the predicted results of the testing dataset were compared with the ground truth, and the IoU of the model on the testing dataset was calculated and divided into two classes according to Section 2.4:

1. When IoU > 0.6, a good prediction is obtained. The target region was successfully segmented, and the borders fit well (the yellow arrows in Figure 8a–d). In a few cases, as indicated by the red arrow on the right in Figure 8b, the output of U-Net was better than the manual ground truth. The borders of the segmentation masks were slightly different from the manual ground truth (the red arrows in Figure 8c,d) at times, but the error was acceptable, with less than one wavelength.

2. When IoU < 0.6, the segmentation results are weak. Some of the segmentation maps were incomplete, and the boundaries were incorrect, as the red arrows indicate in Figure 8e,f.

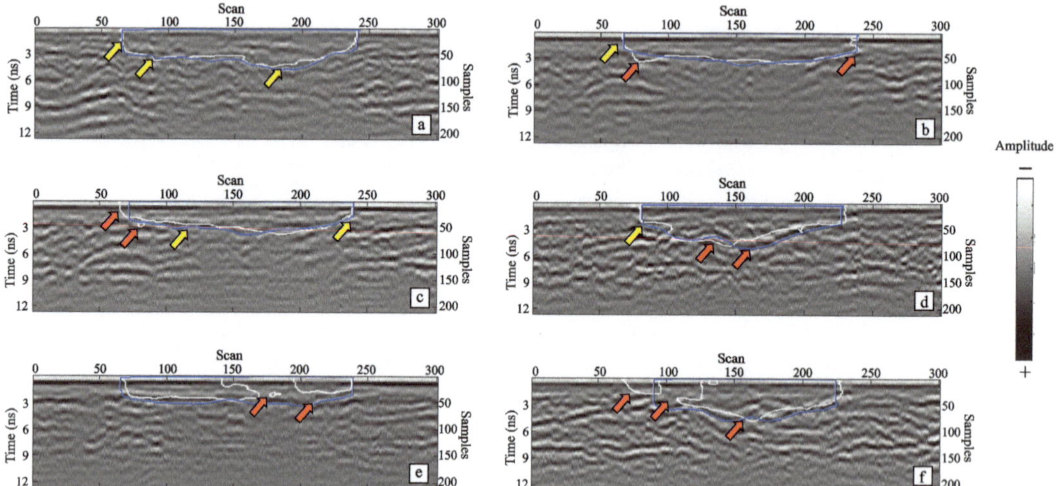

Figure 8. Typical examples of the accuracy of the testing dataset prediction results; the white curves are the boundaries of the segmentation masks, and the blue curves are the boundaries of the manual ground truth. (**a**) Profile P87, IoU = 0.89; (**b**) profile P62, IoU = 0.89; (**c**) profile P61, IoU = 0.87; (**d**) profile P50, IoU = 0.80; (**e**) profile P69, IoU = 0.57; (**f**) profile P55, IoU = 0.64.

The average IoU of the testing dataset was 0.83, indicating that the model could segment the target. The segmentation results of the data collected on 20 May were arranged according to the sequence of profile numbers from small to large, as shown in Figure 9, corresponding to west to east. In addition, the direction of each mask is from south to north.

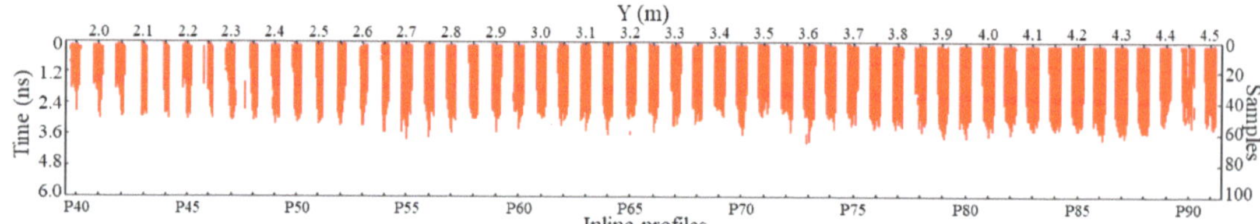

Figure 9. Segmentation results of data collected on 20 May. The horizontal axis represents the y-coordinate and the number of inline profiles, and the segmentation masks of the backfill pit are in red.

3.2. Monitoring

P50 profiles were used as an example to show the changes in the data. As shown in Figure 10a–g, the borders were approximately "U"-shaped. However, a gradual change to a "V" shape of the borders (yellow arrows) was observed, as shown in Figure 10h–j.

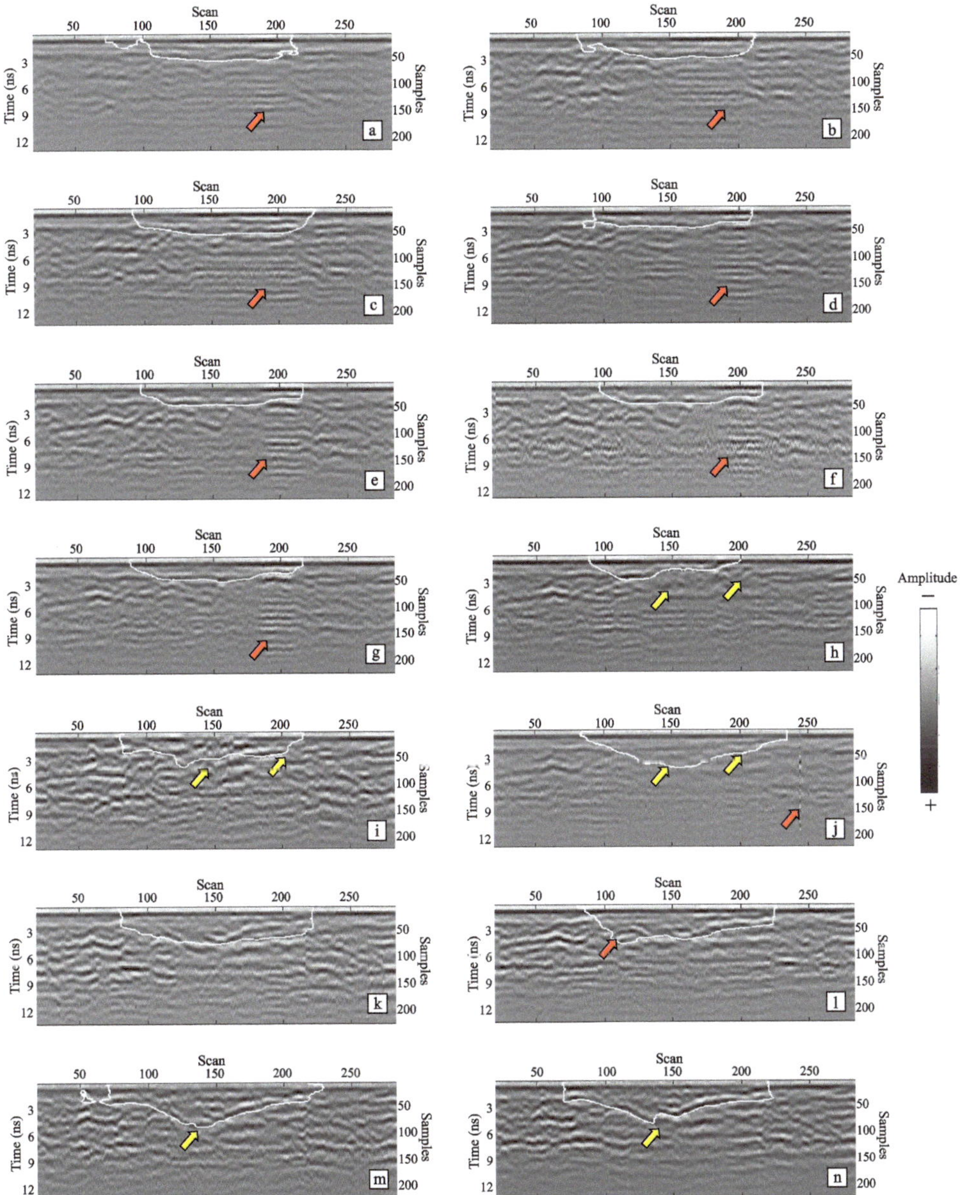

Figure 10. Segmentation results of time-lapse data of the P50 profiles. (**a–n**) are the P50 profiles from 20 May to 26 September. The date and subplot number correspond to those in Table A1. The segmentation was not affected by interferences from multiple reflections (red arrow in (**a–g**)) or a dead signal (red arrow in (**j**)). In a few cases, the segmentation slightly deviated from an ideal performance (red arrow in (**l**)).

To further determine the changes in the 3D data, the overlapped masks on 4 June, 22 August, and 26 September are listed in a crossline direction, as shown in Figure 11, in which changes are shown in blue and yellow, and unchanged areas in red.

1. June 4: As shown in Figure 11a, changes in the left and right borders are noted on the blue masks from Y = 1.95 m to Y = 2.25 m, and changes in the bottom borders can be seen on the yellow masks from Y = 2.35 m to Y = 3.35 m.
2. August 8: As shown in Figure 11b, a significant change in the lateral boundaries can be seen from Y = 2.10 m to Y = 2.65 m and from Y = 3.00 m to Y = 4.50 m (yellow masks in Figure 11b). The bottom borders also change from Y = 2.10 m to Y = 2.65 m (yellow masks) and from Y = 3.90 m to Y = 4.40 m (blue masks).
3. September 26: Significant changes are noted on the bottom borders from Y = 1.95 m to Y = 3.70 m (yellow masks in Figure 11c). Apparently, the change in the left borders from Y = 2.10 m to Y = 3.00 m is clearer compared to that in Figure 11a,b.

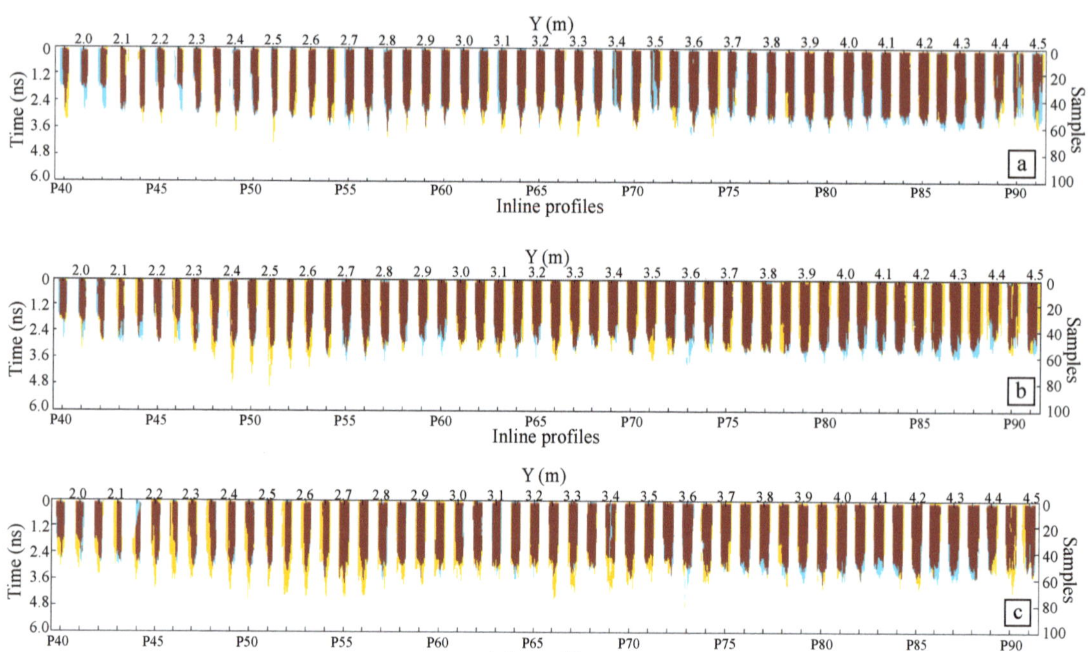

Figure 11. Comparison between segmentation masks: (**a**) 4 June, at Y = 4.45 m, yellow and blue masks within the red mask indicate the incompleteness of the reference mask rather than the change in the target; (**b**) 22 August; (**c**) 26 September, the blue masks in the upper left at Y = 2.15 m and in the middle at Y = 3.40 m are caused by mis-segmentation in the 26 September data, rather than changes.

4. Discussion

4.1. Did the Backfill Pit Really Change?

The time-lapse changes in this study refer to the segmentation masks in the preprocessed radar data, as mentioned in Section 2.5. They provide a sense of lateral variation in the shallow targets, but the changes cannot be accurately calculated because the two-way travel time is not linked to depth, which is crucial information for interpreting GPR data, as the propagation velocity is unknown. For a two-layer model with a horizontal interface, the propagation velocity of the first layer can be calculated using the common middle point (CMP). However, the bottom of this backfill pit was neither irregular nor clear; thus, the speed cannot be calculated using this nondestructive detection method. Moreover, experimental conditions prevented the sampling and laboratory analysis of the material in the backfill pit. Given that the goal of this study was to discuss the fast segmentation of GPR data using U-Net and provide an idea of data comparisons, the study of velocity is not the focus of this study.

For all time-lapse detections, we attempted to avoid rainfall and select similar dry weather for data collection to avoid the influence of water content on speed as much as possible. However, for the rainy season, such as August, the water content and dielectric constant of the subgrade could be higher than during other times; thus, the main frequency and speed of the radar signal may be reduced. On the radar profile, the position of the upper border may be seen further downward, creating a "deepening" illusion, such as the significant expansion of bottom borders from Y = 1.95 m to Y = 3.70 m in Figure 11c. However, if the propagation velocity has not changed dramatically, this phenomenon must be caused by the deepening of the backfill pit. In this case, the backfill pit may have been refilled in August, because it expanded significantly more in September than in May, unlike in August or earlier. Settlement may have occurred in the backfill pit after being rolled, leading to the deformation of the road, and the subgrade may have been filled to facilitate the passage of engineering vehicles. Nonetheless, lateral changes are still valuable since the positions of survey lines were kept unchanged among the time-lapse detections.

4.2. Outlook

Previous studies have used neural networks to recognize targets [5] instead of segmented images [15,16]. The goal of GPR is to identify the range of a target; thus, the network should be capable of both recognition and segmentation. However, most studies remain focused on recognizing strong reflected signals, such as diffraction waves and antenna ringing [9,11]. This is because the result of semi-supervised segmentation is highly dependent on ground truths, which are manually annotated. The complex signals and variable features of GPR data introduce difficulties in annotation and segmentation; thus, methods are expected to be applied to identify the features. For instance, the texture feature coding method was used to extract texture-based features from GPR B-scans [9], and the detection results from the neural network may be more satisfactory with these strong features.

In further research, a dynamic 3D underground database should be realized using TLFC 3D GPR. In this study, TLFC 3D GPR was successfully used to monitor a small field, but the samples and data used to train U-Net were only from the backfill pit in this study. That is, the neural network trained in this study may be inapplicable to targets with other characteristics. If the time-lapse detection can be conducted in a large area to collect more target samples, a more universal neural network model can be established to achieve the segmentation and monitoring of multiple targets.

5. Conclusions

In this study, U-Net was used to segment the inline GPR profiles and divide the boundary of the target backfill pit. After 432 training sessions, an effective and available network model was obtained with IoUs of 0.96 and 0.83 on the training and testing datasets, respectively. This network was used to segment all radar data collected from 20 May to 26 September, and the comparison of segmentation masks with the same location but different acquisition dates provides a method to determine the change in the target.

Author Contributions: Conceptualization, K.S., R.Q. and J.L.; methodology, K.S., J.L., A.S. and J.X.; validation, K.S., R.Q., F.Z., A.S., J.X. and Z.Z.; formal analysis, R.Q., F.Z. and Z.Z.; investigation, K.S. and A.S.; data processing, K.S., J.L., A.S. and J.X.; writing—original draft preparation, K.S. and J.L.; writing—review and editing, K.S., J.L., A.S., F.Z., J.X., Z.Z. and R.Q.; supervision, F.Z. and Z.Z.; project administration, J.L. and J.X.; funding acquisition, R.Q. All authors have read and agreed to the published version of the manuscript.

Funding: This research was funded by Natural Science Foundation of Beijing Municipality (8212016)

Data Availability Statement: Data associated with this research are available and can be obtained by contacting the corresponding author.

Acknowledgments: Zhenning Ma, Jun Zhang, Wenyuan Jin, Xu Liu, Yuchen Wang from China University of Geosciences Beijing are appreciated for their assistance in field experiment.

Conflicts of Interest: The authors declare no conflict of interest.

Appendix A

Table A1. Dataset and acquisition date (year: 2019; date: MM. DD).

Datasets	Date	Datasets	Date	Datasets	Date
(a)	05.20	(f)	08.02	(k)	08.30
(b)	06.04	(g)	08.05	(l)	09.05
(c)	06.18	(h)	08.08	(m)	09.12
(d)	07.22	(i)	08.14	(n)	09.26
(e)	07.30	(j)	08.22	(o)	10.25

Appendix A.1. Raw Data and Misalignments

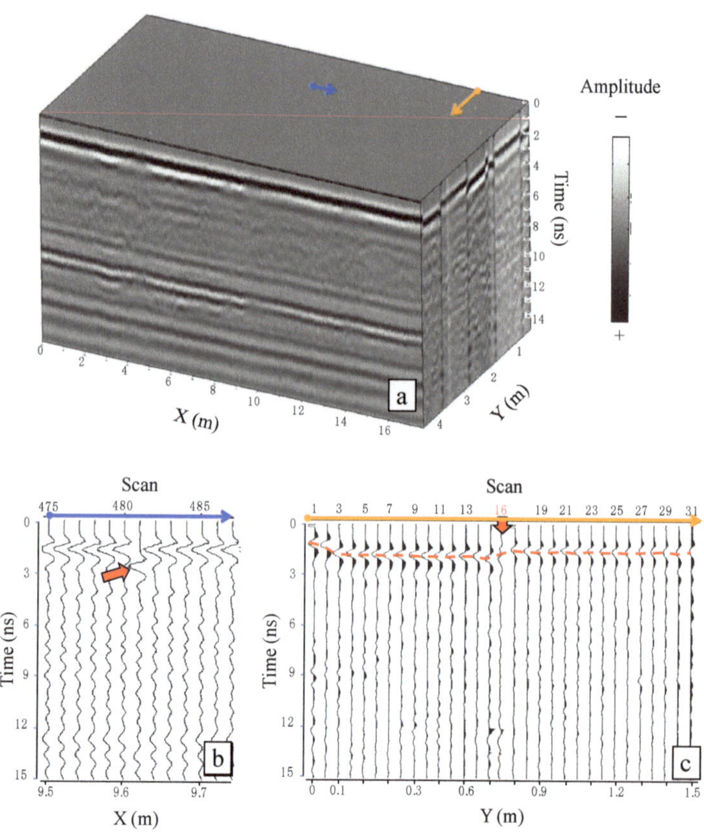

Figure A1. Three-dimensional display and time zero misalignments of GPR data. (**a**) The data cube is displayed as an intensity cube. The colored arrows in (**b**,**c**) indicate the position and orientation of the profiles in the data cube, corresponding to that in (**a**). (**b**) The red arrow indicates the time zero misalignments of the inline, where scans were collected by one antenna. (**c**) The red dash line indicates the time zero misalignments of the crossline, where scans were collected by an antenna array.

Appendix A.2. Antenna Lifting Test

This test was performed in the field prior to each survey by raising the antenna slowly from the ground surface and then returning it to the ground while observing the reflected waveforms as a V-shaped return on a B-scan (Figure A2a). A horizontal BGR filter was applied (Figure A2b) to distinguish positive and negative polarity reflections. Using this method, the direct wave on the ground was separated from the direct wave in the air by picking the first positive peak in the air. Note that there was a slight shift to an earlier time position in the direct wave when the antenna was on the ground surface, as the green dashed line shows on the right side of Figure A2b. The antenna lifting test was conducted several times, and the average reflection time was set as the time zero.

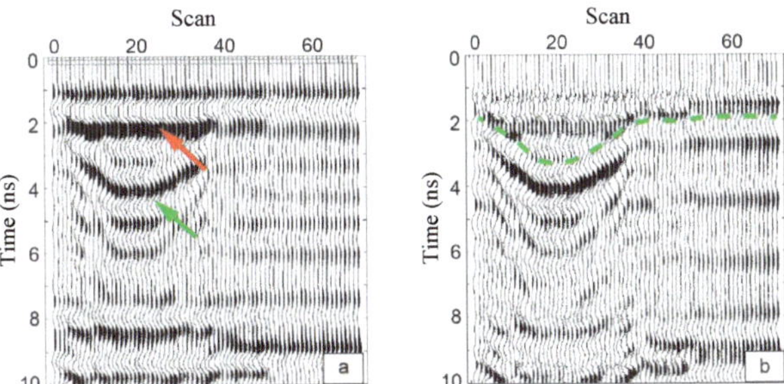

Figure A2. Antenna lifting test. (**a**) Direct wave in air (red arrow) and reflected wave on the ground (green arrow) in the raw data. (**b**) BGR filter and time zero (green dashed line).

Appendix A.3. Preprocessing on Inline Profiles

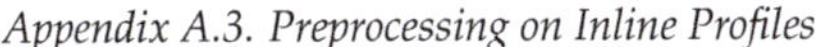

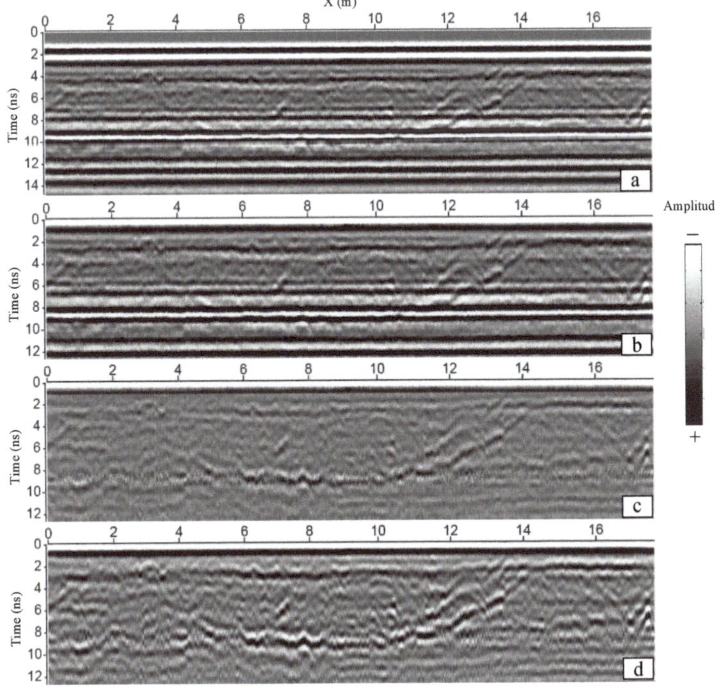

Figure A3. Raw data and preprocessed data of inline profile. (**a**) Raw data. (**b**) After time zero correction. (**c**) After BGR filter. (**d**) After frequency filter.

Appendix A.4. Three-Dimensional Time Zero Normalization

The cross-correlation sequence of the direct wave between $x_{ref}(n)$ and $x_{mat}(n)$ was calculated according to Equation (A1), where the value K is obtained when r_{rm} is maximum.

$$r_{rm}(m) = \sum_{n=1}^{N} x_{ref}(n)x_{mat}(n-m) \tag{A1}$$

where $x_{ref}(n)$ is a reference GPR signal with a known direct wave travel time, and $x_{mat}(n)$ is the signal to be matched.

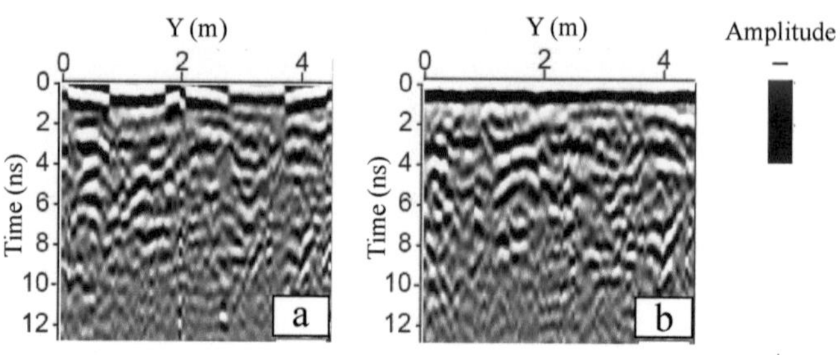

Figure A4. Raw data and preprocessed data of crossline profile. (**a**) Raw data. (**b**) After 3D time zero correction.

Appendix A.5. Three-Dimensional Data Combination and Imaging

The correlation coefficient of all the overlapping scans was calculated according to Equation (A2).

$$\rho_{rc}(m_r, m_c) = \frac{\sum_{n=0}^{N} x_r(m_r, n)x_c(m_c, n)}{\left[\sum_{n=0}^{N} x_r{}^2(m_r, n) \sum_{n=0}^{N} x_c{}^2(m_c, n)\right]^{1/2}} \tag{A2}$$

where m_r and m_c are the scan numbers, N is the number of sampling points of each scan, and $x_r(m_r, n)$ and $x_c(m_c, n)$ denote the reference signals and the signals to be spliced, respectively. When the maximum value of $\rho_{rc}(m_r, m_c)$ is obtained, it is considered that the scan m_c in x_c corresponds to the same position of the trace m_r in x_r, and the scan numbers of m_c and m_r are recorded as M_i and i, respectively. The scan numbers corresponding to the same position were subtracted and averaged according to Equation (A3) to obtain the relative offset Δn.

$$\Delta n = \frac{1}{M} \sum_{i=1}^{M} (M_i - i) \tag{A3}$$

The offset between adjacent group lines was successively calculated and moved along the Y direction to obtain the data cube after combination.

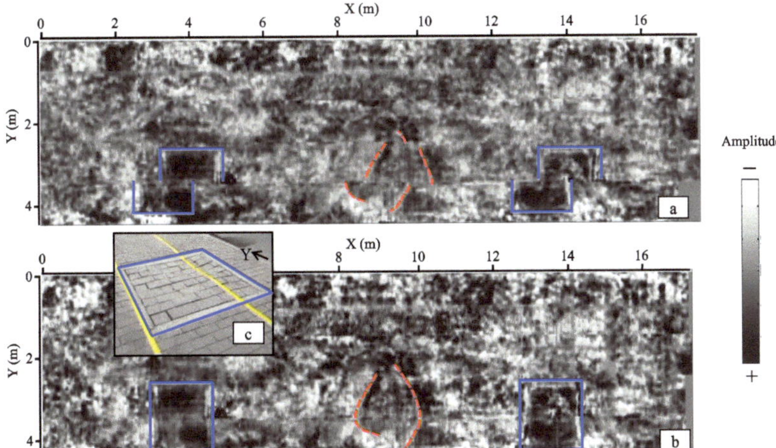

Figure A5. Time slice of 3D GPR data at t = 1.6 ns. The blue boxes indicate the boundaries of two square backfill pits, and the red dash lines indicate the borders of the backfill pit which was segmented in this study. (**a**) Time slice before data combination. There is a significant misplacement at Y ≈ 2.5 m and Y ≈ 3.5 m, as data were collected from different group lines. (**b**) Time slice after data combination. The misplacement was rectified by the correlation applied to the overlapping data of adjacent group lines. The backfill pits marked by the blue boxes are clearly visible and easy to be located on the radar slice and can be used to match the position of the time-lapse data. (**c**) The square backfill pit.

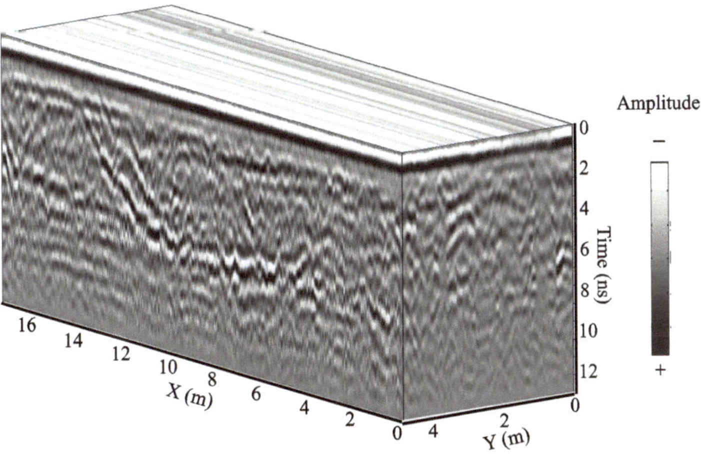

Figure A6. Three-dimensional GPR data cube after preprocessing.

References

1. Reeves, B.A.; Muller, W.B. Traffic-speed 3-D noise modulated ground penetrating radar (NM-GPR). In Proceedings of the 2012 14th International Conference on Ground Penetrating Radar, Shanghai, China, 4–8 June 2012; pp. 165–171. [CrossRef]
2. Simonin, J.M.; Baltazart, V.; Hornych, P.; Derobert, X.; Thibaut, E.; Sala, J.; Utsi, V. Case study of detection of artificial defects in an experimental pavement structure using 3D GPR systems. In Proceedings of the 15th International Conference on Ground Penetrating Radar, Brussels, Belgium, 30 June–4 July 2014; pp. 847–851. [CrossRef]
3. Linjie, W.; Jianyi, Y.; Ning, D.; Changjun, W.; Gao, G.; Haiyang, X. Analysis of interference signal of ground penetrating radar in road detection. *J. Hubei Polytech. Univ.* **2017**, *33*, 35–40. [CrossRef]
4. Simonyan, K.; Zisserman, A. Very deep convolutional networks for large-scale image recognition. *arXiv* **2014**, arXiv:1409.1556.

5. Russakovsky, O.; Deng, J.; Su, H.; Krause, J.; Satheesh, S.; Ma, S.; Huang, Z.; Karpathy, A.; Khosla, A.; Bernstein, M.; et al. ImageNet large scale visual recognition challenge. *Int. J. Comput. Vis.* **2014**, *45*, 211–252. [CrossRef]
6. Tompson, J.; Jain, A.; Lecun, Y.; Bregler, C. Joint training of a convolutional network and a graphical model for human pose estimation. *arXiv* **2014**, arXiv:1406.2984.
7. Girshick, R.; Donahue, J.; Darrell, T.; Malik, J. Region-based convolutional networks for accurate object detection and segmentation. *IEEE Trans. Pattern Anal. Mach. Intell.* **2016**, *38*, 142–158. [CrossRef] [PubMed]
8. Shan, K.; Guo, J.; You, W.; Lu, D.; Bie, R. Automatic facial expression recognition based on a deep convolutional-neural-network structure. In Proceedings of the IEEE International Conference on Software Engineering Research, London, UK, 7–9 June 2017; pp. 123–128. [CrossRef]
9. Besaw, L.E.; Stimac, P.J. Deep convolutional neural networks for classifying GPR B-Scans. In Proceedings of the SPIE the International Society for Optical Engineering, San Diego, CA, USA, 9–13 August 2015; SPIE: Bellingham, WA, USA, 2015; Volume 9454. [CrossRef]
10. Lameri, S.; Lombardi, F.; Bestagini, P.; Lualdi, M.; Tubaro, S. Landmine detection from GPR data using convolutional neural networks. In Proceedings of the 2017 25th European Signal Processing Conference (EUSIPCO), Kos, Greece, 26 October 2017; pp. 508–512. [CrossRef]
11. Kim, N.; Kim, S.; An, Y.K.; Lee, H.J.; Lee, J.J. Deep learning-based underground object detection for urban road pavement. *Int. J. Pavement Eng.* **2020**, *21*, 1638–1650. [CrossRef]
12. Kim, N.; Kim, S.; An, Y.K.; Lee, J.J. A novel 3D GPR image arrangement for deep learning-based underground object classification. *Int. J. Pavement Eng.* **2021**, *22*, 740–751. [CrossRef]
13. Klesk, P.; Godziuk, A.; Kapruziak, M.; Olech, B. Fast analysis of C-scans from ground penetrating radar via 3-D haar-like features with application to landmine detection. *IEEE Trans. Geosci. Remote Sens.* **2015**, *53*, 3996–4009. [CrossRef]
14. Jing, H.; Vladimirova, T. Novel algorithm for landmine detection using C-scan ground penetrating radar signals. In Proceedings of the 2017 Seventh International Conference on Emerging Security Technologies (EST), Canterbury, UK, 6–8 September 2017; pp. 68–73. [CrossRef]
15. Ronneberger, O.; Fischer, P.; Brox, T. U-net: Convolutional networks for biomedical image segmentation. In *Medical Image Computing and Computer-Assisted Intervention–MICCAI 2015, Proceedings of the 8th International Conference, Munich, Germany, 5–9 October 2015*; Lecture Notes in Computer Science; Springer: Cham, Switzerland, 2015; Volume 9351, p. 9351. [CrossRef]
16. Berezsky, O.; Pitsun, O.; Derysh, B.; Pazdriy, I.; Melnyk, G.; Batko, Y. Automatic segmentation of immunohistochemical images based on U-net architecture. In Proceedings of the 2021 IEEE 16th International Conference on Computer Sciences and Information Technologies (CSIT), Lviv, Ukraine, 22–25 September 2021; pp. 29–32. [CrossRef]
17. Ling, J.; Qian, R.; Shang, K.; Guo, L.; Zhao, Y.; Liu, D. Research on the dynamic monitoring technology of road subgrades with time-lapse full-coverage 3D ground penetrating radar (GPR). *Remote Sens.* **2022**, *14*, 1593. [CrossRef]

 *remote sensing*

Technical Note

Research on the Dynamic Monitoring Technology of Road Subgrades with Time-Lapse Full-Coverage 3D Ground Penetrating Radar (GPR)

Jianyu Ling [1], Rongyi Qian [1,*], Ke Shang [2], Linyan Guo [1], Yu Zhao [1] and Dongyi Liu [1]

[1] School of Geophysics and Information Technology, China University of Geosciences, Beijing 100083, China; jianyu@email.cugb.edu.cn (J.L.); guoly@cugb.edu.cn (L.G.); 2110210041@email.cugb.edu.cn (Y.Z.); 2010200050@cugb.edu.cn (D.L.)
[2] School of Earth and Space Sciences, Peking University, Beijing 100871, China; shangke@stu.pku.edu.cn
* Correspondence: rongyiqian@cugb.edu.cn

Abstract: Road safety is important for the rapid development of the economy and society. Thus, it is of great significance to monitor the dynamic changing processes of road diseases, such as cavities, to provide a basis for the daily maintenance of roads and prevent any possible car accidents. The ground penetrating radar (GPR) technology is widely used in road disease detection due to its advantages of nondestructiveness, rapidness, and high resolution. Traditionally, one-time 2D GPR detection cannot obtain the 3D spatial changes of subgrades. Thus, we developed a road subgrade monitoring method based on the time-lapse full-coverage (TLFC) 3D GPR technique by focusing on solving the key problems of time and spatial position mismatches in experimental data. Moreover, we used the time zero consistency correction, 3D data combination, and spatial position matching methods, as they greatly improve the 3D imaging quality of underground spaces. Finally, the time-lapse attribute analysis method was used in the TLFC 3D GPR data to obtain detailed characteristics and an overall rule of the dynamic subgrade change. Overall, this research proves that TLFC 3D GPR is an optimal choice for road subgrade monitoring.

Keywords: time-lapse full-coverage (TLFC) 3D GPR; road disease monitoring; data processing; 3D GPR imaging; attribute interpretation

Citation: Ling, J.; Qian, R.; Shang, K.; Guo, L.; Zhao, Y.; Liu, D. Research on the Dynamic Monitoring Technology of Road Subgrades with Time-Lapse Full-Coverage 3D Ground Penetrating Radar (GPR). *Remote Sens.* **2022**, *14*, 1593. https://doi.org/10.3390/rs14071593

Academic Editors: Alessandro Mei, Xianfeng Zhang and Valerio Baiocchi

Received: 11 March 2022
Accepted: 25 March 2022
Published: 26 March 2022

Publisher's Note: MDPI stays neutral with regard to jurisdictional claims in published maps and institutional affiliations.

1. Introduction

In recent years, road collapses, which are usually caused by the development of cavities in road subgrades, have frequently occurred. They may affect traffic or cause casualties and serious property losses. Therefore, it is necessary to dynamically monitor road subgrade diseases.

Due to the GPR advantages of rapidness, high resolution, and nondestructiveness, it is widely used in road layer thickness measurements [1–3], cavity and crack detection [4–6], asphalt layer quality assessment [7–9], the detection of moisture and water content changes [10–12]. However, most of these studies were performed based on 2D GPR detection, which has the problems of false detection and missing detection, as the survey lines were sparse and the acquired information was inadequate. Using 3D GPR for data acquisition and imaging can well reveal underground structures [13]. Full-resolution 3D GPR imaging can be realized by arranging 2D grids with a line spacing of less than one-quarter of the wavelength [14–16]. Unfortunately, the main limitation of this method is that it is time-consuming. As a result, it is hard to perform the full-coverage detection of roads without affecting traffic. With the appearance of multi-channel array antenna 3D GPR, the detection efficiency and accuracy of the combination of survey lines have greatly been improved, making this method suitable for the full-coverage detection of roads [17].

Furthermore, road subgrades have the characteristics of dynamic change under the action of vehicle rolling and underground engineering. Therefore, it is urgent to perform short-interval periodic detection on roads based on 3D GPR so as to repair the areas where road disasters may occur. At present, the application of GPR in the monitoring field is mainly based on underground hydrological process monitoring [18–23], pollutant monitoring [24,25], gas migration monitoring [26], road stripping monitoring [27], underground facility monitoring [28], and railway monitoring [29].

Compared with the existing studies mentioned above, the full-coverage monitoring of roads has the characteristics of long-term, less-known information, dense survey lines, strong interference, and other uncertain factors. In addition, the differences between time-lapse GPR data are weak, and some of them are caused by nonunderground changes, which greatly increase the difficulty of data interpretation. Hence, it is particularly important to study data processing and interpretation methods. At present, the research on time-lapse GPR data processing mainly focuses on amplitude normalization, time zero correction, spatial position matching, four-dimensional grid interpolation, sampling rate fluctuation correction, and abnormal amplitude correction [23,28,30–32]. Furthermore, the interpretation methods of time-lapse data include difference [24,32], attribute analysis [20,23], K-means clustering [28], amplitude analysis [33], and electromagnetic wave velocity analysis [22,34]. In short, eliminating the differences caused by nonunderground changes and searching the changed areas in time-lapse GPR data are key problems of data processing and interpretation.

In this paper, we present a road subgrade monitoring technology based on time-lapse full-coverage (TLFC) 3D GPR. It relies on eliminating the differences caused by nonunderground changes and analyzing the differences in the attributes between time-lapse data. Through the design of a full-coverage road monitoring experiment, TLFC 3D GPR data were acquired. Then, time zero consistency correction, 3D data combination, spatial position matching, and interpretation based on a time-lapse attribute were performed on the acquired data. The 3D imaging quality of the TLFC 3D GPR data to underground space was significantly improved, and the changing area was highlighted. Finally, we summarized the dynamic changing rule of subgrade and its causes are analyzed.

2. Experimental Design and Data Acquisition

2.1. Experimental Area

We designed the periodic detection experiment on a sidewalk whose subgrade may change. This area was rolled by engineering vehicles outside a subway construction site, and it covers an area of 85 (17 × 5) m^2 and has three backfill pits, as shown in Figure 1. The red shaded area faces the gate of the construction site and belongs to the key rolling area.

Figure 1. Overview of the experimental area.

2.2. TLFC 3D GPR Data Acquisition

Mobyscan-V 3D GPR, which was produced by DECOD of Singapore, was used in this monitoring experiment. A 2-GHz high-frequency array antenna that includes 16 pairs of receiving and transmitting antennas was selected, forming 30 data channels with a spacing of 5 cm (Figure 2a). Single group lines could cover a width of 1.5 m, and there was a 10 cm gap between System A and System B. By designing four group lines on the sidewalk, full-coverage of the rolling area was realized. There were 11 overlapping channels between the adjacent group lines, which provided a guarantee for the accurate combination of different group lines (Figure 2b). Furthermore, due to the frequent passing of engineering vehicles on the sidewalks, we selected a short detection time interval, ranging from 8 to 14 days, to clearly show the change process of the subgrade. Table 1 shows the acquisition parameters.

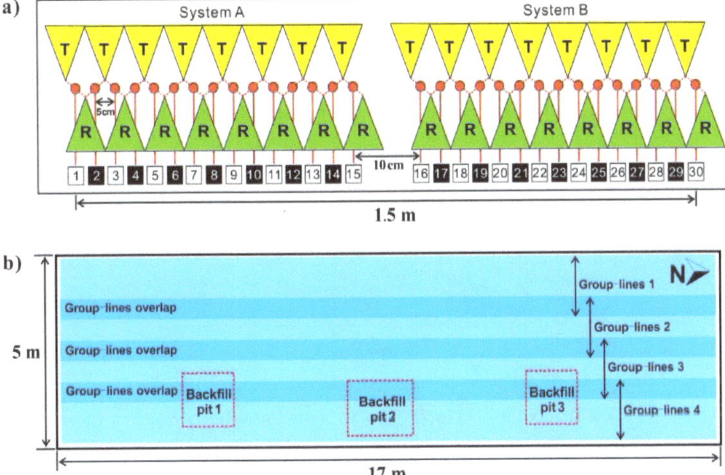

Figure 2. (**a**) Transmitter and receiver arrays of Mobyscan-V. It is divided into two systems, A and B, and each system contains eight pairs of transceiver antennas to form 15 data channels. The spacing between the two systems is 10 cm. (**b**) Layout of the survey lines. Four group lines were used to achieve full-coverage of the experimental area.

Table 1. TLFC 3D GPR data acquisition parameters.

Parameters	Values
Frequency (GHz)	2
Trace spacing (m)	0.02
Sampling points	512
Time window (ns)	15
Number of overlapping channels	11
Time-lapse detection times	5

3. Data Processing and Imaging

Reasonable data processing is crucial for ensuring the reliability of detection results, and it greatly improves the imaging quality of underground 3D spaces. The key problems of TLFC 3D GPR data processing include two aspects. First, for the full-coverage 3D GPR data acquired on the same date, we mainly solve the inconsistency of time zero and the dislocation of multiple group lines. Second, for the time-lapse data acquired on different dates, the main purpose is to eliminate the differences between survey line positions and accurately obtain the repeated parts. Therefore, the designed processing scheme for monitoring experimental data includes (1) preprocessing, (2) time zero consistency correction, (3) 3D data combination of the adjacent group lines, and (4) spatial position matching.

3.1. Preprocessing

Due to the electromagnetic wave characteristics of high frequency and short wavelength, classic processing methods should be selected to ensure the authenticity of reflected waves. Using the ReflexW software, time zero correction, background removal, exponential gain, and bandpass filtering (using a Butterworth bandpass filter with a passband of 400–1500 MHz) are mainly performed. The GPR data profiles before and after preprocessing are shown in Figure 3a,b. The clutter in the processed data was well eliminated, and the events of the reflected wave from the underground interface became clearer.

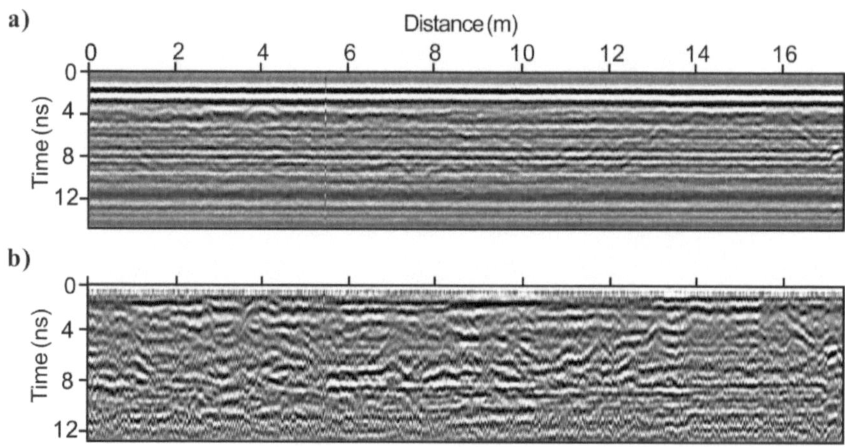

Figure 3. Inline profile of the 3D GPR data: (**a**) before preprocessing; (**b**) after preprocessing.

3.2. Time Zero Consistency Correction and Imaging

After preprocessing, we noticed that the events were rough and fuzzy, as shown in Figure 3b. This is due to the shaking of the antenna during movement and the system stability fluctuation. This problem introduces false anomalies into 3D GPR data, especially in horizontal slices, which greatly reduces the reliability of interpretation results. To realize high-quality imaging of the underground 3D structure and provide interpretation guarantees, the time zero needed to be corrected to the same position before the preprocessing process. Since the ground direct wave has a good correlation and belongs to strong reflection, the most relevant position on the time axis could be found by calculating the correlation sequence of each trace's direct wave. The time zero correction includes the following three procedures.

(1) Determining the reference trace

The reference trace credibility directly determines the accuracy of time zero position normalization and correction. Hence, the one trace $x_{ref}(n)$ with a known direct wave travel time is selected as a reference trace, and the other traces $x_{mat}(n)$ are corrected.

(2) Correction time calculation

The cross-correlation sequence of the direct wave between $x_{ref}(n)$ and $x_{mat}(n)$ is calculated according to Equation 1, and the correction amount K is obtained when r_{rm} is maximum.

$$r_{rm}(m) = \sum_{n=1}^{N} x_{ref}(n)x_{mat}(n-m) \tag{1}$$

(3) Correction

If $K = 0$, there is no need to move the trace to be corrected. Otherwise ($K > 0$ or $K < 0$), the trace to be corrected needs to move K sampling points, positive or negative, along the time axis.

The time zero consistency correction could be completed by performing the above method so that all the traces could be corrected. In the inline profile, after the time zero correction and preprocessing steps, the jitter interference on both sides of the events was eliminated. Also, the imaging quality was greatly improved, as shown in Figure 4.

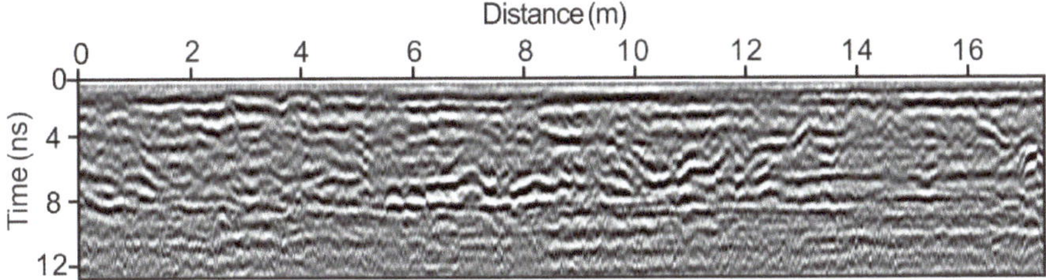

Figure 4. Inline profile of the 3D GPR data after time zero correction and preprocessing.

3.3. 3D Data Combination and Imaging

The data interpretation accuracy could be significantly improved by using multiple parallel-group lines to perform full-coverage detection of roads, especially when abnormalities appear at the edge of group lines.

In this experiment, due to the location, differences among the adjacent group lines in the inline direction exist. If the adjacent group lines are directly combined, imaging dislocation occurs. As shown in the yellow marked-area in Figure 5a, the two backfill pit areas in the horizontal slice are misplaced. Therefore, the 3D GPR data combination algorithm that is based on cross-correlation was studied. This algorithm includes three aspects as follows.

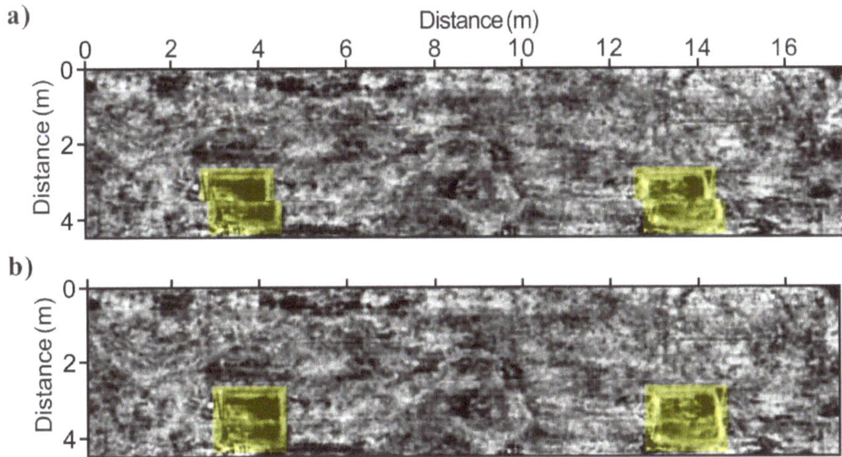

Figure 5. Horizontal slice of the full-coverage 3D GPR data: (**a**) direct combination; (**b**) using the proposed method in this paper. The yellow shaded areas indicate the locations of the backfill pits.

(1)   Determining the reference lines from the overlapping area

Select the profile of one group lines in the overlapping area as a reference profile. Then, the profile corresponding to the position in the other group lines is the profile to be spliced.

(2)   Calculating the relative offset of different group lines

Calculate the correlation coefficient of all the traces between the profile to be spliced and the reference profile according to Equation (2).

$$\rho_{rc}(m_r, m_c) = \frac{\sum_{n=0}^{N} x_r(m_r, n)x_c(m_c, n)}{\left[\sum_{n=0}^{N} x_r^2(m_r, n) \sum_{n=0}^{N} x_c^2(m_c, n)\right]^{1/2}} \tag{2}$$

where m_r and m_c are the trace numbers, N is the number of sampling points of each trace, and $x_r(m_r, n)$ and $x_c(m_c, n)$ denotes the trace of the reference profile and the profile to be spliced, respectively. When the maximum value of $\rho_{rc}(m_r, m_c)$ is obtained, it is considered that the trace m_c in x_c corresponds to the same position of the trace m_r in x_r, and the trace number of m_c and m_r are recorded as M_i and i, respectively. Subtract and average the trace numbers corresponding to the same position according to Equation (3) to obtain the relative offset.

$$\Delta n = \frac{1}{M} \sum_{i=1}^{M} (M_i - i) \tag{3}$$

(3)   3D data volume combination and imaging

When $\Delta n = 0$, data combination can be performed without moving the profiles. When $\Delta n > 0$ or $\Delta n < 0$, the profile is moved to be spliced along with the inline positive or negative direction by Δn traces. Then, the data is combined.

The horizontal slice after the accurate combination of the adjacent group lines is shown in Figure 5b. The misplaced backfill pits were restored to the rectangular and the accurate 3D imaging of the underground space was realized. This shows that the proposed data combination algorithm in this paper is effective and accurate when there is enough overlapping data between adjacent group lines.

3.4. Spatial Location Matching

In large-area road time-lapse monitoring, it is hard to ensure the complete consistency of survey line positions. The location differences between the TLFC 3D GPR data caused the reflection from the same underground object to appear at different positions in the time-lapse GPR profiles (Figure 6a,b). Similar to a previous processing technique, the cross-correlation method was used to calculate the relative offset of the start and end positions of the time-lapse data. Then, the overlapping part of the time-lapse data was extracted to realize spatial position matching. Specifically, it includes the following three procedures.

(1)   Determining the reference data

The position differences between group lines at different acquisition dates are within 40 cm. To ensure the universal applicability of the reference data, the data after cutting 40 cm from the head and tail in the inline profile of the first time acquired are selected as reference data x_r. Then, the subsequent acquired time-lapse data is used as the data to be matched x_c.

(2)   Calculating the position offset

Use Equation (2) to find the trace most similar to x_r in x_c and then obtain the position offset of the time-lapse data according to Equation (3).

(3)   Extract the data with the same survey line position

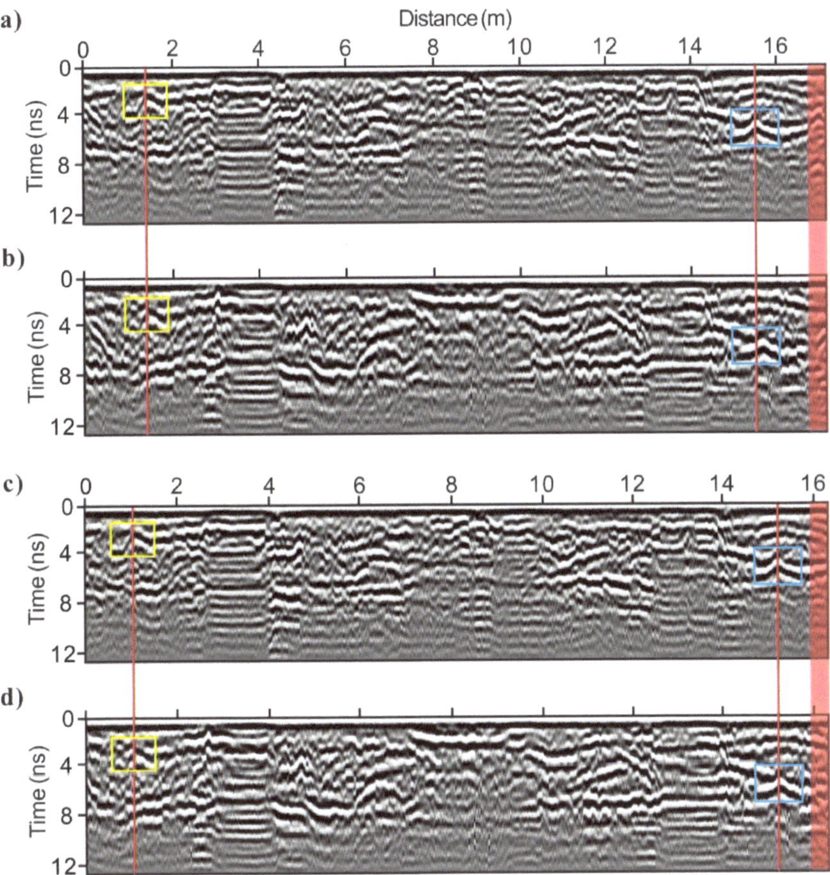

Figure 6. Spatial position matching of the time-lapse GPR data: (**a,b**) before matching; (**c,d**) after matching.

The spatial position matching of time-lapse GPR data can be realized by moving the data to be matched according to the relative offset and deleting the remaining traces that have no correspondence with the reference data.

The time-lapse GPR data after the matching process is shown in Figure 6c,d. The reflection feature with good correspondence on the left remained, and the poor correspondence on the right was improved after the spatial position matching process. In addition, the red shaded area on the right of Figure 6c,d indicates that the time-lapse data have the same survey line terminus; however, this was inconsistent before the data matching process.

4. Data Interpretation

4.1. Time-Lapse Attribute Analysis Method

Although the TLFC 3D GPR data were acquired and processed with great consideration, there were still some problems caused by acquisition and processing. If we simply performed difference processing on the time-lapse data, the wavelet characterization of the data could be preserved in the difference profile. Thus, it would be difficult to clearly show the underground changing area (Figure 7). The time-lapse attribute analysis method proposed by Allroggen et al. has achieved satisfactory results in the time-lapse 2D GPR data of subsurface flow process monitoring [20,23]. Therefore, we used it for the interpretation of the TLFC 3D GPR road subgrade monitoring data. It includes a contrast similarity (CS) attribute and a structural similarity (SS) attribute. The former mainly reflects the areas where the reflected wave energy is different, and the latter mainly highlights the areas where the shape of the events changes. The CS and SS attributes are respectively defined as follows:

$$CS(x,y) = \frac{2\sigma_x\sigma_y + a}{\sigma_x^2 + \sigma_y^2 + a} \tag{4}$$

$$SS(x,y) = \frac{2c_{xy} + a}{\sigma_x\sigma_y + a} \tag{5}$$

where σ_x and σ_y denote the standard deviations within the selected window, c_{xy} denotes a normalized zero-lag cross-correlation after subtracting the window means, and a denotes a stable term for avoiding numerical instability when σ approaches 0. The specific expression is as follows:

$$\sigma_x = (\frac{1}{N-1}\sum_{i=0}^{N}(x_i - \mu_x))^{\frac{1}{2}} \tag{6}$$

$$\mu_x = \frac{1}{N}\sum_{i=1}^{N}x_i \tag{7}$$

$$c_{xy} = \frac{1}{N-1}\sum_{i=1}^{N}(x_i - \mu_x)(y_i - \mu_y) \tag{8}$$

where x and y are the windowed sequences. The normalized similarity attribute (NSA) of the time-lapse data could be obtained by multiplying the CS and SS attributes [23].

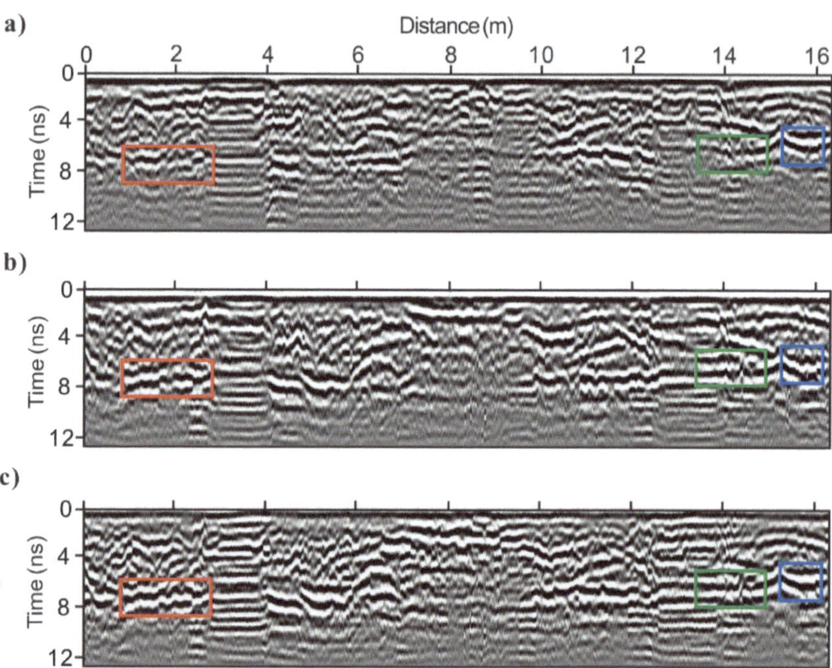

Figure 7. Inline profile of the time-lapse 3D GPR data on (**a**) 22 Jul., and (**b**) 30 Jul., (**c**) differenced profile.

Before calculating the attributes, it is necessary to normalize the amplitude in the selected window, and the value of a is usually 10% of the amplitude variation range in the window after amplitude normalization. During attribute calculation, the window size is the key parameter. With a too-small window, many wavelet features can be retained. Otherwise, it will excessively smooth out the areas with small attribute values. Through parameter experiments, we concluded that the optimal window size for the time-lapse data in this experimental area is 30 × 30. That is, a window contains 30 traces horizontally and 30 sampling points vertically. Figure 8 shows the attribute calculation results of the time-lapse data on 22 Jul. and 30 Jul. in Figure 7. A small attribute value denotes low similarity of the time-lapse data, that is, underground changes may have occurred. The SS attribute highlights the areas with different event shapes, and the regions shown in the red and blue boxes in Figure 7 are highlighted in Figure 8a. The CS attribute highlights the areas with different reflected wave energies, and the regions shown in the red and green boxes in Figure 7 are highlighted in Figure 8b. NSA realizes the fusion of the regions with low attribute values in Figure 8a,b, and it has a comprehensive display of the underground changing areas.

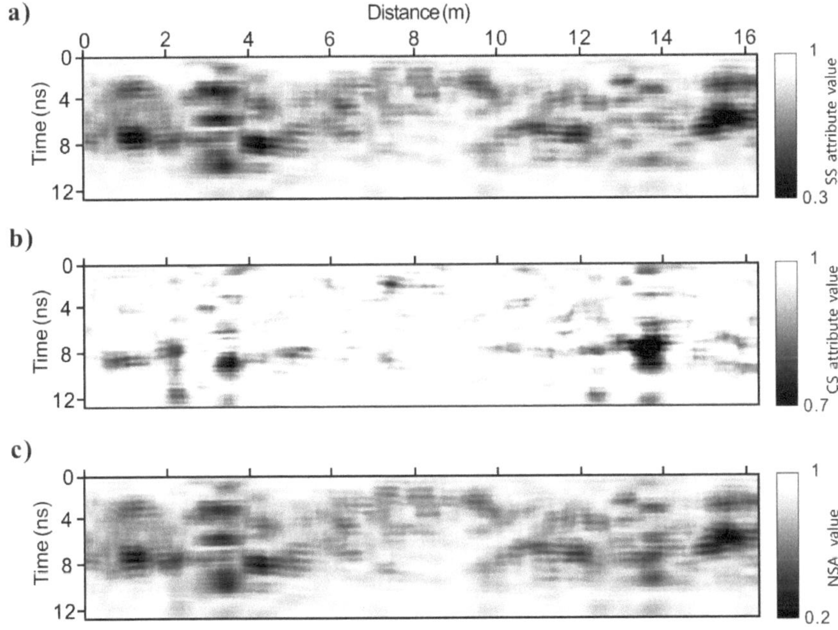

Figure 8. Time-lapse attribute value of the GPR data on 22 Jul. and 30 Jul. (**a**) SS attribute; (**b**) CS attribute; (**c**) NSA.

4.2. Interpretation Based on Attribute Differences

The NSA of the TLFC 3D GPR data was calculated, and 3D images of the results are shown in Figure 9. Since there were still a few differences caused by nonunderground changes after data processing, when the values of the CS and SS attributes were both less than 0.6 or when any one of them was less than 0.4, we regarded the changes in the data as from underground. Hence, only the parts with NSA values lower than 0.4 were demonstrated, as shown in Figure 9. The red dotted circle in Figure 9 corresponds to the near area of the backfill pit 2. Since this area faces the gate of the construction site, it belongs to the key rolling area of the engineering vehicles. The changes in this area are explained as follows.

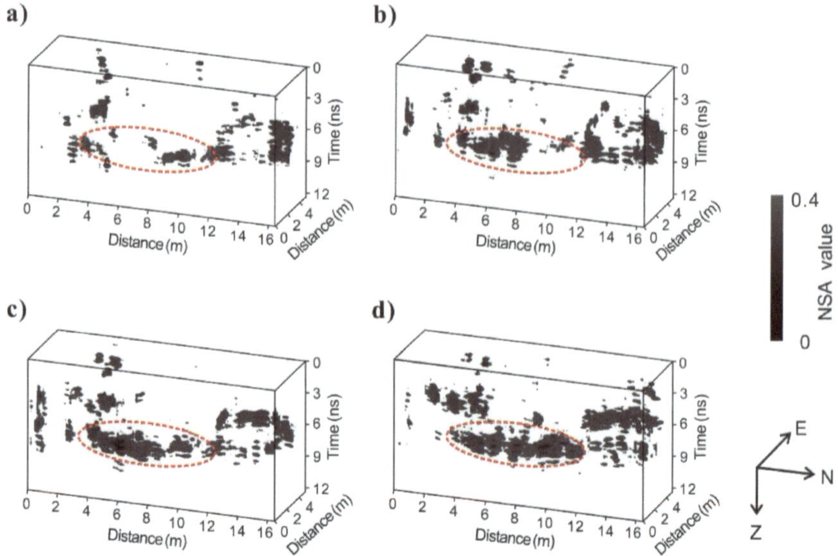

Figure 9. NSA value of the TLFC 3D GPR data: (**a**) 22 Jul. to 30 Jul.; (**b**) 22 Jul. to 8 Aug.; (**c**) 22 Jul. to 22 Aug.; (**d**) 22 Jul. to 30 Aug. The red dotted circle area faces the gate of the construction site, which belongs to the key rolling area of the engineering vehicles.

As time went by, the area with a low NSA value gradually expanded, indicating that the passing of engineering vehicles on this area was continuous during the monitoring period. In addition, according to the changing trend of the attribute value, this area was almost invariable from 22 July to 30 July, and the subgrade was relatively stable. From 30 July to 8 August, the subgrade was dramatically changed, and changes occurred in the southern part. The changing speed of the subgrade tended to ease from 8 August to 22 August, and mainly the northern region has changed. From 22 August to 30 August, the subgrade greatly changed again, mainly in the northern part. It is speculated that the reason for the above changes in the subgrade is that the flow of engineering vehicles was rare before 30 July Thus, the subgrade could maintain its original structure to a certain extent. From 30 July to 8 August, the original structure of the subgrade was damaged and greatly changed due to the frequent passing of engineering vehicles. From 8 August to 22 August, the damaged subgrade was compacted to form a new structure. It was difficult for the engineering vehicles to make great changes in the subgrade structure again. Therefore, although the time interval between Figure 9b,c is two weeks, the change in the low attribute value area was small. From 22 Aug. to 30 Aug., relatively large changes took place in the low attribute value area, which may be because the subgrade was damaged again due to the passing of a large number of engineering vehicles.

5. Discussion

To overcome the disadvantages of 2D GPR (missing detection and inability to obtain underground dynamic changing information), we proposed a new road subgrade monitoring technology (TLFC 3D GPR), which transformed the traditional detection strategy into a monitoring strategy. The main achievements of this paper are as follows. (1) We presented a method of time zero consistency correction, thus eliminating jitter interference on both sides of events and improving imaging quality. (2) 3D data combination was used in multiple group lines to realize the full-coverage imaging of underground spaces. (3) Thorough spatial position matching well solved the mismatch problem in the time-lapse data. (4) The time-lapse attribute analysis method was applied to interpret the road subgrade monitoring data of TLFC 3D GPR, the wavelet characterization influence was eliminated, and subgrade dynamic changing information was obtained.

Compared with the existing research, this paper mainly has three innovations. First, most road disease studies are based on one-time detection [5,17]. Thus, we presented a full-coverage monitoring method and proved its feasibility. Second, the previously proposed time-lapse processing methods are usually based on 2D GPR data [28,30–32]. We realized the data processing of TLFC 3D GPR and improved the imaging quality. Third, the time-lapse attribute has achieved satisfactory results in monitoring the data of 2D GPR for fast subsurface processes [20,23]. Thus, we applied it to long-term road subgrade monitoring data with TLFC 3D GPR, and equally excellent results were obtained.

It is worth mentioning that there were still a few differences caused by the nonunderground changes in the processed data, and they may be because of the following reasons. (1) The sampling rate of the instrument fluctuates during data acquisition, resulting in inconsistent time intervals of adjacent sampling points in each trace. (2) The trigger of the ranging wheel may be affected by multiple factors, such as inaccurate encoder positioning, slight topographic relief, and tire pressure changes, resulting in uneven trace spacing. Because the specific impact of the above problems cannot be determined at present, there is no targeted processing, which has caused some interferences to the time-lapse attribute analysis. We could note that some low attribute value areas in Figure 9 only appeared at an earlier time. Then, the range became smaller or even disappeared.

Moreover, by using TLFC 3D GPR for road monitoring, a massive amount of data can be obtained. At present, manual data processing and interpretation need a lot of time. Therefore, when developing data processing and interpretation methods for TLFC 3D GPR in the future, we should not only pay attention to method accuracy but also to calculation efficiency and automation.

To sum up, in the upcoming research, we will focus on eliminating the influence of the irrelevant variables to ensure that the difference between time-lapse GPR data comes from underground changes to realize the quick and high-precision processing of TLFC 3D GPR data and the accurate characterization of underground 3D spatial changes.

6. Conclusions

In this study, we first put forward a road subgrade monitoring technology based on TLFC 3D GPR. It is an optimal choice for road disease monitoring because of its advantages, which include high data acquisition efficiency, wide-coverage, and the ability to obtain dynamic changing information of underground 3D spaces.

Through time-zero consistency correction, 3D data combination, and spatial position matching of TLFC 3D GPR data, full-coverage and accurate imaging of underground 3D space could be achieved. Furthermore, a time-lapse attribute analysis was performed for the TLFC 3D GPR data. Then, the change rule of the subgrade was comprehensively mastered, and the rapid change period was determined and interpreted.

This paper proves the feasibility of applying TLFC 3D GPR to large-scale road subgrade monitoring, provides a reference for relevant data acquisition, processing, and interpretation, and points out necessary directions for further research.

Author Contributions: Conceptualization, J.L., R.Q., and K.S.; methodology, J.L. and K.S.; validation, J.L., R.Q., K.S., L.G., Y.Z., and D.L.; formal analysis, J.L., R.Q., K.S., L.G., Y.Z., and D.L.; investigation, J.L., K.S., and L.G.; data curation, J.L., R.Q., K.S., and L.G.; writing—original draft preparation, J.L. and Y.Z.; writing—review and editing, R.Q. and K.S.; visualization, J.L., R.Q., and K.S.; supervision, J.L., R.Q., and K.S.; project administration, R.Q.; funding acquisition, R.Q. All authors have read and agreed to the published version of the manuscript.

Funding: This research was funded by the National Natural Science Foundation of China (41974159), Natural Science Foundation of Beijing Municipality (8212016), and 2021 Graduate Innovation Fund Project of China University of Geosciences, Beijing (640221003).

Institutional Review Board Statement: Not applicable.

Informed Consent Statement: Not applicable.

Data Availability Statement: Data associated with this research are available and can be obtained by contacting the corresponding author.

Acknowledgments: The authors especially thank Yuchen Wang and Xu Liu for their field assistance.

Conflicts of Interest: The authors declare no conflict of interest.

References

1. Al-Qadi, I.L.; Lahouar, S. Measuring layer thicknesses with GPR-Theory to practice. *Constr. Build. Mater.* **2005**, *19*, 763–772. [CrossRef]
2. Bastard, C.L.; Baltazart, V.; Wang, Y.; Saillard, J. Thin-pavement thickness estimation using GPR with high-resolution and superresolution methods. *IEEE Trans. Geosci. Remote Sens.* **2007**, *45*, 2511–2519. [CrossRef]
3. Liu, H.; Sato, M. In situ measurement of pavement thickness and dielectric permittivity by GPR using an antenna array. *NDT E Int.* **2014**, *64*, 65–71. [CrossRef]
4. Solla, M.; Lagüela, S.; González-Jorge, H.; Arias, P. Approach to identify cracking in asphalt pavement using GPR and infrared thermographic methods: Preliminary findings. *NDT E Int.* **2014**, *62*, 55–65. [CrossRef]
5. Kim, N.; Kim, S.; An, Y.; Lee, J. A novel 3D GPR image arrangement for deep learning-based underground object classification. *Int. J. Pavement Eng.* **2019**, *22*, 740–751. [CrossRef]
6. Torbaghan, M.E.; Li, W.; Metje, N.; Burrow, M.; Chapman, D.N.; Rogers, C.D.F. Automated detection of cracks in roads using ground penetrating radar. *J. Appl. Geophys.* **2020**, *179*, 104118. [CrossRef]
7. Poikajärvi, J.; Peisa, K.; Herronen, T.; Aursand, P.O.; Maijala, P.; Narbro, A. GPR in road investigations-equipment tests and quality assurance of new asphalt pavement. *Nondestr. Test. Eval.* **2012**, *27*, 293–303. [CrossRef]
8. Rodés, J.P.; Reguero, A.M.; Pérez-Gracia, V. GPR spectra for monitoring asphalt pavements. *Remote Sens.* **2020**, *12*, 1749. [CrossRef]
9. Bezina, Š.; Stančerić, I.; Domitrović, J.; Rukavina, T. Spatial representation of GPR data-accuracy of asphalt layers thickness mapping. *Remote Sens.* **2021**, *13*, 864. [CrossRef]
10. Grote, K.; Hubbard, S.; Rubin, Y. GPR monitoring of volumetric water content in soils applied to highway construction and maintenance. *Lead. Edge* **2002**, *21*, 482–485. [CrossRef]
11. Benedetto, A.; Tosti, F.; Ortuani, B.; Giudici, M.; Mele, M. Mapping the spatial variation of soil moisture at the large scale using GPR for pavement applications. *Near Surf. Geophys.* **2015**, *13*, 269–278. [CrossRef]
12. Rodés, J.P.; Pérez-Gracia, V.; Martínez-Reguero, A. Evaluation of the GPR frequency spectra in asphalt pavement assessment. *Constr. Build. Mater.* **2015**, *96*, 181–188. [CrossRef]
13. Solla, M.; Pérez-Gracia, V.; Fontul, S. A review of GPR application on transport infrastructures: Troubleshooting and best practices. *Remote Sens.* **2021**, *13*, 672. [CrossRef]
14. Grasmueck, M.; Weger, R.; Horstmeyer, H. Full-resolution 3D GPR imaging. *Geophysics* **2005**, *70*, K12–K19. [CrossRef]
15. Gaballah, M.; Grasmueck, M.; Sato, M. Characterizing subsurface archaeological structures with full resolution 3D GPR at the early dynastic foundations of saqqara necropolis, Egypt. *Sens. Imaging* **2018**, *19*, 23. [CrossRef]
16. Dinh, K.; Gucunski, N.; Tran, K.; Novo, A.; Nguyen, T. Full-resolution 3D imaging for concrete structures with dual-polarization GPR. *Autom. Constr.* **2021**, *125*, 103652. [CrossRef]
17. Liu, Z.; Wu, W.; Gu, X.; Li, S.; Wang, L.; Zhang, T. Application of combining YOLO models and 3D GPR images in road detection and maintenance. *Remote Sens.* **2021**, *13*, 1081. [CrossRef]
18. Mangel, A.R.; Lytle, B.A.; Moysey, S.M.J. Automated high-resolution GPR data collection for monitoring dynamic hydrologic processes in two and three dimensions. *Lead. Edge* **2015**, *34*, 190–196. [CrossRef]
19. Allroggen, N.; van Schaik, N.L.M.B.; Tronicke, J. 4D ground-penetrating radar during a plot scale dye tracer experiment. *J. Appl. Geophys.* **2015**, *118*, 139–144. [CrossRef]
20. Allroggen, N.; Tronicke, J. Attribute-based analysis of time-lapse ground-penetrating radar data. *Geophysics* **2016**, *81*, H1–H8. [CrossRef]
21. Pan, X.; Jaumann, S.; Zhang, J.; Roth, K. Efficient estimation of effective hydraulic properties of stratal undulating surface layer using time-lapse multi-channel GPR. *Hydrol. Earth Syst. Sci.* **2019**, *23*, 3653–3663. [CrossRef]
22. Saito, H.; Kuroda, S.; Iwasaki, T.; Sala, J.; Fujimaki, H. Estimating infiltration front depth using time-lapse multioffset gathers obtained from ground-penetrating-radar antenna array. *Geophysics* **2021**, *86*, WB109–WB117. [CrossRef]
23. Allroggen, N.; Beiter, D.; Tronicke, J. Ground-penetrating radar monitoring of fast subsurface processes. *Geophysics* **2020**, *85*, A19–A23. [CrossRef]
24. Johnson, T.C.; Routh, P.S.; Barrash, W.; Knoll, M.D. A field comparison of fresnel zone and ray-based GPR attenuation-difference tomography for time-lapse imaging of electrically anomalous tracer or contaminant plumes. *Geophysics* **2007**, *72*, G21–G29. [CrossRef]
25. Steelman, C.M.; Klazinga, D.R.; Cahill, A.G.; Endres, A.L.; Parker, B.L. Monitoring the evolution and migration of a methane gas plume in an unconfined sandy aquifer using time-lapse GPR and ERT. *J. Contam. Hydrol.* **2017**, *205*, 12–24. [CrossRef] [PubMed]

26. Yuan, H.; Looms, M.C.; Nielsen, L. On the usage of diffractions in GPR reflection data-implications for time-lapse gas migration monitoring. *Geophysics* **2020**, *85*, H83–H95. [CrossRef]
27. Dérobert, X.; Baltazart, V.; Simonin, J.; Todkar, S.S.; Norgeot, C.; Hui, H. GPR monitoring of artificial debonded pavement structures throughout its life cycle during accelerated pavement testing. *Remote Sens.* **2021**, *13*, 1474. [CrossRef]
28. Luo, T.X.H.; Lai, W.L.W. Subsurface diagnosis with time-lapse GPR slices and change detection algorithms. *IEEE J. Sel. Top. Appl. Earth Observ. Remote Sens.* **2020**, *13*, 935–940. [CrossRef]
29. Ciampoli, L.B.; Gagliardi, V.; Clementini, C.; Latini, D.; Frate, F.D.; Benedetto, A. Transport infrastructure monitoring by InSAR and GPR data fusion. *Surv. Geophys.* **2020**, *41*, 371–394. [CrossRef]
30. Friedt, J. Passive cooperative targets for subsurface physical and chemical measurements: A systems perspective. *IEEE Geosci. Remote Sens. Lett.* **2017**, *14*, 821–825. [CrossRef]
31. Allroggen, N.; Jackisch, C.; Tronicke, J. Four-Dimensional Gridding of Time-Lapse GPR Data. In Proceedings of the 9th International Workshop on Advanced Ground Penetrating Radar (IWAGPR), Edinburgh, UK, 28–30 June 2017.
32. Giertzuch, P.; Doetsch, J.; Jalali, M.; Alexis, S.; Schmelzbach, C.; Maurer, H. Time-lapse GPR difference reflection imaging of saline tracer flow in fractured rock. *Geophysics* **2020**, *85*, H25–H37. [CrossRef]
33. Haarder, E.B.; Looms, M.C.; Jensen, K.H.; Nielsen, L. Visualizing unsaturated flow phenomena using high-resolution reflection ground penetrating radar. *Vadose Zone J.* **2011**, *10*, 84–97. [CrossRef]
34. Cheung, W.Y.B.; Lai, W.L.W. Field validation of water-pipe leakage detection through spatial and time-lapse analysis of GPR wave velocity. *Near Surf. Geophys.* **2019**, *17*, 231–246. [CrossRef]

MDPI AG
Grosspeteranlage 5
4052 Basel
Switzerland
Tel.: +41 61 683 77 34

Remote Sensing Editorial Office
E-mail: remotesensing@mdpi.com
www.mdpi.com/journal/remotesensing